Molecular Cloning

A LABORATORY MANUAL

T. Maniatis Harvard University

E. F. Fritsch Michigan State University

J. Sambrook Cold Spring Harbor Laboratory

Cold Spring Harbor Laboratory
1982

Molecular Cloning
A LABORATORY MANUAL

Front cover: The electron micrograph of bacteriophage λ particles stained with uranyl acetate was digitized and assigned false color by computer. *Thomas R. Broker, Louise T. Chow, and James I. Garrels*

Back cover: E. coli VL361 with fimbriae was negatively stained with phosphotungstic acid and the electron micrograph was digitized and assigned false color by computer. *Jeffrey A. Engler, Thomas R. Broker, and James I. Garrels*

Cataloging in Publications data

Maniatis, T.
    Molecular cloning.

    (A laboratory manual)
    Bibliography: p.
    Includes index.
    1. Molecular cloning.   2. Eukaryotic
cells.   I. Fritsch, Edward F.   II. Sambrook,
Joseph.   III. Title.   IV. Series.
QH442.2.F74   574.87'3224   81-68891
ISBN 0-87969-136-0   AACR2

Other manuals available from Cold Spring Harbor Laboratory

Hybridoma Techniques
Advanced Bacterial Genetics,
    A Manual for Genetic Engineering
    (Strain Kit available)

Experiments with Normal and Transformed Cells
Experiments in Molecular Genetics
    (Strain Kit available)
Methods in Yeast Genetics

All Cold Spring Harbor Laboratory publications are available through booksellers or may be ordered directly from Cold Spring Harbor Laboratory, Box 100, Cold Spring Harbor, New York.

SAN 203-6185

Preface

This manual began as a collection of laboratory protocols that were used during the 1980 Cold Spring Harbor course on the Molecular Cloning of Eukaryotic Genes. These procedures had been in use in our laboratories at that time but were scattered throughout the notebooks of many different people. In 1981 we decided to produce a more complete and up-to-date manual not only for use in the next Cold Spring Harbor course, but also for eventual publication. Out of the many permutations of the methods being used, we assembled a set of "consensus protocols," which were photocopied and widely distributed to many laboratories even as the 1981 course was underway. Then in the winter of 1981-1982, the manual was substantially rewritten, and new or revised protocols and figures, as well as entirely new chapters, were added.

Even since this last rewriting, however, the field has progressed: New methods are constantly being invented and existing techniques are altered in response to changing needs. Although we have included in this manual only those protocols that have been thoroughly tested and used successfully in our laboratories, we make no claim that they are inviolable or perfect. We would welcome suggestions for improvements, and we would be grateful to be told about any new procedures that are devised.

The evolution of protocols poses the difficult problem of attribution. We have tried to give credit at appropriate places in the text to the people who originally developed the procedures presented here, but in many cases tracing a particular method to its undisputed roots has proved to be impossible. We therefore wish to apologize—and to express gratitude—to those we have been unable to acknowledge for an idea, procedure, or recipe. Our major function has been to compile, to verify, and, we hope, to clarify; less frequently we have introduced modifications, and only in rare instances have we devised new protocols. In large part, then, the manual is based on procedures developed by others, and it is to them that any credit belongs.

Because the manual was originally written to serve as a guide to those who had little experience in molecular cloning, it contains much basic material. However, the current version also deals in detail with almost every laboratory task currently used in molecular cloning. We therefore hope that newcomers to cloning and veterans alike will find material of value in this book.

Although molecular cloning seems straightforward on paper, it is more difficult to put into practice. Most protocols involve a large number of individual steps and a problem with any one of them can lead the experimenter into difficulty. To deal with these problems, a well-founded understanding of the principles underlying each procedure is essential. We have therefore provided background information and references that may be useful if trouble should arise. We also suggest that, as a matter of course, the products of each step in a protocol be tested to verify that the reaction was successful.

This manual could not have been written without the help and advice of members of our laboratories and contributions from many others. We therefore wish to thank Joan Brooks, John Fiddes, Mary-Jane Gething, Tom Gingeras, David Goldberg, Steve Hughes, David Ish-Horowicz, Mike Mathews, Patty Reichel, Joe Sorge, Jim Stringer, Richard Treisman, and Nigel Whittle. We wish particularly to thank Arg Efstratiadis for his helpful discussions and criticisms of Chapter 7; Brian Seed for permission to include a description of his unpublished procedure for screening libraries by recombination (Chapter 10) and many other useful suggestions; Doug Hanahan for advice on transformation (Chapter 8); Bryan Roberts for suggestions on methods of hybrid-selection and cDNA cloning; Doug Melton for providing a protocol for injection of *Xenopus* oocytes; Ronni Greene for suggesting improvements to many protocols; Nina Irwin for providing a critical anthology of methods available for expressing eukaryotic proteins in bacteria (Chapter 12); Rich Roberts for supplying the computer analysis of the sequence of pBR322; Barbara Bachmann and Ahmad Bukhari for reviewing and correcting the list of *E. coli* strains; and Tom Broker, Louise Chow, Jeff Engler, and Jim Garrels for producing the elegant photographs used for the front and back covers.

We also thank all those who participated in the Cold Spring Harbor Molecular Cloning courses of 1980 and 1981. They were an excellent group of students, who struggled through the first two drafts of the manual and made many useful suggestions. We also thank Nancy Hopkins, who helped us to teach the course the first year and convinced us that producing a manual would be a worthwhile task. In 1981 Doug Engel helped teach the course and suggested many improvements to the manual. Contributing to the success of both courses were the efforts of the teaching assistants, who were Catherine O'Connell and Helen Doris Keller in the summer of 1980 and Susan Vande-Woude, Paul Bates, and Michael Weiss in 1981.

We wish to thank Patti Barkley and Marilyn Goodwin for their cheerfulness and forbearance during the typing of successive revisions of the manuscript. Our artists, Fran Cefalu and Mike Ockler, worked with great dedication and perseverance to produce the drawings for the manual. Joan Ebert kept track of the many references added to and deleted from the text and assembled the reference list. We are also grateful to Nancy Ford, Director of Publications, Cold Spring Harbor Laboratory, for her encouragement and support. Finally, without the patience, skill, and diplomacy of Doug Owen, who prepared the manuscript for the printer and helped us in many other ways, this book would not exist.

Tom Maniatis
Ed Fritsch
Joe Sambrook

Contents

10 The Identification of Recombinant Clones 309

11 Analysis of Recombinant DNA Clones 363

1

Vector-Host Systems

Four types of vector can be used to clone fragments of foreign DNA and propagate them in *Escherichia coli:*

plasmids
bacteriophage λ
cosmids
bacteriophage M13

Although very different in size and structure, these four types of vector share the following properties:

1. They can replicate autonomously in *E. coli* (i.e., they are replicons in their own right), even when joined covalently to a foreign DNA fragment.

2. They can be easily separated from bacterial nucleic acids and purified.

3. They contain regions of DNA that are not essential for propagation in bacteria. Foreign DNA inserted into these regions is replicated and propagated as if it were a normal component of the vector.

Each type of vector has particular biological features that make it useful for different purposes. In this chapter we describe these cloning vectors and discuss the principles of their application to problems of molecular cloning.

PLASMIDS

Plasmids are extrachromosomal genetic elements found in a variety of bacterial species. They are double-stranded, closed circular DNA molecules that range in size from 1 kb to greater than 200 kb. Often, plasmids contain genes coding for enzymes that, under certain circumstances, are advantageous to the bacterial host. Among the phenotypes conferred by different plasmids are:

> resistance to antibiotics
> production of antibiotics
> degradation of complex organic compounds
> production of colicins
> production of enterotoxins
> production of restriction and modification enzymes

Under natural conditions, many plasmids are transmitted to new hosts by a process similar to bacterial conjugation. In the laboratory, however, plasmids can be transferred to bacteria by an artificial process, known as transformation, in which they are introduced into bacteria that have been treated in ways that make some of the cells temporarily permeable to small DNA molecules. The new phenotype conferred upon the recipients by the plasmid (e.g., resistance to an antibiotic) allows simple selection of bacteria that have been successfully transformed.

For the most part, replication of plasmid DNA is carried out by the same set of enzymes used to duplicate the bacterial chromosome. Some plasmids are under "stringent control," which means that their replication is coupled to that of the host so that only one or at most a few copies of the plasmid will be present in each bacterial cell (for review, see Novick et al. 1976). Plasmids under "relaxed control," on the other hand, have copy numbers of 10–200. More importantly, the copy number of "relaxed" plasmids can be increased to several thousand per cell if host protein synthesis is stopped (e.g., by treatment with chloramphenicol) (Clewell 1972). In the absence of protein synthesis, replication of relaxed plasmids continues, whereas replication of chromosomal DNA and of stringent plasmids ceases.

To be useful as a cloning vector, a plasmid should possess several properties. It should be relatively small and should replicate in a relaxed fashion. It should carry one or more selectable markers to allow identification of transformants and to maintain the plasmid in the bacterial population. Finally, it should contain a single recognition site for one or more restriction enzymes in regions of the plasmid that are not essential for replication. Preferably, these restriction sites, into which foreign DNA can be inserted, should be located within the genes coding for selectable markers so that insertion of a foreign DNA fragment inactivates the gene.

Below we describe a number of versatile cloning vectors that embody many of these properties (for reviews, see Bolivar and Backman 1979 and Bernard and Helinski 1980). The most widely used vector is pBR322, a plasmid under relaxed control that contains both ampicillin- and tetracycline-resistance genes and a number of convenient restriction sites (Bolivar et al. 1977) (see page 5). The complete nucleotide sequence of pBR322 is known (Sutcliffe 1978) and is given in Appendix B.

Recently, two derivatives of pBR322 have become available that have the advantage of replicating to even higher copy numbers. One derivative, pAT153, was constructed (Twigg and Sherratt 1980) by deleting the *Hae*IIB and G (A. Cowie and E. Ruley, pers. comm.) fragments, which span a region of the plasmid genome involved in control of copy number (see page 6). About 1.5 to 3.0 times as many copies of pAT153 are present per cell than pBR322. The second derivative, pXf3 (D. Hanahan, unpubl.), is even smaller than pAT153 (see page 6).

The advantages of small size are manifold: The plasmid DNA is easier to handle in that it is less susceptible to physical damage, and it has a simpler restriction map. Furthermore, the fact that smaller plasmids generally have higher copy numbers increases the sensitivity with which bacteria carrying foreign DNA sequences can be identified using radiolabeled hybridization probes. However, reduction in size may lead to the elimination of useful cloning sites. pXf3, for example, lacks the *Bal*I and *Ava*I sites that are present in pBR322 and pAT153. To extend the range of useful cloning sites, polylinkers have been inserted into several small plasmids. Polylinkers are segments of DNA that contain closely spaced sites for several restriction enzymes. An example of a plasmid, plink322 (B. Seed, unpubl.), carrying a polylinker is shown on page 7.

Also shown are diagrams of plasmids that, although not as widely used as pBR322 and its derivatives, are extremely useful for particular cloning purposes. pMK16 (see page 7), for example, contains single *Sma*I and *Xho*I restriction sites in a gene coding for kanamycin resistance. pKC7 and pACYC184 (see page 8) carry useful cloning sites within a gene coding for chloramphenicol resistance. The large plasmids pCRl (11.4 kb) and pSC101 (9.9 kb; see page 9) are not routinely used in cloning experiments but have the unique property of not hybridizing to one another. Thus, it is possible to clone DNA fragments from one source in pSC101 and from another source in pCRl and then carry out cross-hybridization experiments between the inserted DNAs without interference from plasmid DNA sequences. Perhaps the most specialized of all plasmids is πVX, which is used to select from a population of recombinant λ bacteriophages those phages that contain DNA sequences homologous to a foreign DNA segment inserted into the plasmid (B. Seed, unpubl.) (see page 10).

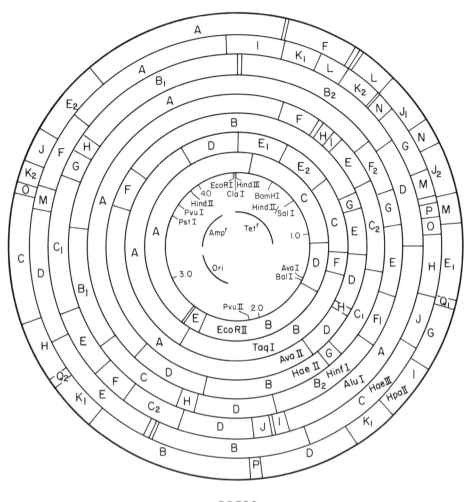

pBR322

Size: 4.3 kb

Replicon: Col E1, relaxed

Selective markers: Amp^r, Tet^r

Single sites: Ava I, Pst I, BamHI, Pvu II, Cla I, Sal I, EcoRI, Hind III

Insertional inactivation: Amp^r - Pst I

Tet^r-BamHI, Hind III (variable), Sal I

References: Bolivar et al. (1977); Sutcliffe (1978, 1979).

Comments: pBR322 is the most versatile of the plasmid cloning vectors.
Its complete nucleotide sequence is known (Sutcliffe 1979).

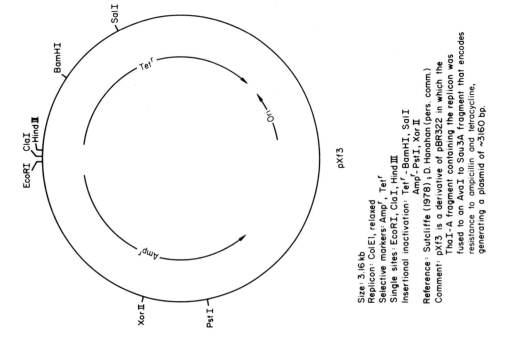

pXf3

Size: 3.16 kb
Replicon: ColEl, relaxed
Selective markers: Amp^r, Tet^r
Single sites: EcoRI, ClaI, HindIII
Insertional inactivation: Tet^r-BamHI, SalI
    Amp^r-PstI, XorII
Reference: Sutcliffe (1978); D. Hanahan (pers. comm.)
Comment: pXf3 is a derivative of pBR322 in which the
    ThaI-A fragment containing the replicon was
    fused to an AvaI to Sau3A fragment that encodes
    resistance to ampicillin and tetracycline,
    generating a plasmid of ≈3160 bp.

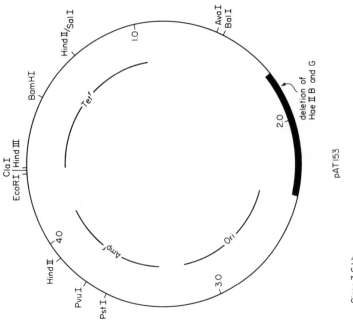

pATI53

Size: 3.6 kb
Replicon: ColEl, relaxed
Selective markers: Amp^r, Tet^r
Single sites: AvaI, PstI, BamHI, ClaI, SalI, EcoRI, HindIII
Insertional inactivation: Amp^r-PstI
    Tet^r-BamHI, HindIII (variable), SalI
Reference: Twigg and Sherratt (1980)
Comment: A high-copy variant of pBR322.

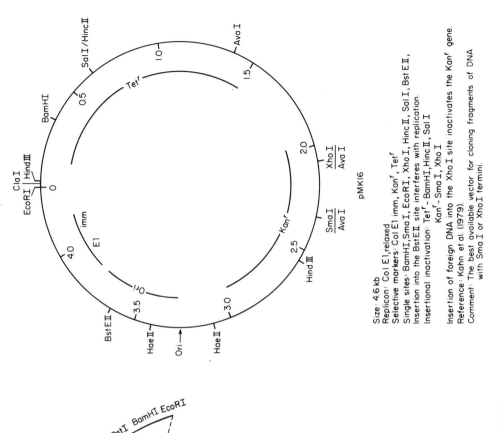

pMKl6

Size: 4.6 kb
Replicon: Col E1, relaxed
Selective markers: Col E1 imm, Kan^r, Tet^r
Single sites: BamHI, SmaI, EcoRI, XhoI, HincII, SalI, BstEII,
Insertion into the BstEII site interferes with replication.
Insertional inactivation: Tet^r - BamHI, HincII, SalI
                          Kan^r - SmaI, XhoI
Insertion of foreign DNA into the XhoI site inactivates the Kan^r gene.
Reference: Kahn et al. (1979).
Comment: The best available vector for cloning fragments of DNA
         with SmaI or XhoI termini.

plink 322

Size: 3.8 kb
Replicon: ColE1, relaxed
Selection: Amp^r
Single sites: ClaI, HindIII, XbaI, BglII, BamI
plink 322, a variant of pBR322, contains a polylinker which
increases the number of useful cloning sites.
Reference: B. Seed (unpubl.)
Polylinker sequence:

```
GAATTCTCATGTTTGACAGCTTATCATCGATAAGCTTCTAGAGATCT
EcoRI                      ClaI  HindIII XbaI  BglII

TCCATACCTACCAGTTCTCCGCCTGCAGCAATGGCAACAACGTTGCC
                              PstI

CGGATCCGGTCGCGCGAATTC
BamHI              EcoRI
```

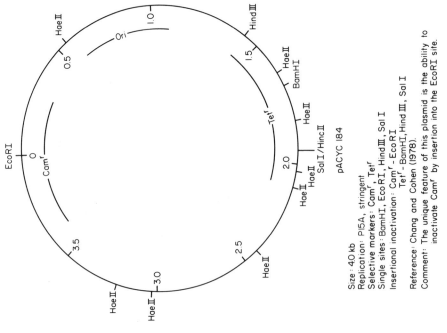

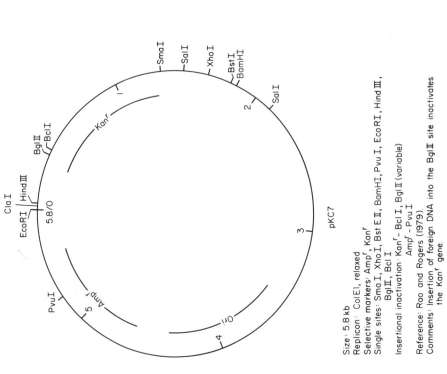

Size: 4.0 kb
Replication: Pl5A, stringent
Selective markers: Cam^r, Tet^r
Single sites: BamHI, EcoRI, HindIII, SalI
Insertional inactivation: Cam^r – EcoRI
Tet^r – BamHI, HindIII, SalI
Reference: Chang and Cohen (1978).
Comment: The unique feature of this plasmid is the ability to
inactivate Cam^r by insertion into the EcoRI site.

Size: 5.8 kb
Replicon: ColEI, relaxed
Selective markers: Amp^r, Kan^r
Single sites: SmaI, XhoI, BstEII, BamHI, PvuI, EcoRI, HindIII,
BglII, BclI
Insertional inactivation: Kan^r – BclI, BglII (variable)
Amp^r – PvuI
Reference: Rao and Rogers (1979).
Comments: Insertion of foreign DNA into the BglII site inactivates
the Kan^r gene.

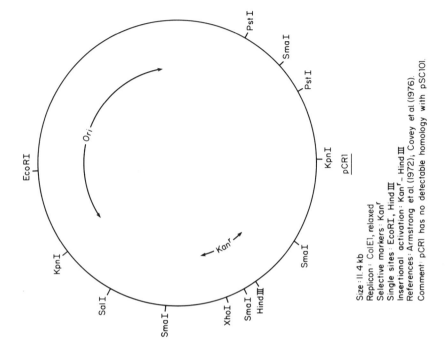

pCRI

Size: 11.4 kb
Replicon: ColEI, relaxed
Selective markers: Kan^r
Single sites: EcoRI, Hind III
Insertional activation: Kan^r – Hind III
References: Armstrong et al (1972), Covey et al (1976).
Comment: pCRI has no detectable homology with pSCIOI.

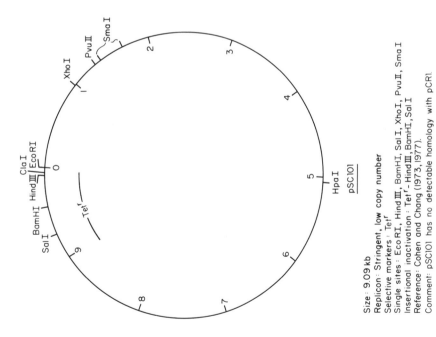

pSCIOI

Size: 9.09 kb
Replicon: Stringent, low copy number
Selective markers: Tet^r
Single sites: EcoRI, Hind III, BamHI, Sal I, XhoI, PvuII, Sma I
Insertional inactivation: Tet^r – Hind III, BamHI, Sal I
Reference: Cohen and Chang (1973, 1977).
Comment: pSCIOI has no detectable homology with pCRI.

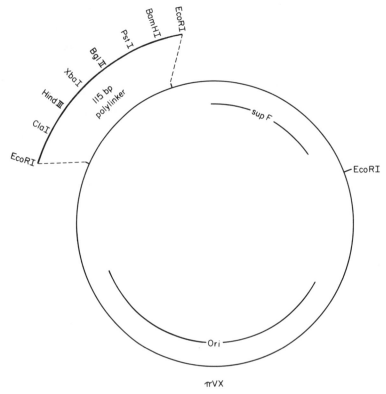

πVX

Size : 902 bp
Replication : relaxed ?
Selection : sup F
Single sites : Cla I, Hind III, Bgl II, Pst I, Bam I
Reference : B. Seed (unpubl.)
Comment : This plasmid consists of 3 Eco RI fragments : a 508 bp
          fragment derived from pMB1, a 207 bp synthetic tyrosine
          tRNA suppressor gene and a 115 bp "polylinker" composed
          of the sites listed above.
Polylinker sequence : See map of plink 322.

Cloning in Plasmids

In principle, cloning in plasmid vectors is very straightforward. The plasmid DNA is cleaved with a restriction endonuclease and joined in vitro to foreign DNA. The resulting recombinant plasmids are then used to transform bacteria. In practice, however, the plasmid vector must be carefully chosen to minimize the effort required to identify and characterize recombinants. The major difficulty is to distinguish between plasmids that contain sequences of foreign DNA and vector DNA molecules that have recircularized without insertion of foreign sequences. Recircularization of the plasmid can be limited to some extent by adjusting the concentrations of the foreign DNA and vector DNA during the ligation reaction. However, a number of procedures, described below, have been developed either to reduce recircularization of the plasmid still further or to distinguish recombinants from nonrecombinants by genetic techniques.

Insertional Inactivation

This method can be used with plasmids that carry two or more antibiotic-resistance markers (see Fig. 1.1). The DNA to be inserted and the purified plasmid DNA are digested with a restriction enzyme that, in this example,

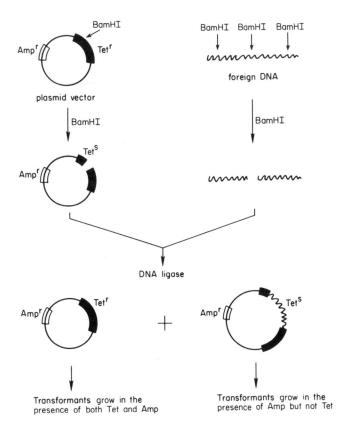

Figure 1.1

Insertional inactivation.

recognizes a unique site located in the plasmid within the tetracycline-resistance gene. After ligating the two DNAs at the appropriate concentrations, the ligation mixture is used to transform, for example, ampicillin-sensitive *E. coli* to ampicillin resistance. Some of the colonies that grow in the presence of ampicillin will contain recombinant plasmids; others will contain plasmid DNA that has recircularized during ligation without insertion of foreign DNA. To discriminate between the two kinds of transformants, a number of colonies are streaked in identical locations on plates containing ampicillin or tetracycline (see Fig. 1.2). The colonies that survive and grow in the presence of tetracycline contain plasmids with an active tetracycline-resistance gene; such plasmids are unlikely to carry insertions of foreign DNA. The colonies that grow only in the presence of ampicillin contain plasmids with inactive tetracycline-resistance genes; these plasmids are likely to carry foreign DNA sequences.

In a few cases, methods have been developed to apply positive selection to obtain bacteria that are sensitive to an antibiotic from populations that are predominantly resistant. In this way, it is possible to select for recombinant plasmids that carry an inactivated antibiotic-resistance gene as a consequence of insertion of a foreign DNA sequence. The most useful of these

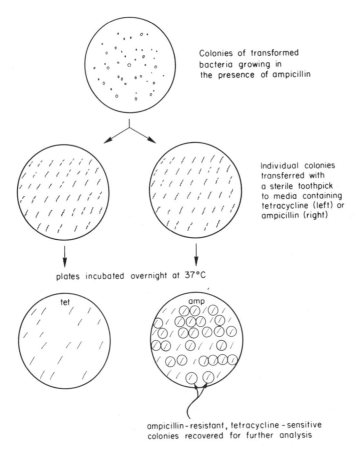

Colonies of transformed bacteria growing in the presence of ampicillin

Individual colonies transferred with a sterile toothpick to media containing tetracycline (left) or ampicillin (right)

plates incubated overnight at 37°C

tet

amp

ampicillin-resistant, tetracycline-sensitive colonies recovered for further analysis

Figure 1.2

Screening for insertions of foreign DNA by inactivation of plasmid-borne, antibiotic-resistance genes.

systems is that described by Bochner et al. (1980) and Maloy and Nunn (1981), who developed media containing the lipophilic, chelating agents fusaric acid or quinaldic acid, which allow the direct positive selection of Tet^s clones from a population of Tet^s and Tet^r bacteria. For most strains of *E. coli*, approximately 90% of the colonies obtained on media containing tetracycline and fusaric acid were found to be Tet^s when plated onto media containing tetracycline alone. It is therefore possible to select from a population of bacteria transformed with pBR322 or pAT153 those cells that carry plasmids with insertions at the *Bam*HI and *Sal*I sites.

A similar technique has been developed to select for bacteria sensitive to paromomycin (Slutsky et al. 1980). This should allow the selection of derivatives of pMK16 that contain insertions at the *Sma*I or *Xho*I site (Kahn et al. 1979).

Although insertion of foreign DNA sequences within an antibiotic-resistance gene almost always leads to inactivation of that gene, at least one case is known where an insertion left the gene in a functional state. Villa-Komaroff et al. (1978) found that insertion of a segment of rat preproinsulin cDNA into the *Pst*I site of pBR322 did not inactivate the ampicillin-resistance gene. Presumably, a small piece of foreign DNA had been inserted that did not alter the reading-frame of the ampicillin-resistance gene, so that a fusion protein was formed which retained β-lactamase activity.

Directional Cloning

Most plasmid vectors carry two or more unique restriction enzyme recognition sites. For example, the plasmid pBR322 contains single *Hind*III and *Bam*HI sites (see Fig. 1.3). After cleavage by both enzymes, the larger fragment of plasmid DNA can be purified by gel electrophoresis and ligated to a segment of foreign DNA containing cohesive ends compatible with those generated by *Bam*HI and *Hind*III. The resulting circular recombinant is then used to transform *E. coli* to ampicillin resistance. Because of the lack of complementarity between the *Hind*III and *Bam*HI protruding ends, the larger vector fragment cannot circularize efficiently; it therefore transforms *E. coli* very poorly. Therefore, most of the colonies resistant to ampicillin contain plasmids that carry foreign DNA segments forming a bridge between the *Hind*III and *Bam*HI sites. Of course, different combinations of enzymes can be used, depending on the locations of restriction sites within vector and the segment of foreign DNA.

Phosphatase Treatment of Linear, Plasmid Vector DNA

During ligation, DNA ligase will catalyze the formation of a phosphodiester bond between adjacent nucleotides only if one nucleotide contains a 5′-phosphate group and the other a 3′-hydroxyl group. Recircularization of plasmid DNA can therefore be minimized by removing the 5′ phosphates from both ends of the linear DNA with bacterial alkaline phosphatase or calf intestinal phosphatase (Seeburg et al. 1977; Ullrich et al. 1977). As a result, neither strand of the duplex can form a phosphodiester bond. However, a

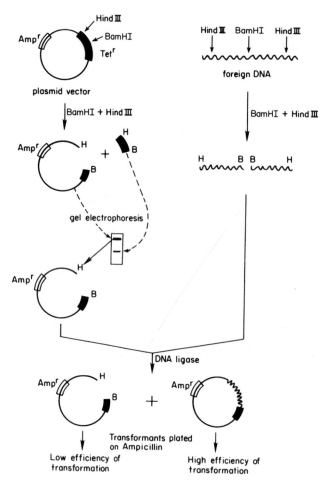

Figure 1.3

Directional cloning.

foreign DNA segment with 5'-terminal phosphates can be ligated efficiently to the dephosphorylated plasmid DNA to give an open circular molecule containing two nicks (see Fig. 1.4). Because circular DNA (even nicked circular DNA) transforms much more efficiently than linear plasmid DNA, most of the transformants will contain recombinant plasmids. A protocol for phosphatase treatment of plasmid DNA is given on page 133.

Problems in Cloning Large DNA Fragments in Plasmids

Finally, the size of the foreign DNA to be inserted can also affect the ratio of transformants containing recombinant plasmids to those containing recircularized vectors. In general, the larger the insertion of foreign DNA, the lower the efficiency of transformation. Thus, when cloning large DNA fragments ($>$10 kb), it is especially important to take all possible measures to keep the number of recircularized vector molecules to a minimum. Even so, the background is relatively high, and it is usually necessary to use an in situ hybridization procedure (Grunstein and Hogness 1975; Hanahan and Meselson 1980) to identify recombinant transformants.

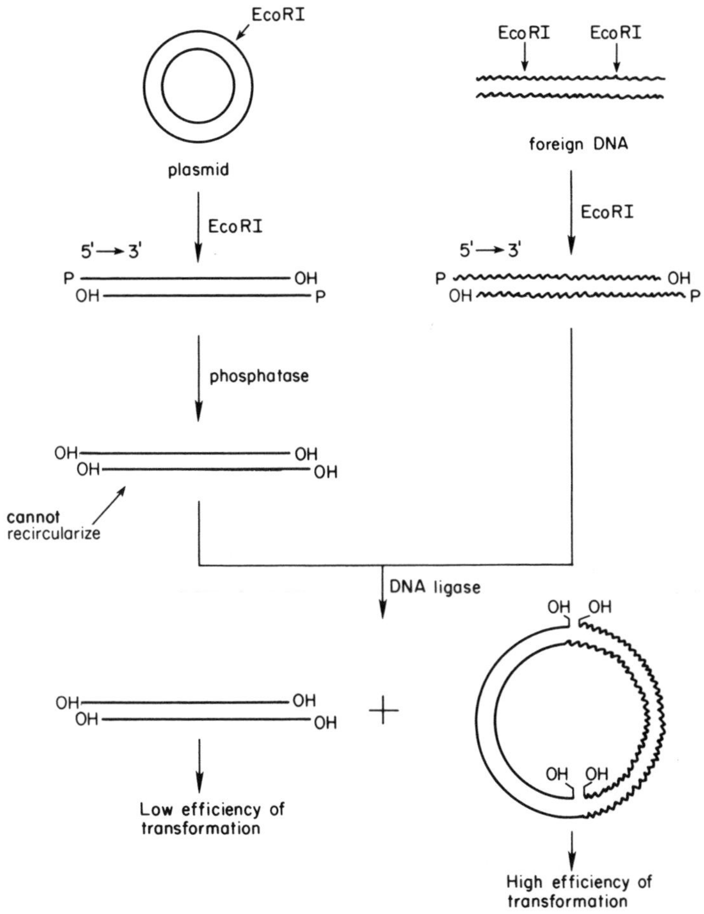

Figure 1.4

Use of phosphatase to prevent recircularization of vector DNA.

BACTERIOPHAGE λ

Since the first demonstration of the feasibility of using bacteriophage λ as a cloning vehicle (Murray and Murray 1974; Rambach and Tiollais 1974; Thomas et al. 1974), a large variety of vectors has been constructed (for review, see Williams and Blattner 1980). The effective use of these vectors requires a basic understanding of the molecular biology of bacteriophage λ (for a more exhaustive discussion, see Herskowitz 1973; Hendrix et al. 1982).

Bacteriophage λ is a double-stranded DNA virus with a genome size of approximately 50 kb (see Fig. 1.5). In bacteriophage λ particles, the DNA is in the form of a linear duplex molecule with single-stranded complementary ends 12 nucleotides in length (cohesive ends). Soon after entering a host bacterium, the DNA circularizes through pairing of the cohesive ends and is transcribed as a circular molecule during the early phase of infection. During this phase, one of two alternate pathways of replication is chosen (see Fig. 1.6): (1) During lytic growth, the circular DNA is replicated manyfold in the cell, a large number of bacteriophage gene products are synthesized, progeny bacteriophage particles are formed and mature, and the cell eventually lyses, releasing many new infectious virus particles. (2) Alternatively, during lysogenic growth, the infecting bacteriophage genome is integrated into the bacterial host DNA and is subsequently replicated and transmitted to progeny bacteria like any other chromosomal gene.

In the discussion that follows, the genes and gene products involved in each of the two pathways of bacteriophage λ growth are described in more detail.

The Lytic Cycle

Immediate Early Transcription

Early in infection, transcription is initiated at two promoters, p_L and p_R, located just to the left and right of the cI (repressor) gene (see Fig. 1.7). The resulting "immediate early" RNAs terminate at the ends of the N and cro genes, at sites t_L and t_{Rl}, respectively, although some rightward transcripts continue through genes O and P (which code for proteins involved in DNA replication) and terminate at t_{R2}. The leftward transcript codes for the N protein, whose action is essential for the next phase of infection.

Delayed Early Transcription

The N protein, by neutralizing the activity of the host termination factor ρ, allows transcription to proceed through the early terminators t_L, t_{Rl}, and t_{R2} into the remainder of the early gene region. N protein is therefore a positive regulatory element whose activity is necessary for the lytic growth of bacte-

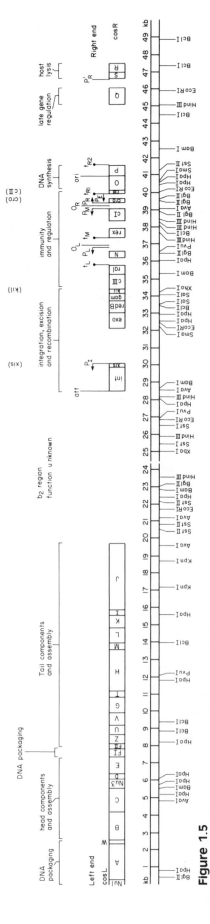

Figure 1.5

Physical and genetic map of wild-type bacteriophage λ. The general locations of genes that encode various lytic and lysogenic functions are indicated at the top of the figure by brackets. The genetic map below the brackets shows the locations of specific bacteriophage λ genes (for a description of the function of each of these genes, see Hendrix et al. 1982). The physical map of bacteriophage λ DNA is presented at the bottom of the figure. Distances from the left end are given in kilobase (kb) pairs, and the locations of restriction endonuclease cleavage sites are indicated below the double line. See Williams and Blattner (1980) for a detailed description of the restriction map.

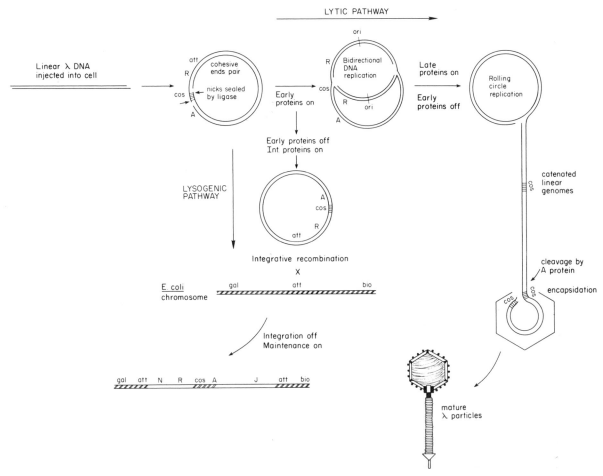

Figure 1.6

A schematic diagram of the bacteriophage λ life cycle. Double-stranded λ DNA, which enters the bacterial host as a linear duplex, is indicated by double lines. The lytic pathway, which is indicated by the horizontal arrow, requires early and late gene functions, whereas the lysogenic pathway, which is indicated by the vertical arrow, requires only early gene function.

riophage carrying t_{R2}. However, N^- mutants can grow (albeit relatively poorly) if t_{R2} is removed by deletion. Such phages are know as *nin* (*N*-independent) mutants.

DNA Replication

During the early phase of infection, circular λ DNA replicates bidirectionally as a Cairns or θ form from a single origin (*ori*) (see Fig. 1.7), whose activation requires the two virus-coded proteins, O and P. Later, rolling circle replication begins (what determines the shift from Cairns structures to rolling circle is unknown), and catenated, linear DNA molecules are generated. During packaging, these catenates are cleaved at the *cos*L and *cos*R

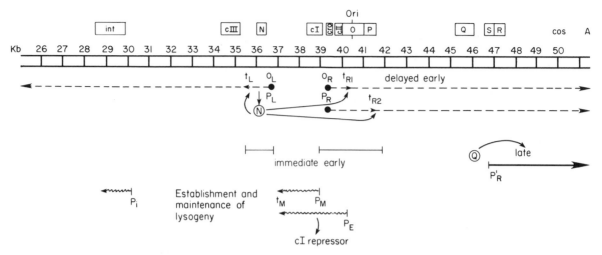

Figure 1.7

A schematic diagram of bacteriophage λ gene regulation. The boxes above the map indicate the locations of genes or DNA sites involved in λ gene regulation (see text for a discussion of the function of these genes.) (●) Promoter sites; (- - -→) transcription products; (t_L, t_{R1}, and t_{R2}) the locations of ρ-mediated termination sites; (N) the product of the N gene, which prevents termination at t_L, t_{R1}, and t_{R2}; (Q) the product of the Q gene, a positive regulator of late transcription at p'_R; (←⌇⌇) transcripts that encode gene products involved in the establishment and maintenance of lysogeny.

sites by the product of the λ *A* gene to form monomer-length linear DNAs with cohesive ends that are found in mature bacteriophage particles.

Late Transcription

One of the products of early transcription from p_R codes for a protein called *cro*. As it accumulates during infection, *cro* binds to o_L and o_R, inhibiting attachment of RNA polymerase and thereby switching off transcription from p_L and p_R. By this time, however, sufficient amounts of another control protein, called *Q*, have been synthesized to cause transcription of the late genes to begin. *Q* acts as a positive regulator of RNA synthesis from p'_R, the promoter used for transcription of the entire late region, which contains many genes involved in head and tail assembly and cell lysis.

Assembly

The two major units of the mature particles—the heads and tails—are assembled separately.

The earliest precursor identified in the pathway leading to head assembly is the scaffolded prehead (see Fig. 1.8). Further maturation of the scaffolded prehead, which involves removal of the scaffolding protein and proteolytic processing of other components, depends on a function (*groE*) supplied by the host. The resulting structures are known as preheads. The first stage in

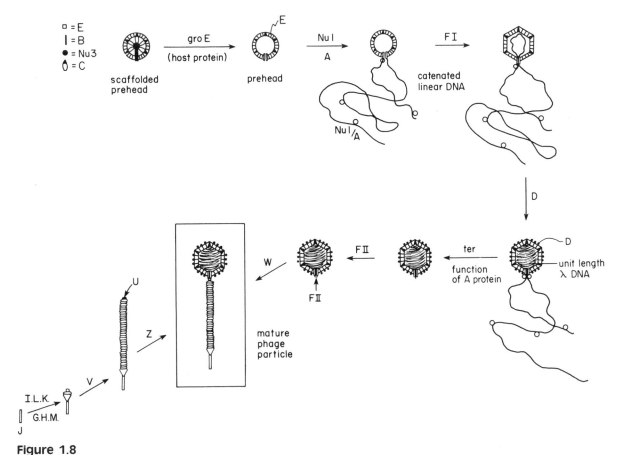

Figure 1.8

A schematic diagram of the assembly pathway of bacteriophage λ. The gene products involved in each step are indicated (for details, see text and Hohn 1979).

packaging DNA into the prehead involves two proteins, *Nu1* and *A*, that bind to catenated, linear DNA close to but not at the left *cos* site. This complex then becomes attached to a defined area on the prehead. In the presence of protein *F*I, the DNA becomes wound into the prehead, which expands in size by some 20%. When the head is filled, the *D* or "decoration" protein attaches to the outside of the capsid, locking the head in place around the DNA. During packaging, the left and right *cos* sites on the catenated, linear DNA are brought close together at the entrance to the head, where they are cleaved in a staggered fashion by the *ter* function of the *A* protein to generate the 12-nucleotide cohesive ends. Finally, the filled heads associate with preformed tail units, which have been assembled by a separate pathway.

Lysis

Two phage proteins, *R* and *S*, are required for lysis of the host cell and liberation of the progeny phage. Mutants in the *S* protein are particularly useful because they allow large numbers of intracellular particles to accumulate and thereby improve the yield of phage. The particles synthesized by S^- mutants may be liberated by lysing the cell with chloroform.

Lysogeny

In a small proportion of infected cells, the lytic cycle is aborted and an alternative pathway is activated that leads to the integration of λ DNA into the genome of its bacterial host. The choice between the lytic and lysogenic pathways depends on an intricate balance of a number of host and bacteriophage factors that are present in the infected cell. To establish lysogeny, proteins coded by genes cII and cIII (whose transcription is controlled by p_R and p_L, respectively) activate leftward transcription from promoters p_E and p_I so that genes cI and *int* are expressed. The cI-gene product is a repressor of early transcription and consequently blocks late gene expression. The *int* protein recognizes sequences at the *att* sites in the bacterial and bacteriophage genomes and catalyzes a breaking and joining event that leads to insertion of viral DNA into the host chromosome.

Almost all viral transcription is repressed in the integrated state; only the cI gene is expressed. The resulting cI-gene product represses transcription of early genes and, in addition, regulates its own synthesis. Low concentrations of the cI-gene product cause the activity of the maintenance promoter, p_m, to increase; high concentrations inhibit transcription from p_m. All of these effects are mediated by binding of the cI-gene product to three 17-nucleotide-long sites in o_L and o_R (Johnson et al. 1979).

Bacteriophages mutated in genes cI, cII, or cIII are unable to lysogenize; they therefore form clear plaques. Bacteriophages carrying a temperature-sensitive mutation in cI are able to establish and maintain the lysogenic state as long as the cells are propagated at a temperature that allows the cI-gene product to retain its ability to repress transcription from p_L and p_R. Prophages containing a wild-type cI repressor gene can also be induced into lytic growth, most commonly by cleavage of the cI-gene product by the host *recA*-gene product, following exposure to agents that damage DNA. If the repressor activity is destroyed, transcription resumes from p_L and p_R. A protein called *xis* is then synthesized, which, by causing the *att* sites to recombine, detaches the prophage DNA from the host chromosome. The lytic cycle then follows its usual course.

Construction of Bacteriophage λ Vectors

At first glance, it would seem a bleak prospect to use bacteriophage λ, with its large and complex genome, as a vector. Its DNA contains several sites for many of the restriction enzymes that are most useful in cloning; and often these sites are located in regions of the genome essential for lytic growth of the virus. Moreover, since particles will not accommodate molecules of DNA that are much longer than the viral genome itself, it would seem that λ might be useful as a vector only for small pieces of foreign DNA.

However, these problems are not as formidable as they might seem. First, the central third of the viral genome, lying between the *J* and *N* genes (see Fig. 1.5) was shown to be nonessential for lytic growth. Studies of specialized transducing phages, carried out several years ago (Campbell 1971), had shown that this region could be replaced by genetic manipulation in vivo with any one of a variety of segments of *E. coli* DNA. Without the consequent wealth of detailed knowledge of the structure and expression of these phages, it is certain that λ would not have been so quickly developed as a vector, and it is quite possible that it would not have been developed at all.

Second, knowledge of the genetics of restriction and modification of DNA provided the opportunity to select, in vivo, derivatives of λ from which all sites for *Eco*RI had been eliminated from the essential portion of the viral genome. By cycling alternately on *E. coli* strain K and on a K strain harboring a plasmid coding for *Eco*RI, strains of λ were isolated having only one or two targets for *Eco*RI located in the nonessential region (Murray and Murray 1974; Rambach and Tiollais 1974; Thomas et al. 1974).

This kind of selection in vivo was possible only when bacteriophage λ could be grown in hosts that synthesized the restriction enzyme of interest, and alternative methods were required to obtain vectors suitable for use with other restriction enzymes. For example, derivatives of λ have been constructed that carry DNA segments from other lambdoid phages (e.g., φ80) that either contain or lack the desired target sites. More recently, improvements in the efficiency with which λ DNA can be packaged into particles have made it possible to eliminate unwanted target sites in vitro. The bacteriophage λ DNA is treated with a restriction enzyme, and the surviving molecules are packed into particles in vitro and propagated in *E. coli*. After several successive cycles, phage genomes are obtained that have lost target sites by mutation while retaining infectivity (Murray 1977).

These approaches, singly and in combination, have led to the development of a large number of vectors that can accept and propagate foreign DNA in a variety of restriction sites (see Williams and Blattner 1980). Those derivatives having a single target site at which foreign DNA is inserted are known as *insertion vectors;* those having a pair of sites spanning a segment that can be removed and replaced by foreign DNA are known as *replacement* or *substitution vectors.*

Cloning with bacteriophage λ vectors involves several steps:

1. The vector DNA is digested to completion with the appropriate restriction enzyme, and in the case of replacement vectors, the left and right arms are separated from the central "stuffer" fragments by velocity gradient centrifugation or gel electrophoresis.

2. The arms are then ligated in the presence of fragments of foreign DNA having termini compatible with those of the arms.

3. The resulting recombinant DNAs are packaged in vitro into viable bacteriophage particles that form plaques on the appropriate hosts.

4. Recombinant phages carrying the desired foreign DNA sequences are then identified, most usually by a procedure involving nucleic acid hybridization.

Choosing the Appropriate Vector

There is no single bacteriophage λ vector suitable for cloning all DNA fragments. It is therefore necessary to choose carefully among the available vectors for the one best suited to the particular task at hand. Two considerations influence this choice: (1) the restriction enzyme(s) that is to be employed, and (2) the size of the fragment of foreign DNA that is to be inserted.

Only about 60% of the viral genome (the left arm, ~ 20 kb in length, including the head and tail genes *A–J*, and the right arm from p_R through the *cos*R site) is necessary for lytic propagation of the phage; the middle one third of the genome, being inessential for lytic growth, can therefore be replaced by foreign DNA. However, the viability of phages decreases dramatically when DNA longer than 105% or shorter than 78% of the wild-type genome is packaged. It is therefore important to choose a combination of vector and foreign DNA such that the size of the recombinant phage falls within acceptable limits. These constraints sometimes can be turned to advantage. After the central "stuffer" portion of some replacement vectors is removed, the deleted genome formed by fusion of the right and left arms is too short to be packaged. With such vectors, therefore, there is a positive selection for recombinant phages. Furthermore, certain vectors contain a stuffer fragment that donates a readily recognizable phenotype to the phage. For example, several vectors are available that carry a segment of *E. coli* DNA coding for β-galactosidase. Such vectors form dark blue plaques when plated on *lac*⁻ hosts in the presence of the chromogenic substrate 5-bromo-4-chloro-3-indolyl-β-D-galactoside (Xgal). Cloning with these vectors results in replacement of most of the β-galactosidase gene with foreign DNA. The resulting phages may be recognized by their ability to form colorless plaques when plated on *lac*⁻ hosts in the presence of Xgal (see Miller 1972).

Finally, some vectors have been designed to take advantage of the fact that growth of wild-type λ is restricted in lysogens carrying prophage P2. This phenotype is called Spi⁺ (sensitive to P2 interference). However, strains of λ lacking two genes involved in recombination (*red* and *gam*) display the Spi⁻ phenotype and grow well in P2 lysogens (Zissler et al. 1971) as long as they carry a *chi* sequence. (As discussed below, the *chi* site is a substrate for recombination catalyzed by the *recA* system.) Cloning in a vector such as λL47.1, λ1059, or λBF101 (see pages 41, 43, and 44, respectively) is accomplished by removal of a stuffer fragment containing *red* and *gam*. The resulting recombinant phages are Spi⁻, and they may be recognized or selected by their ability to grow in *recA*⁺ P2 lysogens of *E. coli*.

chi mutations were originally detected as suppressors of the small-plaque phenotype that is normally seen when *red*⁻ *gam*⁻ bacteriophages are plated on wild-type *E. coli* hosts. Subsequently, Stahl and his coworkers (Stahl et al.

1975; Stahl 1979) found that *chi* mutations are hot spots for *rec*-mediated recombination. The suppression of small-plaque phenotype may therefore be explained in the following way.

Ordinarily, the bacteriophage λ *gam*-gene product inactivates the potent exonuclease coded by the host *recBC* system. In the absence of the *gam*-gene product, the *recBC* exonuclease degrades the catenated, linear λ DNA produced by rolling-circle replication. Under these circumstances, the only DNA molecules that are available as substrates for packaging into bacteriophage particles are the exonuclease-resistant, closed circular dimers generated by θ-form replication and subsequent recombination. In the absence of a *chi* mutation, this recombination process is inefficient and small plaques are therefore formed.

Because most bacteriophage λ vectors are red^- gam^- when the stuffer fragments are removed, the presence of a *chi* mutation in one of the two arms is essential for the construction of representative libraries. Since *chi* sequences have been found in all eukaryotic DNAs so far studied, some recombinant phages will carry insertions that, by chance, contain *chi* sequences. Thus, recombinants constructed in vectors that do not contain a *chi* site and are gam^- as recombinants (e.g., λ Charon 28) exhibit large differences in plaque size. Furthermore, those recombinants that happen to pick up a *chi* site gain a selective advantage over those that do not and become overrepresented during amplification of libraries. This problem can be avoided by using vectors such as λ1059 and λL47.1, which contain *chi* sites in the right and left arms, respectively, and give rise to recombinant plaques of relatively uniform size.

Note that almost all replacement vectors lose the *gam* gene when the stuffer fragment is removed. Recombinants formed with these vectors must therefore be propagated in $recA^+$ bacteria. The one exception is λsep6-lac5, which carries the *red* and *gam* genes in its right arm. Recombinants formed with this vector are Spi^+ and can be propagated in bacteria carrying $recA^-$ mutations.

Maps of Bacteriophage λ Vectors

Each of the maps for the bacteriophage λ vectors is set out in the following format. The top line is a scale in kilobase (kb) pairs. The second line shows the location of restriction enzyme sites and the positions of the major genes in λ wild-type DNA. The third line is a map of the vector. Open boxes represent regions of the vector that are derived from wild-type λ DNA and are drawn directly beneath the corresponding locations of the wild-type map. Specific mutations (such as *W*am or *ts*) are indicated above the open boxes. Restriction sites that have been removed from the wild-type sequence by in vivo or in vitro selection during construction of the vector are indicated by small x's above the open boxes. Deletions from the wild-type λ sequence are indicated by a solid line. Insertions and substitutions are indicated by boxes with diagonal stripes (insertion from *E. coli*), horizontal stripes (substitution from φ434), stippled pattern (substitution or insertion of λ), or complete shading (substitution from $\phi21$). The extents of the concomitant deletions from the wild-type λ sequence are indicated by the dashed lines. Restriction enzyme sites not present in wild-type λ DNA are indicated on the vector map. Below and to the right of the vector map is a compilation of the approximate locations (in base pairs from the left end [L end]) of various useful restriction enzyme sites. The numbers given are the best estimates available and are of limited precision. Below and to the left of the vector map is a table of the various possible combinations of enzymes for which the vector is a useful cloning vehicle and an indication of whether the vector serves as an insertion or substitution vector. To the right of each line in the table is a pictorial representation of the cloning vehicle that is obtained when the vector is prepared by cleavage with various restriction enzymes. Above each line are indicated the maximum (on the left) and minimum (on the right) fragment sizes (in kb) that the cloning vehicle will accommodate.

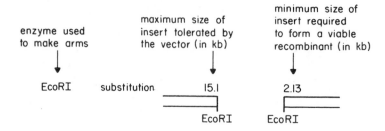

In some cases, where the left and right arms are generated by cleavage with different restriction enzymes, an asterisk (*) appears. The asterisk denotes that an insert of foreign DNA is required, not to bring the recombinant genome to a packageable size, but to allow efficient ligation of the left and right arms.

A description of many of the genetic markers used in bacteriophage λ vectors is presented in Table 1.1.

TABLE 1.1. GENETIC MARKERS USED IN BACTERIOPHAGE λ VECTORS

*lac*5 *bio*1 *bio*256	substitutions from the *lac* and *bio* regions of *E. coli*
att^{80} *imm*^{80} *QSR*^{80}	substitutions from bacteriophage φ80
imm^{21}	a substitution from bacteriophage φ21
int^{29}	a substitution from bacteriophage φ29
KH100	an insertion containing an *Eco*RI site
pacl29	ColEl plasmids with cloned bacteriophage λ *att* sites
BWl	a deletion resulting in loss of the *Eco*RI site in gene *O*
*b*1007	a *b*-region deletion that damages the *att* site and therefore prevents lysogenization
KH53	a deletion in the region of bacteriophage φ80, analogous to λ *c*I, that effectively prevents lysogenization
KH54	a deletion in the *rex-c*I region that effectively prevents lysogenization
*nin*5	a deletion that removes the transcription termination site t_{R2} and thereby renders delayed-early transcription independent of the *N*-gene product
*nin*L44	a deletion removing the *Bam*HI site near the *ral* gene
*c*I857	a *ts* mutation that renders the *c*I-gene product thermolabile
chi	specialized sites in bacteriophage λ that promote directional recombination

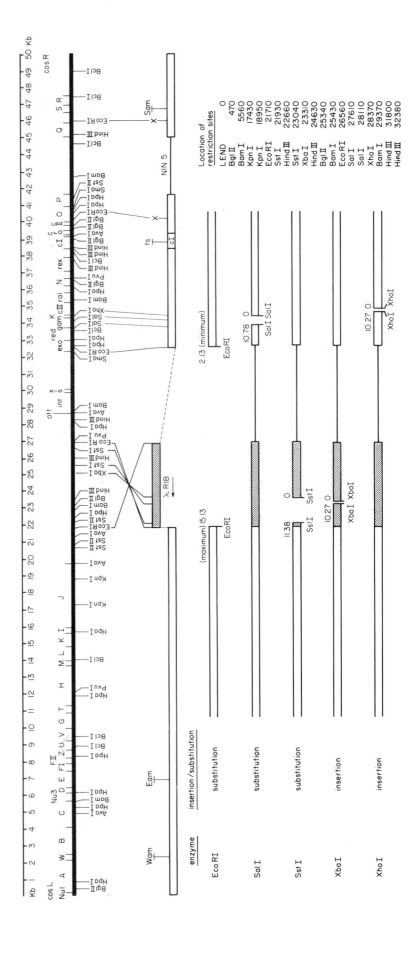

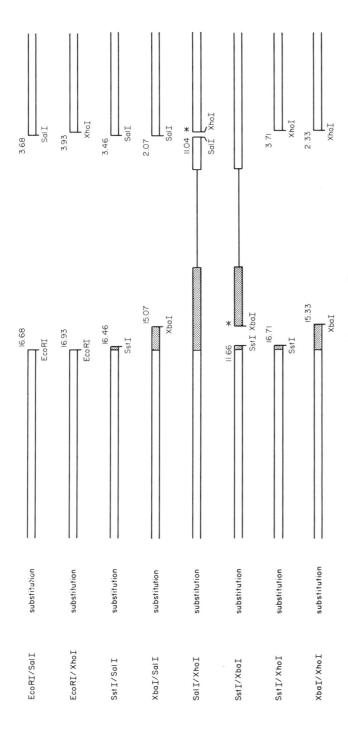

λgt wes·λB : Total length: 40.4 Kb

Genetic markers: Wam 403, Eam 1100 and Sam 100: CI 857 ts

Requires host bacteria carrying sup F

Reference: Leder, P., Tiemeier, D. and Enquist, L. (1977) Science 196: 175

Comment: This vector has been used extensively to clone intermediate-sized EcoRI fragments. It was derived from the original λgt·λB of Struhl et al (1976).

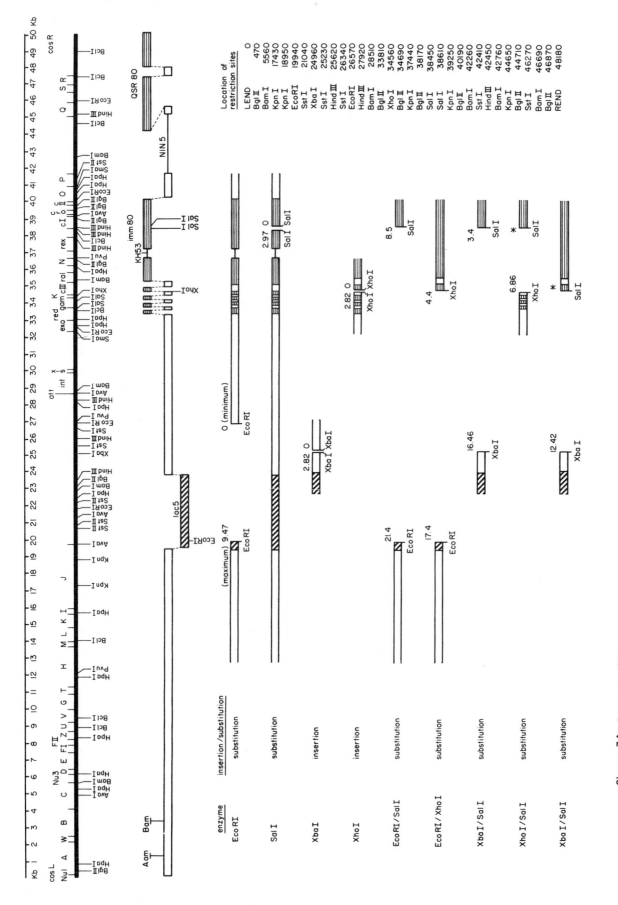

Charon 3A: Total length: 48.3 Kb

Genetic markers: Aam32, Bam1, lac⁺, imm 80, Requires host bacteria carrying suⅡ.

Forms blue plaques on Xgal plates.

References: Blattner, F. et al. (1977) Science 196:161; Williams, B. and Blattner, F. (1979) J. Virol. 29: 555
de Wet, J. et al. (1980) J. Virol.33: 401

Comment: This vector has been used to construct libraries of complete EcoRI digests of eukaryotic DNA.

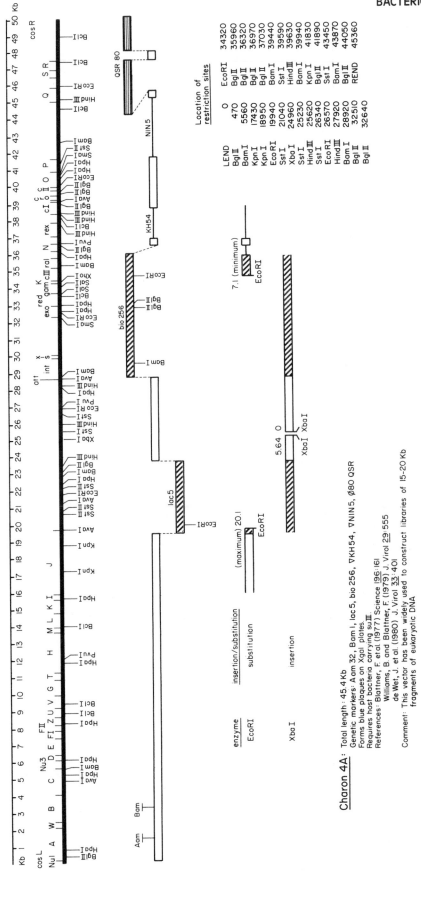

Location of restriction sites

LEND	BglII	0
	BglII	470
	BamI	5560
	KpnI	17430
	KpnI	18950
	EcoRI	19940
	SstI	21040
	XbaI	24960
	SstI	25230
	HindIII	25620
	SstI	26340
	EcoRI	26570
	HindIII	27920
	BamI	28920
	BglII	32510
	BglII	32640
	EcoRI	34320
	BglII	35960
	BglII	36320
	BglII	36970
	BamI	37030
	BamI	39440
	SstI	39590
	HindIII	39630
	BamI	39940
	KpnI	41830
	BglII	41890
	SstI	43450
	BamI	43870
	BglII	44050
	REND	45360

enzyme	insertion/substitution
EcoRI	substitution
XbaI	insertion

Charon 4A: Total length: 45.4 Kb
Genetic markers: Aam 32, Bam I, lac 5, bio 256, ∇KH54, ∇NIN5, Ø80 QSR
Forms blue plaques on Xgal plates.
Requires host bacteria carrying su III.
References: Blattner, F. et al. (1977) Science 196:161
Williams, B. and Blattner, F. (1979) J. Virol 29:555
de Wet, J. et al. (1980) J. Virol 33:401
Comment: This vector has been widely used to construct libraries of 15-20 Kb
fragments of eukaryotic DNA

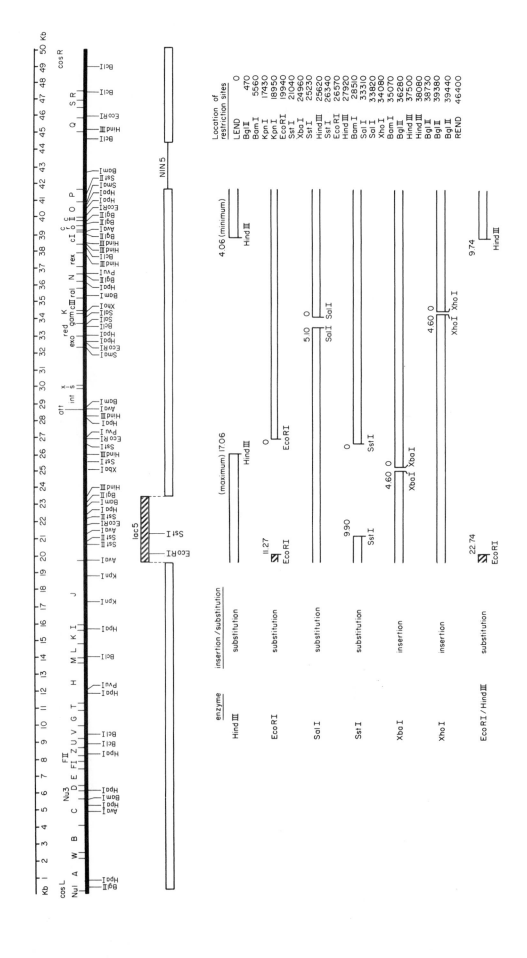

Charon 17 : Total length: 46.4 Kb
Genetic markers: lac5, NIN5, cIam, Fec⁻ Forms blue plaques on plates containing Xgal.
Forms lysogens in hosts carrying suⅡ or suⅢ
Reference: Williams, B.G. and Blattner, F.R. (1979) J. Virol. 29:555

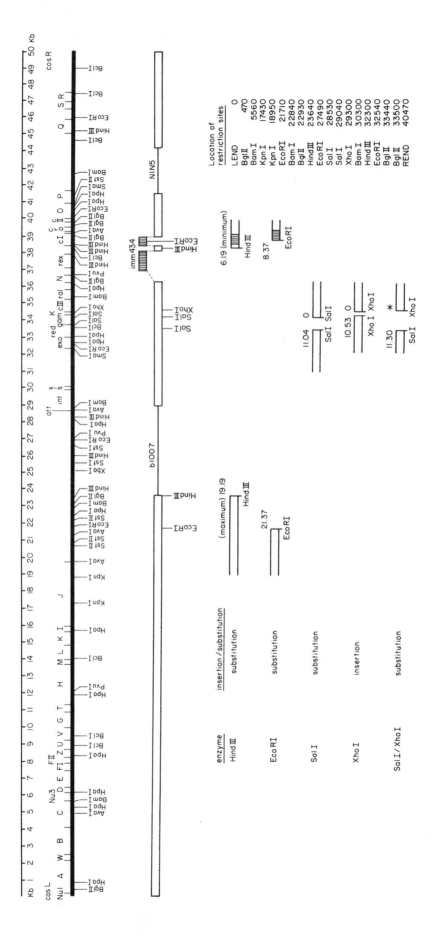

Charon 20 : Total length : 40.36 Kb
Genetic markers : b1007, NIN 5, imm 434
Reference : Williams, B. and Blattner, F. (1979) J Virology 29 : 555

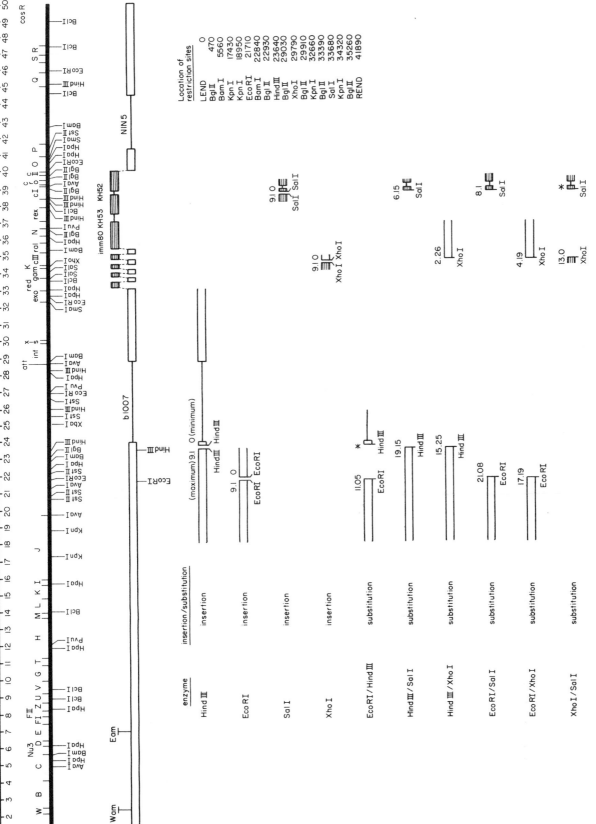

Charon 21A: Total length : 41.7 Kb

Genetic markers: Wam 403, Eam 1100, imm 80, ▽b1007, ▽KH53, ▽KH52, ▽NIN5

Requires host bacterium carrying supE or supF

Reference: Blattner, F. et al. Application for certification of a host-vector system for DNA cloning supplement IX data on Charon 21A.

Comment: This bacteriophage has been widely used as a vector for Hind III fragments. Because it contains amber mutations, it is suitable for use in recombination screening with πVX.

Location of restriction sites

LEND	0
Bgl II	470
Bam I	5560
Kpn I	17430
Kpn I	18950
EcoRI	21710
Bam I	22840
Bgl II	22930
Bgl II	23640
Hind III	29030
Xho I	29790
Bgl II	29910
Kpn I	32660
Bgl II	33390
Sal I	33680
Kpn I	34320
Bgl II	35260
REND	41890

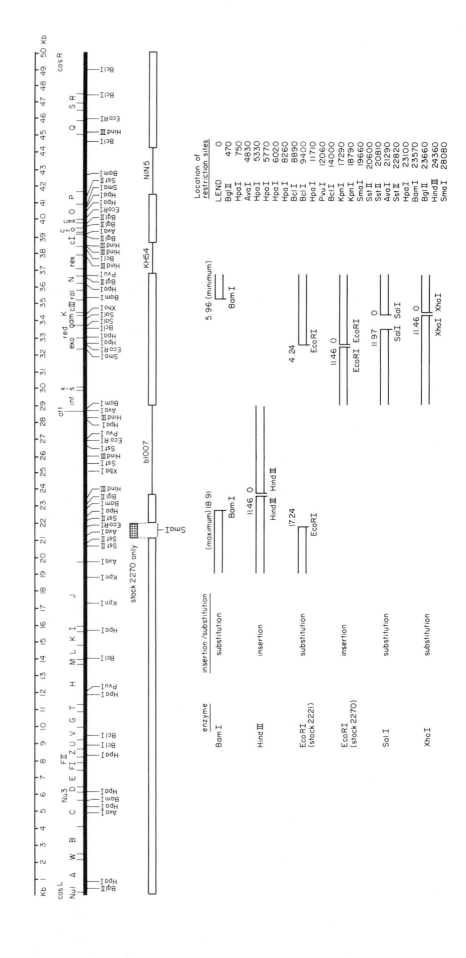

EcoRI	28210	
HpaI	28280	
HpaI	26680	
BclI	29200	
SalI	29240	
SalI	29740	
XhoI	30000	
BamI	31000	
HpaI	31760	
BglII	32220	
PvuI	32280	
BglII	32500	
AvaI	32620	
BglII	33160	
BglII	33220	
HpaI	34020	
HpaI	34240	
SmaI	34300	
SstII	34830	
BclI	35310	
BclI	38080	
BclI	39650	
REND	40230	

EcoRI/BamI (stock 2221) substitution 20.05 EcoRI 7.05 BamI

EcoRI/BamI (stock 2270) substitution 14.30 EcoRI ~0.5 BamI

HindIII/SalI substitution 16.89 HindIII 3.86 SalI

EcoRI/SalI (stock 2221) substitution 18.75 EcoRI 5.75 SalI

EcoRI/SalI (stock 2270) substitution 13.21 EcoRI SalI *

EcoRI/XhoI (stock 2221) substitution 17.91 EcoRI 4.96 XhoI

EcoRI/XhoI (stock 2270) substitution 13.61 EcoRI XhoI *

SalI/XhoI substitution 12.30 SalI XhoI *

Charon 28 : Total length: stock 2221 - 39.39 Kb (contains an insertion of ~900 bp of unknown origin in the left arm)
stock 2270 - 40.29 Kb

Genetic markers: bl007, KH54, NIN5
Reference: Rimm, D et al. (1980) Gene 12: 301

Comment: Until recently, this was the only bacteriophage vector that could be used to clone fragments compatible with BamI termini. It has been used to construct libraries of partial MboI digestion products of eukaryotic DNA. Lin et al.(1980). The absence of chi sites in this vector leads to large variations in plaque size in recombinants (see text for discussion). More recently derived BamI vectors (eg. L47I or BF101) have largely superseded the use of Ch28 as a BamI vector.

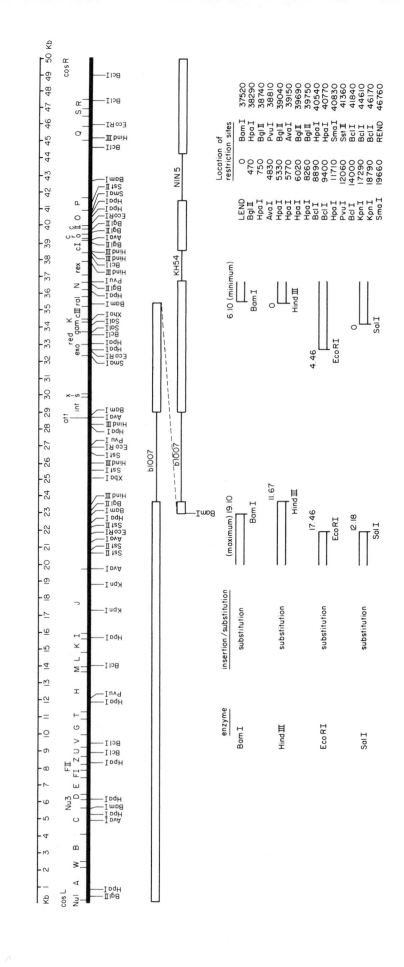

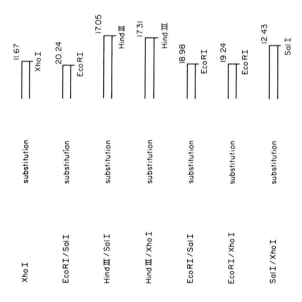

			Sst II	20600
			Sst II	20810
			Ava I	21290
			EcoRI	21520
			Sst II	21920
			Hpa I	22200
			Bam I	22670
			Bgl II	22760
			Hind III	23460
			Sma I	27180
			EcoRI	27310
			Hpa I	27380
			Hpa I	27780
			Bcl I	28300
			Sal I	28340
			Sal I	28840
			Xho I	29100
			Bam I	30100
			Bgl II	30190
			Hind III	30900
			Sma I	34600
			EcoRI	34740
			Hpa I	34810
			Hpa I	35210
			Bcl I	35730
			Sal I	35760
			Sal I	36270
			Xho I	36530

Charon 30 : Total length : 46.76 Kb
Genetic markers : bI007, KH54, NIN5, dup1 (sbh 2-3)
Reference : Rimm, D. et al (1980) Gene 12:301

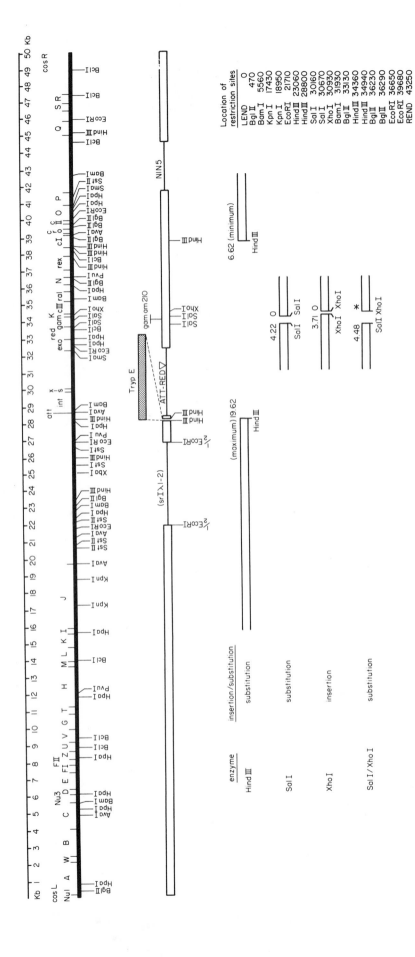

λ709 : Total length : 43.25 Kb
Genetic markers : gam am 210, tryp E, NIN 5 (sr Iλ1-2) ∇
Reference : Murray, N. et al. (1977) Molec. Gen. Genet. <u>150</u> : 53

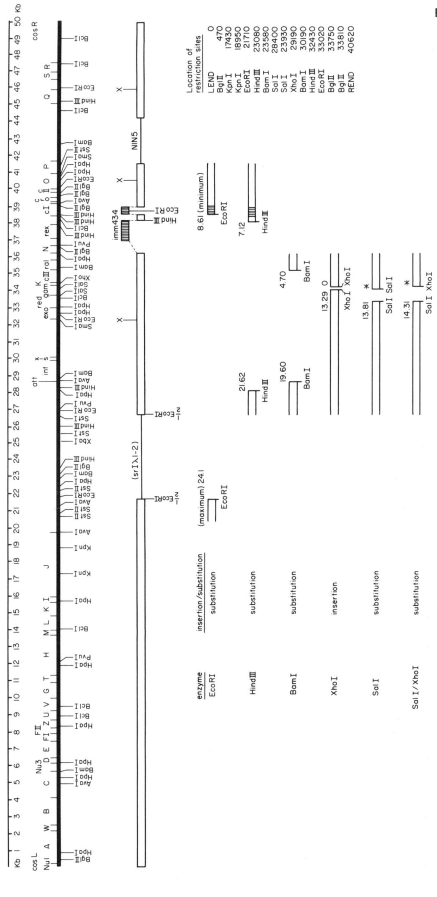

Location of
restriction sites

LEND	0
BglⅡ	470
KpnI	17430
KpnI	18950
EcoRI	21710
HindⅢ	23080
BamI	23580
SalI	28400
SalI	23930
XhoI	29190
BamI	30190
HindⅢ	32430
EcoRI	33020
BglⅡ	33750
BglⅡ	33810
REND	40620

enzyme	insertion/substitution
EcoRI	substitution
HindⅢ	substitution
BamI	substitution
XhoI	insertion
SalI	substitution
SalI / XhoI	substitution

λL47.1 : Total length : 40.6 Kb
Genetic markers : (srIλ1-2)▽, imm434 cI⁻, NIN5, chiA13I
Reference: Loenen, W.A. and Brammar, W.J. (1980) Gene 10 : 249
Comment: A highly versatile vector that can be used to clone large
EcoRI, HindⅢ, and BamHI fragments.

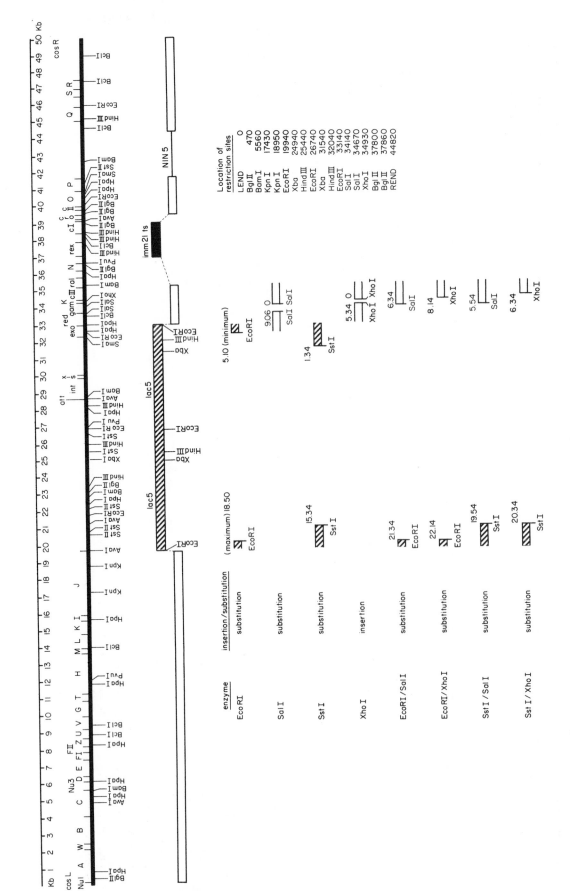

λ sep6 - lac5 : Total length : 44.3 Kb
Genetic markers : lac5, imm 21 ts, NIN 5
Reference : Meyerowitz and Rambach, unpublished
Comment : This vector has been used extensively by Hogness and his
coworkers to construct libraries of Drosophila DNA. A derivative
exists (λ sep6-lac5A) that carries amber mutations in the A and B genes (D. Goldberg, unpubl.).
Recombinants of both phages will grow on recA[+] hosts.

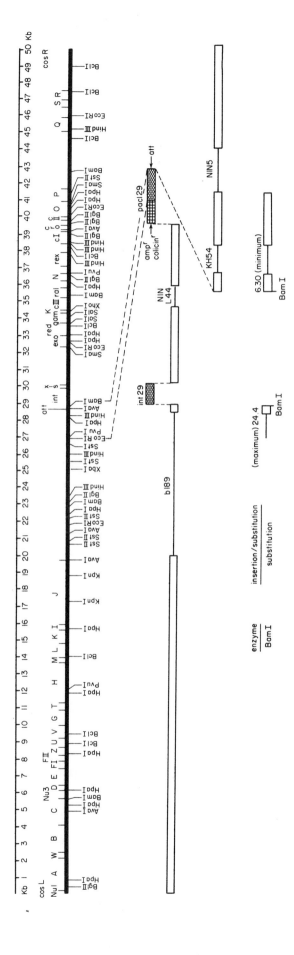

λ 1059 : Total length : 44.0 Kb

Genetic markers : hλ sBαm I° bl89 (int29, NIN L44, cI857, pacI 29) Δ (int–cIII) KH54 sRI 4° NIN5 chi3

Replacement of the stuffer fragment (which carries red and gam) with foreign DNA causes the
phenotype of the phage to change from spi⁺ to spi⁻

Reference : Karn et al. Proc. Natl. Acad. Sci. US 71 : 5172 (1980)

Map of λ1059

This is a vector for large *Bam*HI fragments, in which replacement of the stuffer fragment (which carries the *red* and *gam* genes) with foreign DNA causes the phenotype of the phage to change from Spi⁻ to Spi⁻. The presence of the pBR322 origin of DNA replication allows the vector to be propagated as a plasmid (phasmid) in λ lysogens. However, the presence of this pBR322 sequence makes it difficult to screen libraries by using as probes fragments derived from recombinant plasmids. Even after gel purification, such fragments inevitably contain small amounts of pBR322 sequences that hybridize to plaques formed by nonrecombinant phages.

Another difficulty with this vector is the occurrence of spontaneous rearrangements, which lead to the appearance in the parental phage stock of Spi⁻ derivatives (0.1%).

The Spi⁻ derivatives are proposed to be the result of *int*-mediated deletions from the *att* site into the *spi* region. The absence of spontaneous Spi⁻ phage in stocks of the *int*⁻ derivative of λ1059 (λBF101) supports this interpretation.

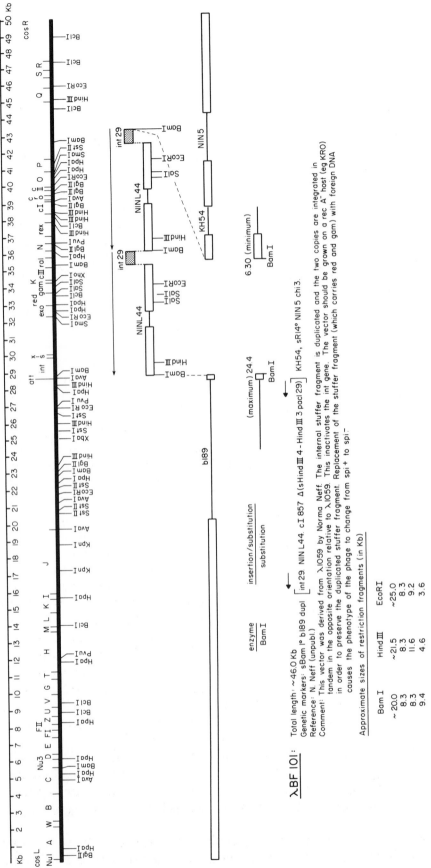

λBF 101:

Total length: ~46.0 Kb

Genetic markers: sBam I° b189 dupl [int 29 NIN L44. cI 857 Δ(sHind III 4 - Hind III 3 pac 29]

Reference: N. Neff (unpubl.)

Comment: This vector was derived from λ1059 by Norma Neff. The internal stuffer fragment is duplicated and the two copies are integrated in tandem in the opposite orientation relative to λ1059. This inactivates the int gene. The vector should be grown on a rec A⁻ host (eg KRO) in order to preserve the duplicated stuffer fragment. Replacement of the stuffer fragment (which carries red and gam) with foreign DNA causes the phenotype of the phage to change from spi⁺ to spi⁻

Approximate sizes of restriction fragments (in Kb)

BamI	HindIII	EcoRI
~20.0	~21.5	~25.0
8.3	8.3	8.3
8.3	11.6	9.2
9.4	4.6	3.6

Map of λBF101

This phage is a derivative of λ1059, which was altered in two ways. First, the *Hind*III fragment containing the pBR322 sequence was removed to overcome the problem of detection by hybridization of nonrecombinant phages (see λ1059 map). Second, the resulting *Bam*HI stuffer fragment was inverted and duplicated, thereby inactivating the *int* gene (compare the maps of λ1059 and λBF101). This inactivation of the *int* gene was carried out to overcome the spontaneous background of Spi⁻ phages in the parental vector stock (see λ1059 map). Stocks of this phage should be grown on the *recA*⁻ strain KRO to prevent homologous but unequal crossing over between the duplicated *Bam*HI fragments.

COSMIDS

The construction of libraries in bacteriophage λ vectors has proven to be an effective means of isolating specific segments of DNA from complex eukaryotic genomes. However, the limited size capacity (<23 kb) of λ cloning vectors is a disadvantage in some situations: Some genes are too large to be cloned in a single piece of DNA. For example, the α-2 collagen gene of chickens, approximately 38 kb in length (Vogeli et al. 1980; Wozney et al. 1981), was isolated as a series of overlapping genomic clones (Ohkubo et al. 1980). Furthermore, the laborious process of walking along a chromosome by stepping from one overlapping clone to another could be facilitated by the ability to clone larger fragments of DNA.

Cosmids, which were first developed by Collins and Hohn (1978), are vectors specifically designed for cloning large fragments of eukaryotic DNA. The essential components of cosmid cloning vectors are: (1) a drug-resistance marker and a plasmid origin of replication; (2) one or more unique restriction sites for cloning; (3) a DNA fragment that carries the ligated cohesive end (cos) site of bacteriophage λ; and (4) a small size, so that eukaryotic DNA fragments up to 45 kb in length can be accommodated.

The basic principles of cloning in cosmids are shown in Figure 1.9. A ligation mixture is set up containing high concentrations of cosmid DNA and foreign DNA that have been digested with a restriction endonuclease. Among the ligation products will be catenates in which the foreign DNA has become linked to cosmid molecules, such that both the cos sites are in the same orientation and the entire complement of plasmid information is contained between the two cos sites. When such molecules are incubated in an in vitro λ packaging reaction (see pages 301, 303), the cos sites flanking the foreign DNA are cleaved by the ter function of the bacteriophage λ A protein, and the intervening DNA is packaged into mature bacteriophage particles. During infection of E. coli by such bacteriophage particles, the recombinant DNA is injected into the cell and circularizes via the phage cohesive ends. Because the recombinant contains a complete copy of the cosmid genome, it replicates as a plasmid and expresses the drug-resistance marker.

Packaging the recombinant molecules into bacteriophage particles selects for recombinants with a total size (i.e., cosmid plus foreign DNA) of 40–50 kb. Optimally designed cosmid vectors are 4–6 kb in length (Hohn and Collins 1980; Meyerowitz et al. 1980; Ish-Horowicz and Burke 1981), and they can therefore accommodate up to 45 kb of foreign DNA.

Despite this advantage in size, cosmids have not yet replaced bacteriophage λ as the vector of choice for the construction of libraries of eukaryotic genomic DNA. This delay in acceptance has stemmed from a series of technical difficulties that have only recently been solved. These include:

1. vector-to-vector ligation; intramolecular recombination between two or more vector molecules within a single cosmid can lead to excision of a

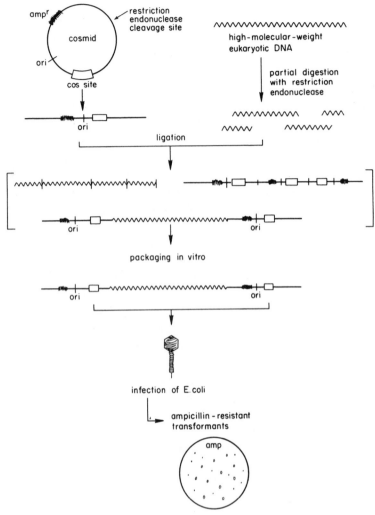

Figure 1.9

Cloning in cosmids.

faster-replicating cosmid vector and eventually to loss of the cosmid by segregation.

2. "scrambling" caused by the insertion into the same vector molecule of two or more fragments that were not adjacent to one another in the original genome;

3. difficulties in screening large numbers of bacterial colonies.

The first problem can be dealt with by treating linearized plasmid DNA with alkaline phosphatase, which prevents vector DNAs from ligating to each other (Meyerowitz et al. 1980; Grosveld et al. 1981). By using this procedure, Meyerowitz et al. (1980) were able to construct a library of *Drosophila melanogaster* DNA in the 4.4-kb cosmid MUA-3.

The second problem can be minimized by selecting fragments of eukaryotic DNA of a certain size (30–45 kb) for ligation to the cosmid vector. Insertion of two or more such fragments into the same cosmid will result in a molecule too large to be packaged into a bacteriophage λ particle.

Recently, Ish-Horowicz and Burke (1981) have described another method for cosmid cloning that overcomes the first two of these problems. In this procedure, which is designed for the vector pJB8 and illustrated in Figure 1.10, the cosmid vector is divided into two aliquots, each of which is cleaved with a different restriction enzyme cutting either to one side or the other of the *cos* sequence. The resulting full-length, linear DNAs are dephosphorylated with alkaline phosphatase. It is this dephosphorylation that prevents formation of tandem vectors and suppresses the background of bacterial colonies containing cosmids lacking inserts. The dephosphorylated DNAs are then both cleaved with *Bam*HI and mixed to generate cohesive ends, which can be ligated to eukaryotic DNA prepared by partial cleavage with *Mbo*I or *Sau*3A followed by dephosphorylation. This protocol gives only one

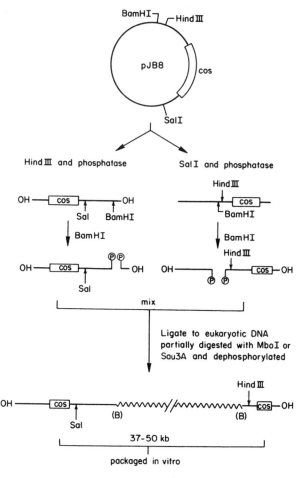

Figure 1.10

Efficient cloning in cosmids (Ish-Horowicz and Burke 1981).

kind of packageable molecule, which contains the "left-hand" *cos* fragment, an insert of 32–42 kb, and the "right-hand" *cos* fragment. The pJB8 vector has the additional advantage of containing two *Eco*RI sites closely flanking the *Bam*HI site, which allow nearly precise excision of the inserted DNA. However, the technique can easily be adapted to other existing cosmid vectors (see pages 49–50). The efficiency of the method is high, yielding more than 5×10^5 clones per microgram of insert *D. melanogaster* DNA (Ish-Horowicz and Burke 1981).

The third problem, the difficulty of screening large numbers of bacterial colonies for the presence of desired DNA sequences, has been circumvented by the high-density, colony-screening technique developed by Hanahan and Meselson (1980) (see page 318).

The feasibility of cloning large fragments of DNA in cosmids was first established by Royal et al. (1979), who were able to isolate from a cosmid library of chicken DNA two overlapping fragments that together span more than 46 kb of DNA and contain the ovalbumin gene and two ovalbuminlike genes. More recently, cosmid DNA libraries of *Drosophila* DNA (Meyerowitz et al. 1980; Ish-Horowicz and Burke 1981), mouse DNA (Cattaneo et al. 1981), and human DNA (Grosveld et al. 1981) have been prepared. Furthermore, the utility of human cosmid libraries has been demonstrated by the isolation of a series of overlapping fragments containing the β-globin gene cluster (Grosveld et al. 1981) and the α-globin gene cluster (P. Charney, unpubl.). The latter recombinants, isolated using the most recent cosmid-cloning procedures, appear to be stable during propagation in bacteria.

In summary, it seems that cloning in cosmids has been developed to a point where it is the method of choice for certain specialized purposes (e.g., isolation of large genes or chromosome-walking experiments). However, it seems likely that bacteriophage λ, because of its greater versatility and demonstrated effectiveness, will continue to be the vector of choice for routine construction of libraries of eukaryotic DNA.

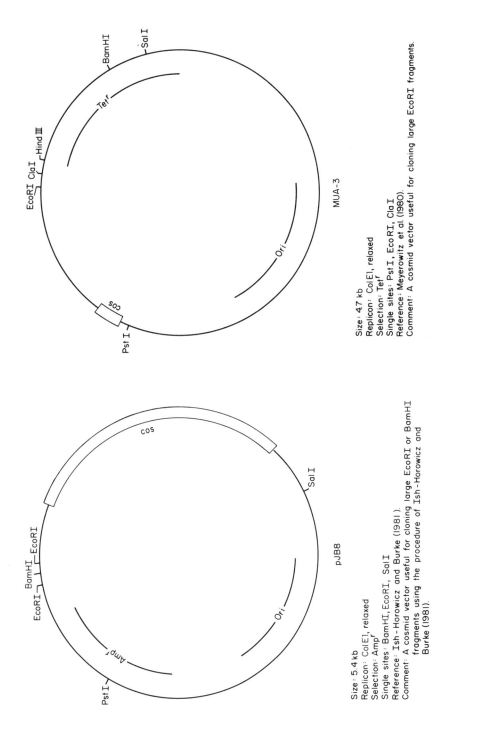

MUA-3

Size: 4.7 kb
Replicon: ColE1, relaxed
Selection: Tet^r
Single sites: Pst I, EcoRI, Cla I
Reference: Meyerowitz et al. (1980).
Comment: A cosmid vector useful for cloning large EcoRI fragments.

pJB8

Size: 5.4 kb
Replicon: ColE1, relaxed
Selection: Amp^r
Single sites: BamHI, EcoRI, Sal I
Reference: Ish-Horowicz and Burke (1981).
Comment: A cosmid vector useful for cloning large EcoRI or BamHI
fragments using the procedure of Ish-Horowicz and
Burke (1981).

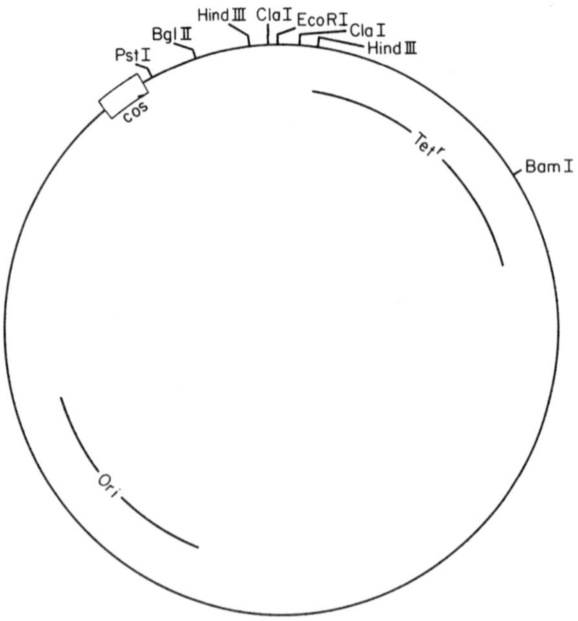

Kos1

Size: 40 kb
Replicon: col El, relaxed
Selection: Tet^r
Single sites: EcoRI, BglII, PstI, HindIII, ClaI
Reference: P.F.R. Little (unpubl.)
Comments: A cosmid vector useful for cloning large restriction fragments into the BglII site. The inverted repeat near to the EcoRI site leads to small deletions at this region in a small percentage (<5%) of the molecules.

SINGLE-STRANDED BACTERIOPHAGES

By far the best-developed, single-stranded DNA bacteriophage vectors are those derived from M13. M13 is a filamentous bacteriophage with a closed circular DNA genome approximately 6500 nucleotides in length (Denhardt et al. 1978). The phages attach to F-pili of *E. coli* and are therefore able to infect only male bacterial cells (F′ or Hfr strains). Following penetration, the single-stranded phage DNA is converted into a double-stranded replicative form (RF), which can be isolated from cells and is used as a double-stranded DNA cloning vector. After 100–200 copies per cell of the RF have accumulated, M13 synthesis becomes assymmetric, producing large amounts of only one of the two DNA strands, which is eventually incorporated into mature bacteriophage particles. The particles are continually extruded from the infected cells. Although the cells are not killed by M13 infection, their growth is somewhat inhibited, and the virus therefore produces turbid plaques on lawns of bacterial cells.

The primary advantage of M13 as a cloning vehicle is that the phage particles released from the cell contain single-stranded DNA that is homologous to only one of the two complementary strands of the cloned DNA and can therefore be used as template for DNA sequencing by the Sanger dideoxy-sequencing method (Sanger et al. 1977). M13 vectors and specific primer DNAs (see below) have been developed to allow sequencing of up to 350 bases from a single clone (Sanger et al. 1977; Anderson et al. 1980). Rapid sequencing of longer stretches of DNA is accomplished by sequencing overlapping cloned fragments (Gingeras et al. 1979; Anderson 1981). Additionally, single-stranded M13 clones can be used to generate single-stranded DNA probes, to select RNA, and to serve as substrates for in vitro mutagenesis (Itakura and Riggs 1980).

The major problem encountered in M13 cloning has been instability of large DNA inserts. In general, cloned sequences greater than 1000 nucleotides are unstable and give rise to deletions during propagation of bacteriophages. However, 300–400-nucleotide inserts are sufficiently stable to carry out the applications described above.

Bacteriophage M13 Vectors

Except for a 507-nucleotide region known as the intergenic sequence (IS), all of the M13 genome contains genetic information essential for virus replication. The IS region has been shown to accept inserts of foreign DNA without affecting phage viability. In the M13 derivatives constructed as cloning vehicles, two types of sequence have been inserted into this region.

(1) The first is a fragment of the *E. coli lac* operon containing the regulatory region and the coding information for the first 146 amino acids of the β-galactosidase (Z) gene. The aminoterminal portion of the β-galactosidase

protein produced in the infected cells is able to complement ("α complementation") a defective β-galactosidase gene present on the *F* episome in the host cell. This complementation produces active β-galactosidase, which gives rise to a blue color when the phage and cells are grown in the presence of the inducer isopropyl-thiogalactoside (IPTG) and the chromogenic substrate Xgal.

(2) The second type of sequence is a small DNA fragment ("polylinker") containing several unique restriction sites for cloning that has been inserted into the aminoterminal portion of the β-galactosidase gene. This insertion does not affect the ability of the β-galactosidase peptide to complement the β-galactosidase mutant. However, insertions of additional DNA into the region destroy the complementation. Phages that contain inserts give rise to colorless plaques when grown in the presence of IPTG and Xgal. The canonical M13 cloning vector is shown in Figure 1.11. M13 cloning vectors differ primarily in the restriction enzyme sites found in the polylinker. Table 1.2 presents the cloning sites available in the vectors presently being used.

Specific, small DNA fragments, which serve as primers for sequencing cloned DNA, have also now been subcloned onto plasmids (Anderson et al. 1980) or synthesized chemically (Rothstein et al. 1980) and are available commercially. These primers are homologous to the M13 vector in the portion of the β-galactosidase gene immediately adjacent to the polylinker region.

SUMMARY

As stated at the beginning of this chapter, the four types of vector—plasmids, bacteriophage λ, cosmids, and the single-stranded bacteriophages—have particular biological and physical properties that make them suitable for different cloning purposes. The salient properties and chief uses of each of the four vectors are summarized in Table 1.3.

The large size of foreign DNA segments accepted by cosmids and various λ bacteriophages make them the vectors of choice for constructing and propagating libraries of eukaryotic genomic DNA. However, they are comparatively clumsy vehicles for detailed analysis and manipulation of small segments of DNA, and as presently designed, they are unsuitable for the construction of cDNA libraries.

Single-stranded DNA cloning vehicles are used chiefly as sources of templates for sequencing by the Sanger dideoxy-sequencing, chain-termination technique and as sources of strand-specific probes for nucleic acid hybridization. The relative instability of DNA inserts larger than about 1 kb effectively eliminates the usefulness of single-stranded bacteriophages for most other cloning purposes.

Plasmids find extensive use in a wide variety of cloning procedures. Their ability to accept moderately sized segments of DNA makes them the vectors of choice for subcloning from libraries of genomic DNA constructed in cosmids or bacteriophage λ. Their great assortment of useful restriction enzyme sites is invaluable during the assembly of novel DNA constructs.

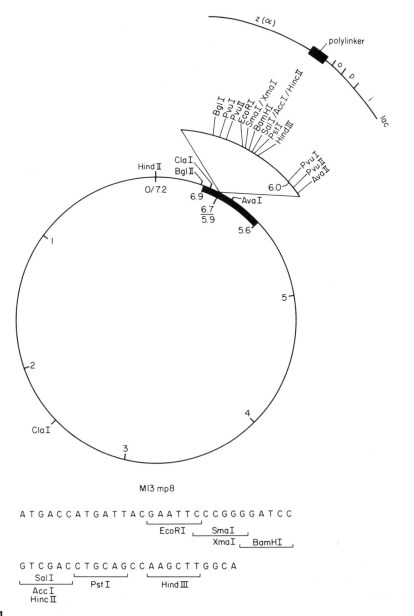

Figure 1.11

This M13 cloning vector (mp8), developed by J. Messing, contains a "polylinker," making it possible to clone a variety of restriction-endonuclease-generated DNA fragments. Insertion of a foreign DNA fragment into the polylinkers inactivates the β-galactosidase gene (Z) and facilitates a simple screening procedure for phage containing inserts (inactivation of α-complementation).

They are the only vectors suitable for cloning of cDNA. Finally, a large number of plasmids have been designed in which coding segments of DNA can be harnessed to powerful bacterial promoters. Such vectors, whose properties are described in detail in Chapter 12, are used to obtain expression of foreign genes in *E. coli*.

A list of the bacterial strains used in this manual is given in Appendix C.

TABLE 1.2. M13 VECTORS AND CLONING SITES

Vector	Cloning sites	Reference
mp2	*Eco*RI[1]	Gronenborn and Messing 1978
mp5	*Eco*RI-*Hind*III-*Eco*RI[2]	J. Messing, pers. comm.
mp7	*Eco*RI-*Bam*HI-*Sal*I-PstI-*Sal*I-*Bam*HI, *Eco*RI	Messing et al. 1981
mp8	*Eco*RI-*Sma*I-*Bam*HI-*Sal*I-*Pst*I-*Hind*III	J. Messing, pers. comm.
mp9	*Hind*III-*Pst*I-*Sal*I-*Bam*HI-*Sma*I-*Eco*RI[3]	J. Messing, pers. comm.

[1]This site is not due to insertion of a linker sequence but due to a chance mutation that created an *Eco*RI site at codon position 5 of the β-galactosidase gene.

[2]The structure of the cloning site in mp5 is more complex than is shown here. It consists of several linkers arranged in tandem and flanked by at least two *Eco*RI linkers on each side.

[3]The polylinker in mp9 is in the opposite orientation to mp8.

TABLE 1.3. CHIEF USES OF DIFFERENT VECTORS IN CLONING

	Plasmids	Bacteriophage λ	Cosmids	Single-stranded bacteriophages
Cloning large DNA fragments	±[1]	+	+	−
Construction of genomic libraries	−	+	+	−
Construction of cDNA libraries	+	−	−	−
Routine subcloning	+	−	−	−
Building new DNA constructs	+	−	−	−
Sequencing	+	−	−	+
Single-stranded probes	−[2]	−	−	+
Expression of foreign genes in *E. coli*	+	−	−	−

[1]Although there is no defined upper limit to the size of DNA that can be cloned in plasmid vectors, the transformation rate and the yield of DNA of recombinant plasmids that contain more than 10 kb of inserted DNA are very low.

[2]Hayashi et al. (1980) have developed a plasmid vector that contains a tract of poly(dA) on one DNA strand and a tract of poly(dT) on the other. After cloning in this vector, the strands of linearized, recombinant DNA can be separated by chromatography on oligo(dT)- and oligo(dA)-cellulose.

2

Propagation and Maintenance of Bacterial Strains and Viruses

The microbiological procedures required for molecular cloning are quite simple and should present no problems to anyone with training in basic sterile techniques. The two difficulties most commonly encountered are cross-contamination of strains and loss or gain of genetic markers. Both of these problems can be minimized by colony or plaque purification, followed by verification of the genotype of the strain. Serial passaging of strains should be avoided by preparing working stocks from verified master cultures kept in long-term storage.

Even in the best-run laboratories, there is always a possibility of contamination, and it is important to check for bacterial colonies with an altered morphology, color, or odor, or for bacteriophage plaques with an unusual appearance or size. Such plates, as well as any that become contaminated with mold or other fungi, should be sealed, autoclaved, and discarded.

STRAIN VERIFICATION

When a new strain of bacteria or bacteriophage is received, it should be plated out and single colonies or plaques picked. Small stocks should then be prepared and the appropriate tests performed to verify the genotype of the strain. A full description of how these tests should be carried out on bacteria may be found in Miller (1972). Appropriate genetic tests for bacteriophages are usually given in the publication describing the development and construction of the particular strain. Table 2.1 contains a list of genetic markers that should be tested frequently. The structure of vector DNAs should be confirmed by digesting small-scale preparations with the appropriate restriction enzymes.

TABLE 2.1. GENETIC MARKERS

Marker	Assay
Auxotrophic markers (e.g., requirement for thymidine, diaminopimelic acid, biotin)	the ability of cells to grow on minimal media with and without the nutrient or supplement
Metabolic markers (e.g., ability to ferment lactose, galactose)	the ability of cells to grow on minimal media containing the appropriate sugar as the sole carbon source; or the ability to form colored colonies on media containing the appropriate dyes or chromogenic substrates
Drug resistance (e.g., tetracycline, ampicillin, kanamycin, streptomycin, nalidixic acid)	the ability of cells to grow on media containing one or more of the drugs
$supE$ and $supF$ (suppressors for amber mutations)	the ability to allow plaque formation by bacteriophages requiring $supE$, $supF$, or both suppressors
Amber mutations in bacteriophages	the ability to form plaques only on bacteria carrying $supE$ or $supF$ suppressors; complementation tests with bacteriophages known to carry amber mutations in particular genes may be used to verify the location of the amber mutation under test

ISOLATION OF SINGLE COLONIES

STREAKING TECHNIQUE

1. Sterilize a platinum transfer loop by flaming until it glows light red. Allow it to cool in air. (The loop may be cooled more rapidly by dipping it into sterile water or medium or by stabbing it into sterile agar medium.)

2. Use the cooled loop to pick up bacteria. Touch the loop to a single, well-isolated bacterial colony growing on the surface of solid medium. Transfer the bacteria that adhere to the loop to a sterile tube containing 1 ml of liquid medium. Vortex briefly to ensure that clumps of bacteria are dispersed.

3. Sterilize the loop by flaming, cool it, and dip it into the bacterial suspension. Streak the bacteria adhering to the loop onto a segment of a plate containing agar medium (see Fig. 2.1). Sterilize the loop by flaming and cool it by stabbing into a region of the agar medium that is free of bacterial cells. Pass the loop once across one end of the primary streak and spread the bacteria that adhere to the loop into a fresh region of the agar medium.

4. Sterilize and cool the loop again and streak from one end of the secondary streak.

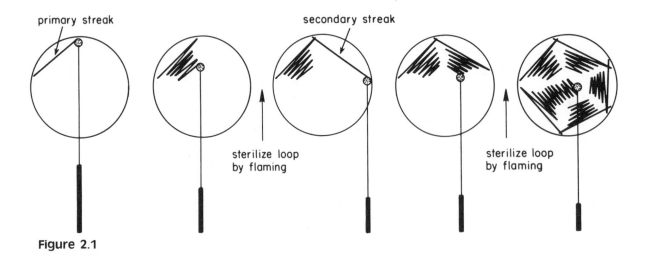

Figure 2.1

5. Repeat step 3 twice more, serially.

6. Replace the lid on the plate. Label the bottom of the plate and incubate it in an inverted position at the appropriate temperature (usually 37°C) for 16–24 hours. Well-separated colonies should be visible in the area of the final streak.

Note

Because each colony is the progeny of a single bacterial cell, genetically pure cultures can be obtained by touching the colony with a sterile loop and then transferring to a plate, stab, or liquid culture. Colonies can also be transferred using a sterile pasteur pipette or a micropipette. The sterile pipette is stabbed through a bacterial colony into the underlying agar medium. The resulting agar plug is drawn together with the colony into the pipette by suction with a pipette bulb. The plug and bacteria are then blown into a sterile flask containing nutrient medium.

POURED-PLATE TECHNIQUE

1. Transfer bacteria with a sterile loop from an agar slant or liquid culture into a sterile tube containing 3.0 ml of sterile medium that contains 0.7% agar. Keep the tube in a heating block at 47°C to prevent the agar from solidifying.

2. Vortex briefly to ensure that the contents are mixed and that no clumps of bacteria remain. Flame and cool the loop and transfer a loopful from the first tube to a second tube containing soft agar medium.

3. Repeat step 2 two or three times, serially, and pour each tube of soft agar medium onto a separate plate containing a layer of hardened, bottom agar medium. Swirl the plate gently to ensure that the molten agar spreads over the entire surface. Replace the lid and label the plate.

4. After allowing the top agar to harden at room temperature, invert the plates and incubate them for 16–24 hours at a temperature appropriate for the bacterial strain.

5. One or more of the plates will contain individual colonies that can be picked as described above and used to start a pure bacterial culture.

SPREADING TECHNIQUE

1. Transfer bacteria from an agar slant or liquid culture to a tube containing 1 ml of liquid medium. Vortex thoroughly to disperse any clumps of bacteria.

2. Dilute the cells by using a sterile micropipette to transfer 10 μl from the first tube to a fresh tube containing 1 ml of medium.

3. Repeat step 2 serially two or three times.

4. Spot a small aliquot (10–50 μl) of each dilution in the center of a plate containing hardened agar medium. Spread the cells over the entire surface of the medium by moving a sterile, bent glass rod back and forth gently over the agar surface and, at the same time, rotating the plate by hand or using a rotating wheel. The glass spreader can be sterilized by dipping it into a beaker containing 95% ethanol and then holding it in the flame of a bunsen burner to ignite the ethanol. The spreader should then be cooled first in air and then by touching the surface of a plate of sterile agar medium.

Caution

Do not place the beaker of ethanol near the lighted bunsen burner. Do not return the hot glass rod to the beaker of ethanol.

GROWTH, MAINTENANCE, AND PRESERVATION OF BACTERIAL STRAINS

GROWTH OF BACTERIA

Under ideal conditions, *Escherichia coli* grows exponentially. The rate of growth is dependent on the medium, the genotype of the strain, the temperature, and the degree of aeration. As the density of the culture increases, the rate of division decreases until the bacteria reach a concentration (saturation density) at which they no longer divide but are viable.

The rate of growth of a bacterial culture is most conveniently monitored by withdrawing aliquots at various times and reading the optical density (OD) at a wavelength of 600 nm. The bacteria may also be grown in flasks fitted with specially designed side arms that fit directly into a densitometer.

As a rough guide,

$$1 \ OD_{600} \simeq 8 \times 10^8 \ \text{cells/ml.}$$

Because the exact relationship between the optical density and the number of bacteria varies a little from strain to strain of *E. coli*, it is advisable, whenever precision is required, to construct a calibration curve that relates the number of viable bacteria to the OD_{600} of the culture.

Aeration of cultures is best achieved by attaching a flask or tube to a rotary shaking platform or a rotating wheel. The volume of the vessel should be at least four times larger than the volume of medium it contains so as to allow vigorous shaking (rotary shaking platform, 250 rpm; rotating wheel, 100 rpm).

SHORT-TERM STORAGE

Colonies of most strains of bacteria can be maintained for periods of a few weeks on the surface of agar media if the plates are tightly wrapped in parafilm and stored inverted at 4°C.

MEDIUM-TERM STORAGE

Most strains of bacteria can be maintained for a year or two in stab cultures. Such cultures are usually prepared in small screw-capped bottles containing 2-3 ml of agar medium. The culture is inoculated using a sterile, straight platinum wire that is dipped into a dense liquid culture of bacteria and then stabbed deep into the agar medium. The cap to the container is kept loosened during overnight incubation of the stab at the appropriate temperature. The cap is then tightened and wrapped tightly in parafilm to prevent desiccation of the medium. The stab culture is stored in the dark at room temperature.

LONG-TERM STORAGE

Bacteria can be stored for many years in media containing 15% glycerol at low temperature without significant loss of viability.

1. Inoculate a culture flask containing 5–10 ml of liquid medium with a single bacterial colony and grow the culture overnight.

2. Transfer 0.85 ml of the overnight culture to a sterile vial containing 0.15 ml of sterile gylcerol. Place the cap on the vial and mix the contents thoroughly by vortexing.

3. The glycerinated cultures can then be stored at −20°C for a few years without loss of viability. Alternatively, the glycerinated suspension can be stored at −70°C. In this case, viable bacteria can be recovered many years after the initial freezing. Viable bacteria can be recovered by simply scratching the surface of the frozen stock with a sterile platinum loop or wire. The frozen suspension can then be returned to the freezer. Several vials of each strain should be frozen.

PLAQUE PURIFICATION OF BACTERIOPHAGE λ

PREPARATION OF PLATING BACTERIA

1. Inoculate a single bacterial colony into 50 ml of rich medium (e.g., NZCYM, LB), supplemented with 0.2% maltose, in a 250-ml flask and grow overnight. Bacteria grown in the presence of maltose adsorb bacteriophage λ more efficiently than if maltose were not present; the sugar induces the maltose operon, which contains the gene *(lamb)* coding for the λ receptor.

2. Centrifuge the cells at $4000g$ for 10 minutes at room temperature.

3. Discard the supernatant and resuspend the cell pellet in sterile 0.01 M $MgSO_4$ (0.4× the volume of the original culture). More consistent results are obtained if the resuspended cells are diluted to an appropriate density (usually $OD_{600} = 2$; i.e., $\sim 1.5 \times 10^9$ cells/ml).

4. Store at 4°C. The bacterial suspension may be used for up to three weeks. However, the highest plating efficiencies are obtained with fresh cells (0–2 days old).

STREAKING PROCEDURE TO OBTAIN BACTERIOPHAGE λ PLAQUES

1. Add 0.1 ml of the plating bacteria to 3.0 ml of soft agar medium at 47°C in a heating block. Pour the soft agar–bacterial suspension onto a plate containing NZCYM or LB bottom agar. To obtain a firmer agar surface, the plates may at this stage be placed at 4°C for 1 hour.

2. Stab a straight platinum wire into a bacteriophage stock or a plaque and then gently streak the wire over the surface of the hardened agar medium.

3. Invert the plate and incubate at 37°C; plaques will appear after about 8 hours of incubation.

PLATING BACTERIOPHAGE λ

1. Prepare 10-fold serial dilutions of bacteriophage stocks in SM. Dispense 0.1 ml of each dilution to be assayed into each of two 13-mm × 100-mm test tubes.

2. Add 0.1 ml of plating bacteria to each tube. Mix by shaking or vortexing. Incubate at 37°C for 20 minutes to allow the bacteriophage particles to adsorb.

3. Add 3.0 ml of medium (47°C) containing melted 0.7% agar or agarose to the first tube, vortex gently, and immediately pour onto a labeled plate containing 30–35 ml of hardened bottom agar medium. Try to avoid air bubbles. Swirl the plate gently to ensure an even distribution of bacteria and top agar/agarose. Repeat with each of the tubes.

4. Close the plates and let them stand for 5 minutes at room temperature to allow the top agar/agarose to harden. Invert the plates and incubate at 37°C. Plaques begin to appear after about 8 hours and should be counted or picked after 12–16 hours of incubation.

Note

Although there can be considerable variability, most plate stocks or liquid culture lysates of bacteriophage λ contain between 10^9 and 10^{11} phages/ml.

PICKING PLAQUES

1. Place 1.0 ml of SM in a 13-mm × 100-mm polypropylene tube. Add 1 drop of chloroform.

2. Using a pasteur pipette equipped with a rubber bulb, or a micropipette, stab through the chosen plaque into the hard agar beneath. Apply mild suction so that the plaque, together with the underlying agar, is drawn into the pipette.

3. Wash out the fragments of agar into the 1.0 ml of SM containing a drop of chloroform. Let stand at room temperature for an hour or two to allow the bacteriophage particles to diffuse out of the agar. An average plaque yields 10^6–10^7 infectious phage particles, which can be stored indefinitely at 4°C in SM/chloroform without loss of viability.

Note

Because bacteriophage λ can diffuse considerable distances through the top agar layer, choose well-separated plaques. For the same reason, it is advisable to pick plaques shortly after the bacterial lawn has grown up and the bacteriophage plaques have first appeared.

PREPARING STOCKS OF BACTERIOPHAGE λ FROM SINGLE PLAQUES

Two techniques are commonly used to prepare phage stocks from single plaques: (1) *plate lysates*, in which the phages are propagated in bacteria grown in soft agar, and (2) *small-scale liquid cultures*, in which the phages are grown in bacteria in liquid medium.

Although the virus yields from the two procedures are approximately equal, the first has the advantage that one can determine, merely by looking at the degree of confluence of the plaques, whether or not the bacteriophage has grown successfully.

Plate Lysate Stocks

To achieve maximum yield, the number of bacteriophages plated should be adjusted so that the outer edges of the expanding plaques just touch after approximately 12 hours of incubation. An inoculum of 10^5 plaque-forming units (pfu) is usually sufficient to produce confluent lysis on an 85-mm plate (567 cm²). By the end of the period of growth, there should be no patches of uninfected bacteria.

Method 1

1. Mix 10^5 pfu of bacteriophage (or $\frac{1}{20}$ of a resuspended plaque) with 0.1 ml of plating bacteria. Incubate at 37°C for 20 minutes.

2. Add 3.0 ml of melted top agar/agarose at 47°C, mix, and pour onto a labeled 85-mm plate containing 30 ml of hardened, bottom NZCYM or LB agar. Freshly poured plates give the best results, but older plates (1–4 days old) give satisfactory yields.

3. Invert and incubate the plate for 8–12 hours, until lysis is confluent.

4. Turn the plate over, add 5 ml of SM, and store the plate at 4°C for several hours with intermittent, gentle shaking.

5. With a pasteur pipette, harvest as much as possible of the SM. Add 1 ml of fresh SM and store the plate for 15 minutes in a tilted position to allow all the fluid to drain into one area. Again remove the SM and combine it with the first harvest. Discard the plate.

6. Add 0.1 ml of chloroform to the pooled SM, vortex briefly, and centrifuge at 4000*g* for 10 minutes at 4°C.

7. Recover the supernatant and add chloroform to 0.3%. The titer of bacteriophage (approximately 10^{10}/ml) usually remains unchanged as long as the stock is stored at 4°C.

Method 2

1. Mix 10^5 bacteriophages (or $^1/_{20}$ of a resuspended plaque) with 0.1 ml of plating bacteria. Incubate at 37°C for 20 minutes.

2. Add 3.0 ml of melted, *0.5%* top agar/agarose. Mix, and pour onto a freshly poured, labeled, 85-mm plate containing 30 ml of bottom agar.

3. Incubate the plates for 8–12 hours at 37°C *without* inversion.

4. When confluent lysis is achieved, gently scrape the soft top agar/agarose into a sterile centrifuge tube by using a sterile, bent glass rod.

5. Add 5 ml of SM to the plate to rinse off any remaining top agar/agarose and add it to the centrifuge tube.

6. Add chloroform (0.1 ml) and mix by rotation or shaking for 15 minutes at 37°C.

7. Centrifuge at 4000g for 10 minutes at 4°C.

8. Recover the supernatant, add chloroform to 0.3%, and store the stock at 4°C. The titer of the stock should be approximately 10^{10}/ml.

Note

Master stocks of important λ bacteriophages (e.g., vectors) whose genotype has been verified should also be stored at −70°C.

Add dimethylsulfoxide (DMSO) to the bacteriophage stock to a final concentration of 7% v/v. Mix gently. Plunge the container into liquid nitrogen. When the liquid has frozen, transfer the container to a freezer at −70°C for long-term storage.

To recover the bacteriophage, scrape the frozen surface of the liquid with a sterile, 18-gauge needle. Streak the needle over the surface of a plate containing indicator bacteria (see page 63) in order to obtain bacteriophage λ plaques.

Small-scale, Liquid Cultures[1]

1. Mix 0.1 ml of a fresh, overnight, bacterial culture with approximately 3×10^6 bacteriophages (about $\frac{1}{2}$–$\frac{1}{3}$ of a resuspended plaque) in a 15-ml, sterile, polypropylene tube.

2. Incubate at 37°C for 20 minutes to allow the bacteriophages to adsorb.

3. Add 4 ml of NZCYM or LB medium and incubate with vigorous shaking at 37°C for 6–12 hours, until lysis occurs. Shaking can be best accomplished by positioning the tube in an incubator shaker or in a rotating wheel.

4. After lysis has occurred, add chloroform to 1% and continue incubation for 15 minutes.

5. Centrifuge at 4000g for 10 minutes at 4°C.

6. Recover the supernatant, add chloroform to 0.3%, and store at 4°C. The titer of the stock should be approximately 10^{10}/ml.

[1]Leder et al. (1977).

MEDIA AND ANTIBIOTICS

LIQUID MEDIA

NZCYM Medium

Per liter:

NZ amine[2]	10 g
NaCl	5 g
casamino acids	1 g
Bacto-yeast extract	5 g
$MgSO_4 \cdot 7H_2O$	2 g

Adjust pH to 7.5 with sodium hydroxide.

NZYM Medium

Identical to NZCYM except that casamino acids are omitted.

LB (Luria-Bertani) Medium

Per liter:

Bacto-tryptone	10 g
Bacto-yeast extract	5 g
NaCl	10 g

Adjust pH to 7.5 with sodium hydroxide.

M9 Medium

Per liter:

Na_2HPO_4	6 g
KH_2PO_4	3 g
NaCl	0.5 g
NH_4Cl	1 g

[2]Type-A hydrolysate of casein from Humko Sheffield Chemical Division of Kraft, Inc., 1099 Wall St. West, Lynnhurst, NJ 07071.

Adjust pH to 7.4, autoclave, and then add:

1 M MgSO₄	2 ml
20% glucose	10 ml
1 M CaCl₂	0.1 ml

The above solutions should be sterilized separately by filtration (glucose) or autoclaving.

M9CA Medium

Identical to M9 medium except that 2.0 g/l of casamino acids are included.

χ1776 Medium

Per liter:

Bacto-tryptone	25 g
Bacto-yeast extract	7.5 g
1 M Tris·Cl (pH 7.5)	20 ml

Autoclave, cool, and then add:

1 M MgCl₂	5 ml
1% diaminopimelic acid	10 ml
0.4% thymidine	10 ml
20% glucose	25 ml

The magnesium chloride should be sterilized separately by autoclaving and the other ingredients should be sterilized separately by filtration.

SOB Medium

Per liter:

Bacto-tryptone	20 g
Yeast extract	5 g
NaCl	0.5 g

Adjust pH to 7.5 with potassium hydroxide and sterilize by autoclaving. Just before use, add 20 ml of 1 M MgSO₄, sterilized separately by autoclaving.

Concentrated Media

Some media (e.g., LB, NZYM, and NZCYM) can be prepared and stored as a 5× concentrate. Medium may then be prepared rapidly by diluting the 5× concentrate with the appropriate volume of sterile water.

Maltose

Maltose (0.2%) is often added to the medium during growth of bacteria that are to be used for plating bacteriophage λ.

maltose	20 g
H$_2$O	to 100 ml

Sterilize by filtration. Add 1 ml of sterile, 20% maltose solution for every 100 ml of medium.

SM

This medium is used for phage storage and dilution.

Per liter:

NaCl	5.8 g
MgSO$_4$ · 7H$_2$O	2 g
1 M Tris · Cl (pH 7.5)	50 ml
2% gelatin	5 ml

Sterilize by autoclaving and store in 50-ml lots.

Note

All of the liquid media above can be stored indefinitely at room temperature following sterilization.

Media Containing Agar or Agarose

Make up liquid media according to the appropriate formula given above. Just before autoclaving, add (per liter) one of the following:

Bacto-agar	15 g	(for plates)
Bacto-agar	7 g	(for top agar)
Agarose (type-1, low EEO)	15 g	(for plates)
Agarose	7 g	(for top agarose)

When preparing plates, media that contain agar or agarose should be sterilized by autoclaving and allowed to cool to 55°C before thermolabile substances (e.g., antibiotics) are added. Plates can then be poured directly from the flask, allowing about 30–35 ml per 85-mm petri dish. If bubbles are a problem, flame the surface of the medium with a bunsen burner before the agar hardens.

For most purposes, the plates should be stored at room temperature for 1–2 days before use to minimize problems of condensation and sweating. Excessive moisture causes streaking of bacteriophage plaques or bacterial colonies.

Top Agarose and Top Agar

There is no difference in the efficiency of plating bacteriophage λ in top agar and top agarose. However, for some purposes (e.g., transferring bacteriophage from plates to nitrocellulose filters for screening by hybridization), top agarose is preferred because it does not tear or stick to the filters as readily as does top agar.

It is convenient to make up a large batch of sterile top agar/agarose, which can be stored indefinitely at room temperature in 50-ml lots in tightly sealed 100-ml bottles. It can be melted just before use, either by immersion for about 10 minutes in a boiling-water bath or by 1–1.5 minutes of treatment in a microwave oven. In either case, loosen the bottle tops before heating. Transfer the bottle to a water bath at 47°C, and allow the top agar/agarose to cool for 20 minutes before use.

ANTIBIOTICS

Ampicillin

Prepare a 25 mg/ml solution of the sodium salt of ampicillin in water. Sterilize by filtration and store in aliquots at −20°C.

For plates: Allow autoclaved media to cool to 55°C before adding ampicillin to a final concentration of 35–50 μg/ml. Plates containing ampicillin may be stored at 4°C for only 1–2 weeks before use.

Chloramphenicol

Dissolve solid chloramphenicol in 100% ethanol at a concentration of 34 mg/ml. Chloramphenicol is stable for at least a year when prepared in this way and stored at −20°C.

For plates: Allow autoclaved media to cool to 55°C and add chloramphenicol to a final concentration of 10 μg/ml just before pouring. Store the plates at 4°C and use within 1–5 days.

Kanamycin

Prepare a 25 mg/ml solution of kanamycin sulfate in water. Sterilize by filtration and store in aliquots at $-20°C$.

For plates: Allow autoclaved media to cool to 55°C before adding kanamycin to a final concentration of 50 μg/ml.

Streptomycin

Prepare a 20 mg/ml solution of streptomycin sulfate in water. Sterilize by filtration and store in aliquots at $-20°C$.

For plates: Allow autoclaved media to cool to 55°C before adding streptomycin to a final concentration of 25 μg/ml.

Tetracycline

Prepare a 12.5 mg/ml solution of tetracycline hydrochloride in ethanol/water (50% v/v). Sterilize by filtration and store in aliquots at $-20°C$ in the dark (e.g., in containers wrapped in aluminum foil).

For plates: Allow autoclaved media to cool to 55°C before adding tetracycline to a final concentration of 12.5–15.0 μg/ml. Because tetracycline is light-sensitive, store plates in the dark at 4°C.

Note

Magnesium ions are antagonists of tetracycline. It is therefore best to use media that do not contain magnesium salts (e.g., LB) for growing bacteria in the presence of the antibiotic. If a richer medium is required, χ1776 medium containing sodium chloride (6 g/l) in place of magnesium chloride may be used.

TABLE 2.2. ANTIBIOTICS: MODE OF ACTION AND MECHANISM OF RESISTANCE

Antibiotic	Mode of action	Mechanism of resistance
Ampicillin (Ap)	a derivative of penicillin that kills growing cells by interfering with a terminal reaction in bacterial cell-wall synthesis	the resistance gene (*bla*) specifies a periplasmic enzyme, β-lactamase, which cleaves the β-lactam ring of the antibiotic
Chloramphenicol (Cm)	a bacteriostatic agent that interferes with bacterial protein synthesis by binding to the 50S subunit of ribosomes and preventing peptide bond formation	the resistance gene (*cat*) specifies an acetyltransferase that acetylates and thereby inactivates the antibiotic
Colicin E1 (ColE1)	a member of the general class of substances known as bacteriocins, which are lethal for strains of bacteria not carrying a resistance gene; colicin E1 effects lethal membrane changes in target bacteria	the resistance gene (*cea*) specifies a product that interferes with the action of the colicin in an unknown manner
Kanamycin (Km)	a bacteriocidal agent that binds to 70S ribosomes and causes misreading of messenger RNA	the resistance gene (*kan*) specifies an enzyme that modifies the antibiotic and prevents its interaction with ribosomes
Streptomycin (Sm)	a bacteriocidal agent that binds to the 30S subunit of ribosomes and causes misreading of the messenger RNA	the resistance gene (*str*) specifies an enzyme that modifies the antibiotic and inhibits its binding to the ribosome
Tetracycline (Tc)	a bacteriostatic agent that prevents bacterial protein synthesis by binding to the 30S subunit of ribosomes	the resistance gene (*tet*) specifies a protein that modifies the bacterial membrane and prevents transport of the antibiotic into the cell

3

Isolation of Bacteriophage λ and Plasmid DNA

Large-scale Preparation of Bacteriophage λ

Two methods are available for the preparation of large quantities of bacteriophage λ. The first involves infection of bacteria at low multiplicity. The infected culture is then inoculated into a large volume of medium. Initially, the concentration of bacteriophage is low, and uninfected cells in the culture continue to divide for several hours. However, successive rounds of infection lead to the production of increasing quantities of bacteriophage; eventually all the bacteria become infected and complete lysis of the culture occurs.

Care is required with this method because small changes in the ratio of cells to bacteriophage in the initial infection greatly affect the final yield of bacteriophage. Furthermore, the optimum ratio varies for different strains of bacteriophage λ and bacteria. However, with a little effort, the method can be adapted for use with most combinations of virus and host cells.

The second method involves induction of a lysogenic bacteriophage. This technique works most efficiently when the bacteriophage encodes a temperature-sensitive repressor (*cIts*). Lytic growth may then be induced by transiently raising the temperature of the growing bacterial culture. The method is straightforward but of course can only be used for bacteriophages that form stable lysogens. Among this group, however, are several useful strains; for example, *cI857ts*Sam7, which yields large amounts of bacteriophage DNA that may be used to prepare molecular weight markers, and λgt-λC, which has been used as a vector. Most other vectors in current use cannot form lysogens.

INFECTION

There are many variants of this method; however, in our hands, the following protocol gives the best results (Blattner et al. 1977; Maniatis et al. 1978).

To prepare bacteriophage from a 2-liter culture of infected bacteria:

1. Inoculate 100 ml of NZCYM in a 500-ml flask with a single colony of an appropriate bacterial host. Incubate overnight at 37°C with vigorous shaking.

2. Read the OD_{600} of the culture. Calculate the cell concentration assuming that 1 $OD_{600} = 8 \times 10^8$ cells/ml.

3. Withdraw four aliquots, each containing 10^{10} cells. Centrifuge at 4000g for 10 minutes at room temperature. Discard the supernatants.

4. Resuspend each of the bacterial pellets in 3 ml of SM.

5. Add bacteriophage and mix rapidly. The number of bacteriophages used is critical. For strains of bacteriophage λ that grow well (e.g., λgtWES-λB), 5×10^7 bacteriophages are added to each suspension of 10^{10} cells; for bacteriophages that grow relatively poorly (e.g., the Charon series), it is better to increase the starting inoculum to 5×10^8. However, there are no hard and fast rules, and you will probably need to experiment to find the multiplicity that gives the best results under your conditions.

6. Incubate at 37°C for 20 minutes with intermittent shaking.

7. Add each infected aliquot of 10^{10} cells to 500 ml of NZCYM, prewarmed to 37°C in a 2-liter flask. Incubate at 37°C with vigorous shaking. Concomitant growth of bacteria and bacteriophage should occur, resulting in lysis of the culture after 9–12 hours.

    A fully lysed culture contains a considerable amount of bacterial debris, which can vary in appearance from a fine splintery precipitate to much larger stringy clumps. If the culture is held up to the light, the Schlieren patterns and silky appearance of a dense, unlysed bacterial culture should not be visible.

8. If lysis is not apparent, check a small sample of the cultures for evidence of bacteriophage growth. Withdraw into glass tubes two aliquots (1 ml) of the infected cultures. Add 1 or 2 drops of chloroform to one of the tubes. Incubate both tubes at 37°C for 5–10 minutes with intermittent shaking. Compare the appearance of the two cultures by holding the tubes to a light. If infection is near completion but the cells have not yet lysed, the chloroform causes the cells to burst and the turbid culture clears to the point where it is translucent. In this case, proceed to step 9.

   If lysis does not occur, the preparation can sometimes be rescued by adding to each of the cultures an additional 500 ml of NZYCM preheated to 37°C. Incubation should be continued for a further 2–3 hours, shaking as vigorously as possible.

9. Add 10 ml of chloroform to each flask and continue incubating and shaking for a further 30 minutes.

10. Proceed to Purification of Bacteriophage λ, step 1 (page 78).

INDUCTION[1]

To prepare bacteriophage λ from a 2-liter culture of induced bacteria:

1. Streak out the appropriate lysogenic bacteria on each of two NZCYM plates. Incubate one at 30°C, the other at 42°C. Colonies should form only on the plate incubated at 30°C.

2. Pick a single colony from the plate incubated at 30°C and inoculate 100 ml of NZCYM in a 500-ml flask. Incubate overnight at 30°C with vigorous agitation.

3. Read the OD_{600} of the overnight culture, and inoculate four 500-ml batches of NZCYM prewarmed to 30°C with sufficient overnight culture to give a starting OD_{600} of 0.05.

4. Incubate at 30°C with vigorous shaking until the OD_{600} reaches 0.5. It is important to induce the culture immediately at this stage, since further growth of the bacteria can result in a significant decrease in the yield of phage.

5. Induce the culture by incubating for 15 minutes in a 45°C water bath with constant shaking.

[1]Pirrotta et al. (1971).

6. Incubate the induced culture at 38°C for 2.5–5 additional hours with vigorous shaking.

7. Check a small sample of the cultures for evidence of bacteriophage growth. Withdraw into glass tubes two aliquots (1 ml) of the induced cultures. Add 1 or 2 drops of chloroform to one of the tubes. Incubate both of the tubes at 37°C for 5–10 minutes with intermittent shaking. Compare the appearance of the two cultures by holding the tubes to a light. If the lysogen was properly induced, the chloroform-treated culture should have become translucent; the untreated culture will remain opaque. If you are satisfied with the result, proceed to step 8. If lysis does not occur, you probably should recheck your lysogen (step 1) and start the experiment over again.

8a. If the phage carries a wild-type S gene, lysis of the induced bacteria should occur spontaneously. In this case, add 10 ml of chloroform to each culture and continue incubation for a further 30 minutes. Then proceed to Purification of Bacteriophage λ, step 1 (page 80).

8b. If the phage carries an amber mutation in the S gene, lysis will not occur spontaneously. The bacteriophage particles remain trapped within the host bacteria until chloroform is added. The infected bacteria can be concentrated by centrifugation and lysed in a small volume of liquid, thereby eliminating the necessity of handling large volumes of phage lysate.

   Collect bacteria by centrifugation at 4000g for 10 minutes at 4°C. Resuspend the cell pellet in:

   10 mM Tris·Cl (pH 7.4)
   50 mM NaCl
   5 mM $MgCl_2$

   Use 10–20 ml per liter of original culture.

   Add 10 drops of chloroform. Vortex well and let stand at room temperature for 30 minutes.

   To remove DNA liberated from the lysed cells, add crystalline pancreatic DNase to a final concentration of 0.2 μg/ml and incubate at room temperature for 5 minutes.

   Remove debris by centrifugation at 12,000g for 15 minutes. Harvest the supernatant, which contains the bacteriophage. Proceed to Purification of Bacteriophage λ (step 8, page 80).

PURIFICATION OF BACTERIOPHAGE λ[2]

1. Chill the lysed cultures to room temperature and add pancreatic DNase and RNase, both to a final concentration of 1 μg/ml. Incubate for 30 minutes at room temperature. Crude commercial preparations of both enzymes are adequate to digest the nucleic acids liberated from the lysed bacteria, which might otherwise entrap bacteriophage particles.

2. Add solid sodium chloride to a final concentration of 1 M (29.2 g/500 ml of culture). Dissolve by swirling. Let stand for 1 hour on ice.

3. Remove debris by centrifugation at 11,000g for 10 minutes at 4°C. Pool the supernatants in a clean flask.

4. Add solid polyethylene glycol (PEG 6000) to a final concentration of 10% w/v (i.e., 50 g/500 ml of supernatant). Dissolve by slow stirring on a magnetic stirrer at room temperature.

   Note. Some workers prefer to add the polyethylene glycol at the same time as the sodium chloride (step 2). The centrifugation step (3) may then be omitted. This procedure works well if the bacteriophage grows well and the titer of bacteriophage in the original lysed culture is greater than 2×10^{10}/ml.

5. Cool in ice water and let stand for at least 1 hour to allow the bacteriophage particles to form a precipitate.

6. Recover the precipitated bacteriophage particles by centrifugation at 11,000g for 10 minutes at 4°C. Discard the supernatant and stand the centrifuge bottles in a tilted position for 5 minutes to allow the remaining fluid to drain away from the pellet. Remove the fluid with a pipette.

7. Using a wide-bore pipette equipped with a rubber bulb, gently resuspend the bacteriophage pellet in SM (8 ml for each 500 ml of supernatant). Wash the walls of the centrifuge bottles thoroughly since the bacteriophage precipitate sticks to them, especially if the bottles are old.

8. Add an equal volume of chloroform to the bacteriophage suspension and vortex for 30 seconds. Separate the organic and aqueous phases by centrifugation at 1600g for 15 minutes at 4°C. Recover the aqueous phase containing the bacteriophage.

9. Measure the volume of the supernatant and add 0.5 g/ml of solid cesium chloride. Mix gently.

[2]Yamamoto et al. (1970).

TABLE 3.1. CESIUM CHLORIDE SOLUTIONS (100 ML) FOR STEP GRADIENTS PREPARED IN SM

Density (ρ)	CsCl (g)	SM (ml)	Refractive index (η)
1.45	60	85	1.3768
1.50	67	82	1.3815
1.70	95	75	1.3990

10. When the cesium chloride has dissolved, carefully layer the bacterio-phage suspension onto cesium chloride step gradients that are pre-formed in Beckman SW41 or SW27 cellulose nitrate centrifuge tubes (or their equivalent) (see Table 3.1).

   The step gradients may be made *either* by layering carefully and sequentially solutions of decreasing density on top of one another *or* by layering solutions of increasing density under one another. Make a mark on the outside of the tube opposite the position of the interface between the ρ 1.50 layer and the ρ 1.45 layer (see Fig. 3.1).

11. Centrifuge in either an SW41 or an SW27 rotor at 22,000 rpm for 2 hours at 4°C.

12. A bluish band of bacteriophage particles should be visible at the inter-face between the 1.45 and 1.50 g/ml layers. If the yield of bacteriophage is low, placing the gradient against a black background and shining a light from above often helps to detect the band of particles.

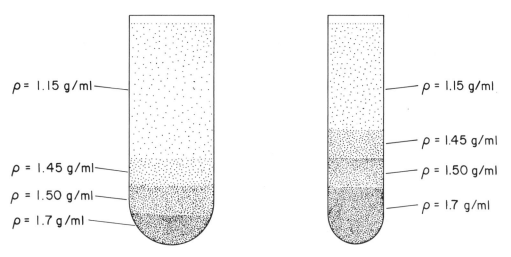

Figure 3.1

Cesium chloride gradients for purifying bacteriophage λ. The bacteriophage will form a visible band at the interface between the 1.45 g/ml and 1.50 g/ml cesium chloride layers.

13. Collect the bacteriophage particles by puncturing the side of the tube. First, place a piece of Scotch tape on the outside of the tube, level with the bacteriophage band. Using a 21-gauge needle, puncture the tube through the tape and collect the band of bacteriophage particles (see Fig. 3.2). Alternatively, the band can be collected from above using a micro-pipette or a pasteur pipette.

   Be careful not to contaminate the bacteriophage with material from other bands that are visible in the gradient. These consist of various types of bacterial debris and unassembled bacteriophage components.

14. Add enough cesium chloride solution (1.5 g/ml in SM) to the bacterio-phage suspension to fill a cellulose nitrate tube that fits either a Type-50Ti rotor or an SW50.1 rotor (or equivalent). Centrifuge at 38,000 rpm for 24 hours at 4°C (Type-50Ti) or at 35,000 rpm for 24 hours at 4°C (SW50.1).

15. Collect the band of bacteriophage particles as described above. Store at 4°C in cesium chloride in a tightly capped tube.

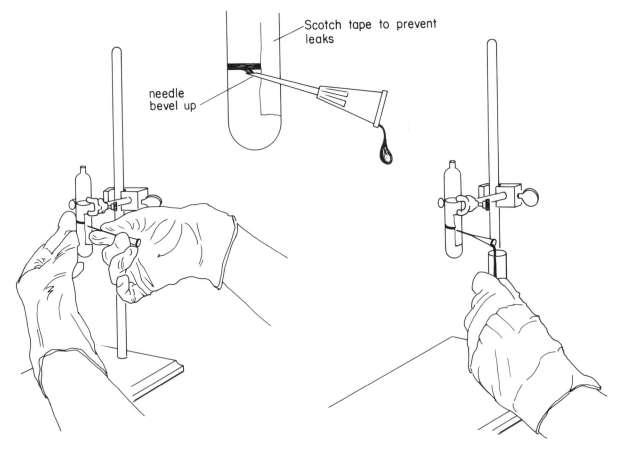

Figure 3.2

Collection of bacteriophage band by side puncture. **Caution:** Keep your fingers out of the path of the needle in case you poke it through the other side of the tube.

Notes

i. Particles of bacteriophage λ are exceedingly sensitive to EDTA, and it is essential that Mg^{++} (10–30 mM) be present at all stages of the purification to prevent disintegration of the particles.

ii. If the yields of purified bacteriophage are low, the number of infectious bacteriophage particles should be determined in samples taken at various stages during the purification in order to determine where losses are occurring.

Alternative Methods of Purification of Bacteriophage λ

The method of purification of bacteriophage particles given on the preceding pages is essentially that of Yamamoto et al. (1970). Over the years many variations and shortcuts have been invented. Three slightly modified procedures are given below.

Pelleting Phage Particles

1. Follow the Yamamoto purification scheme through step 8 (page 80), including the extraction with chloroform.

2. Collect the bacteriophage particles by centrifugation at 25,000 rpm for 2 hours at 4°C in a Beckman SW27 rotor (or its equivalent).

3. Pour off the supernatant. A glassy pellet of bacteriophage should be visible on the bottom of the tube. Add 1–2 ml of SM to each tube and leave it overnight at 4°C, if possible on a rocking platform.

4. The following morning, pipette the solution gently to ensure that all the bacteriophage particles have been resuspended.

Although the preparations obtained in this way are not as clean as those obtained by equilibrium gradient centrifugation, they serve as a useful source of DNA, which can be used in the preparation of bacteriophage λ arms.

Proceed with DNA extraction at step 5 (page 85).

Glycerol Step Gradient[3]

1. Follow the Yamamoto purification scheme through step 6 (page 80).

2. Resuspend the phage pellet in 50 mM Tris · Cl (pH 7.8) and 10 mM $MgSO_4$ (TM buffer) using 5–10 ml per liter of original culture.

[3]Modified from Vande Woude (1979).

3. Extract the bacteriophage suspension once with chloroform.

4. Prepare a glycerol step gradient in a Beckman SW41 cellulose nitrate tube (or its equivalent) as follows:

    a. Pipette 3 ml of a solution consisting of 40% glycerol in TM into the bottom of the tube.

    b. Carefully layer 4 ml of a solution consisting of 5% glycerol in TM over it.

    c. Carefully layer the bacteriophage suspension onto the glycerol solution.

    d. Fill the tube with TM.

5. Centrifuge at 35,000 rpm for 60 minutes at 4°C.

6. Discard the supernatant and resuspend the bacteriophage pellet in 1 ml of TM per liter of original culture.

7. Add pancreatic DNase and RNase to final concentrations of 5 μg/ml and 1 μg/ml, respectively. Digest for 30 minutes at 37°C.

8. Add EDTA from a stock solution (0.5 M, pH 8.0) to a final concentration 20 mM.

9. Proceed with DNA extraction at step 5 (page 85).

Equilibrium Centrifugation in Cesium Chloride

When dealing with small-scale preparations of bacteriophage (1 liter or less), the cesium chloride step gradient can be omitted.

1. After extraction with chloroform (step 8), measure the volume of the aqueous phase and add 0.75 g/ml of solid cesium chloride. Mix gently to dissolve.

2. When the cesium chloride has dissolved, transfer the bacteriophage suspension to an ultracentrifuge tube that fits either an angle or swing-out rotor. Fill the tube with TM to which cesium chloride (0.75 g/ml) has been added.

3. Centrifuge and collect the band of bacteriophage particles as described on page 82.

EXTRACTION OF BACTERIOPHAGE λ DNA

1. Remove cesium chloride from the purified bacteriophage preparation by dialysis at room temperature for 1 hour against a 1000-fold volume of:

   10 mM NaCl
   50 mM Tris · Cl (pH 8.0)
   10 mM MgCl$_2$

2. Transfer the dialysis sac to a fresh flask of buffer and dialyze for an additional hour.

3. Transfer the bacteriophage suspension into a centrifuge tube of a size such that only one third is full.

4. Add EDTA from a stock solution (0.5 M, pH 8.0) to give a final concentration of 20 mM.

5. Add pronase to a final concentration of 0.5 mg/ml, or add proteinase K to a final concentration of 50 μg/ml.

6. Add SDS (stock solution, 20% w/v in water) to a final concentration of 0.5%. Mix by inverting the tube several times.

7. Incubate for 1 hour at 37°C (pronase) or 65°C (proteinase K).

8. Add an equal volume of equilibrated phenol (see page 438). Mix by inverting the tube several times. Separate the phases by centrifugation at 1600g for 5 minutes at room temperature. Use a wide-bore pipette to transfer the aqueous phase to a clean tube.

9. Extract the aqueous phase once with a 50:50 mixture of equilibrated phenol and chloroform.

10. Recover the aqueous phase as described above and extract once with an equal volume of chloroform.

11. Transfer the aqueous phase to a dialysis sac.

12. Dialyze sequentially overnight at 4°C against three 1000-fold volumes of TE (10 mM Tris · Cl [pH 8.0] and 1 mM EDTA).

LARGE-SCALE ISOLATION OF PLASMID DNA

Many methods have been used to isolate plasmid DNA. All of them involve three basic steps: growth of bacteria and amplification of the plasmid; harvesting and lysis of the bacteria; and purification of the plasmid DNA.

These purification procedures exploit in one way or another the two major differences between *Escherichia coli* DNA and plasmid DNA:

1. The *E. coli* chromosome is much larger than the DNA of plasmids commonly used as vectors.

2. The bulk of *E. coli* DNA extracted from cells is obtained as broken, linear molecules. By contrast, most plasmid DNA is extracted in a covalently closed, circular form.

Most of the purification protocols therefore involve a differential precipitation step, in which the long strands of *E. coli* DNA, entangled in the remnants of lysed cells, are preferentially removed. The protocols also take advantage of the distinctive properties of closed circular DNA. Because each of the complementary strands of plasmid DNA is a covalently closed circle, the strands cannot be separated (without breaking one of them) by conditions such as heating or exposure to mild alkali (up to pH 12.5), which break most of the hydrogen bonds in DNA. Closed circular molecules regain their native configuration when cooled or returned to neutral pH. *E. coli* DNA remains in the denatured state.

Plasmid DNA also behaves differently from *E. coli* DNA when the two are centrifuged to equilibrium in cesium chloride gradients containing saturating quantities of an intercalating dye, such as ethidium bromide or propidium diiodide. Covalently closed, circular DNA binds much less of these dyes than linear DNA and therefore bands at a higher density in cesium chloride gradients containing an intercalating agent (Radloff et al. 1967). This method is used when plasmid DNA of high purity is required. However, as recombinant DNA techniques have advanced, it has become unnecessary, for most purposes, to purify large quantities of plasmid DNA to homogeneity. For example, cleavage with restriction endonucleases, ligation, transformation, and even DNA sequencing can be carried out now on relatively crude preparations of plasmid DNA obtained from small-scale (10-ml) cultures. Large-scale preparations are used only when plasmid DNAs are needed in considerable quantity (e.g., in hybridization experiments to select specific mRNAs; or when the 5′ ends of DNA fragments are to be labeled by polynucleotide kinase).

The methods described on the following pages have been used successfully with a wide variety of plasmids carried in several different bacterial strains. In general, the smaller the plasmid, the better the results. As the molecular weight of the plasmid increases, its properties come more and more to resemble those of host DNA. For plasmids greater than 25 kb in size, isolation is very difficult and the yield is poor. However, all plasmids used routinely in cloning are comparatively small and the following methods give good results.

GROWTH OF BACTERIA AND AMPLIFICATION OF THE PLASMID

Amplification in Rich Medium

For many years, it was thought that amplification of plasmids in the presence of chloramphenicol was effective only when the host bacteria were grown in minimal medium. However, the following procedure gives reproducibly high yields (>2 mg of plasmid/liter of culture) with strains of *E. coli* harboring plasmids that carry ColE1 replicons.

1. Inoculate 10 ml of LB medium containing the appropriate antibiotic with a single bacterial colony. Incubate at 37°C overnight with vigorous shaking.

2. The following morning, inoculate 25 ml of LB medium in a 100-ml flask containing the appropriate antibiotic with 0.1 ml of the overnight culture. Incubate at 37°C with vigorous shaking until the culture reaches late log phase ($OD_{600} \simeq 0.6$).

3. Inoculate 25 ml of the late log culture into 500 ml of LB medium prewarmed to 37°C with the appropriate antibiotic in a 2-liter flask. Incubate for exactly 2.5 hours at 37°C with vigorous shaking. The OD_{600} of the culture will be approximately 0.4.

4. Add 2.5 ml of a solution of chloramphenicol (34 mg/ml in ethanol). The final concentration of chloramphenicol in the culture is 170 μg/ml.

5. Incubate at 37°C with vigorous shaking for a further 12–16 hours.

Amplification in Minimal Medium

This procedure (Clewell and Helinski 1972), although now largely superceded by the one above, may still be useful when growing relaxed plasmids that contain replicons other than ColE1.

1. Starting from a single colony grown in the presence of the appropriate antibiotic, inoculate 10 ml of overnight culture in enriched medium (e.g., LB) containing the antibiotic.

2. Inoculate 2.5 ml of the overnight culture into a 2-liter flask containing 500 ml of minimal medium (M9) and the antibiotic. Incubate at 37°C with vigorous shaking until the culture reaches an OD_{600} of 0.4–0.5. It is important not to let the cells grow to a greater density or the efficiency of amplification decreases.

3. Add 2.5 ml of a solution of chloramphenicol (34 mg/ml in ethanol). Continue incubation for a further 12–16 hours.

HARVESTING AND LYSIS OF BACTERIA

Harvesting

1. Harvest the bacterial cells by centrifugation at 4000g for 10 minutes at 4°C. Discard the supernatant.

2. Wash in 100 ml of ice-cold STE (0.1 M NaCl, 10 mM Tris · Cl [pH 7.8], and 1 mM EDTA).

Lysis

Below we give three different lysis procedures. The first two, lysis by boiling or by alkali, are highly efficient and give yields of 2–3 mg/l of small plasmids (<10 kb). The third method is considerably more gentle and is the method of choice for large plasmids (>10 kb).

Lysis by Boiling[4]

1. Resuspend the bacterial pellet from a 500-ml culture in 10 ml of STE (0.1 M NaCl, 10 mM Tris · Cl [pH 8.0], and 0.1 mM EDTA). Transfer to a 50-ml Erlenmeyer flask.

2. Add 1 ml of a freshly made solution of lysozyme (20 mg/ml in 10 mM Tris · Cl, pH 8.0).

   Note. Lysozyme will not work efficiently if the pH of the solution is less than 8.0.

3. Using a clamp, hold the Erlenmeyer flask over the open flame of a Bunsen burner until the liquid *just* starts to boil. Shake the flask constantly.

4. Immediately immerse the flask into a large (2-liter) beaker of boiling water. Hold the flask in the boiling water for 40 seconds.

5. Cool the flask by immersing it in ice-cold water for 5 minutes.

6. Transfer the viscous contents of the flask to an ultracentrifuge tube (Beckman SW41 or equivalent). Centrifuge at 25,000 rpm for 30 minutes at 4°C.

7. Purify the plasmid DNA by centrifugation to equilibrium in cesium chloride–ethidium bromide gradients (see page 93).

[4]Holmes and Quigley (1981).

Lysis by Alkali[5]

1. Resuspend the bacterial pellet from a 500-ml culture in 10 ml of solution I containing 5 mg/ml lysozyme.

    Solution I
    50 mM glucose
    25 mM Tris · Cl (pH 8.0)
    10 mM EDTA

    Solution I can be made up in batches of approximately 100 ml, autoclaved for 15 minutes at 10 lb/in^2, and stored at 4°C. Powdered lysozyme should be dissolved in the solution just before use.

2. Transfer to a Beckman SW27 polyallomer tube (or its equivalent). Let stand at room temperature for 5 minutes.

3. Add 20 ml of freshly made solution II. Cover the top of the tube with parafilm and mix the contents by gently inverting the tube several times. Let stand on ice for 10 minutes.

    Solution II
    0.2 N NaOH
    1% SDS

    Solution II should be made up from stock solutions of 10 N NaOH and 20% SDS.

4. Add 15 ml of an ice-cold solution of a 5 M potassium acetate (pH 4.8) that is prepared as follows: To 60 ml of 5 M potassium acetate, add 11.5 ml of glacial acetic acid and 28.5 ml of H$_2$O. The resulting solution is 3 M with respect to potassium and 5 M with respect to acetate. Cover the top of the tube with parafilm and mix the contents by inverting the tube sharply several times. Let stand on ice for 10 minutes.

5. Centrifuge on a Beckman SW27 (or its equivalent) at 20,000 rpm for 20 minutes at 4°C. The cell DNA and bacterial debris should form a tight pellet on the bottom of the tube.

6. Transfer equal quantities (~ 18 ml) of the supernatant into each of two 30-ml Corex tubes.

7. Add 0.6 volumes (~ 12 ml) of isopropanol to each tube. Mix well and let stand at room temperature for 15 minutes.

[5]This procedure is as described in Birnboim and Doly (1979) and modified by D. Ish-Horowicz (pers. comm.).

8. Recover the DNA by centrifugation in a Sorvall rotor at 12,000g for 30 minutes at room temperature.

    Note. Salt may precipitate if centrifugation is carried out at 4°C.

9. Discard the supernatant. Wash the pellet with 70% ethanol at room temperature. Discard as much ethanol as possible, then dry the nucleic acid pellet briefly in vacuum desiccator.

10. Dissolve the pellets in a total volume of 8 ml of TE (pH 8.0).

11. Purify the plasmid DNA by centrifugation to equilibrium in cesium chloride–ethidium bromide density gradients (see page 93).

Note

Perhaps surprisingly, this method also works well with plasmids that have been amplified in the presence of chloramphenicol. Long exposure to the drug results in the synthesis of plasmids whose DNA has been partially substituted by ribonucleotides. Because these small regions of ribosubstitution will be sensitive to cleavage by alkali, one might expect this method to yield preparations that contain an unusually large amount of nicked, circular molecules. In our hands, however, the DNA prepared in this way consists of less than 5% nicked molecules.

Lysis by SDS[6]

1. Resuspend the bacterial pellet in 10 ml of an ice-cold solution of 10% sucrose in 50 mM Tris · Cl (pH 8.0). Transfer the suspension to a 30-ml Oak Ridge tube.

2. Add 2 ml of a freshly made solution of lysozyme (10 mg/ml in 0.25 M Tris · Cl, pH 8.0).

3. Add 8 ml of 0.25 M EDTA. Mix by inverting the tube several times. Place on ice for 10 minutes at 0°C.

4. Add 4 ml of 10% SDS. Mix *quickly* with a glass rod so as to disperse the SDS evenly through the bacterial suspension and *gently* so as not to shear the liberated bacterial DNA.

5. Immediately add 6.0 ml of 5 M NaCl (final concentration = 1 M). Again, mix gently but thoroughly. Place on ice for at least 1 hour.

6. Centrifuge to remove high-molecular-weight DNA and bacterial debris in a Beckman Type-50Ti rotor (or its equivalent) for 30 minutes at 30,000 rpm at 4°C.

7. Pour off and save the supernatant. Discard the pellet, which should be firm and tight.

8. Extract the supernatant twice with phenol/chloroform and once with chloroform. After each extraction, transfer the aqueous layer to a clean tube.

9. Transfer the aqueous phase to a 250-ml glass centrifuge bottle. Add 2 volumes (~ 60 ml) of ethanol. Mix well. Let stand at -20°C for 1-2 hours, or at -70°C for 15 minutes.

10. Recover the nucleic acids by centrifugation at 1500*g* for 15 minutes at 4°C.

11. Discard the supernatant. Wash the pellet with 70% ethanol at room temperature. Discard as much of the ethanol as possible. Dry the pellet briefly in a vacuum desiccator.

12. Dissolve the DNA in a total volume of 8 ml of TE (pH 8.0).

13. Purify the plasmid DNA by centrifugation to equilibrium in cesium chloride–ethidium bromide gradients (see page 93).

[6]Godson and Vapnek (1973).

PURIFICATION OF CLOSED CIRCULAR DNA BY CENTRIFUGATION TO EQUILIBRIUM IN CESIUM CHLORIDE–ETHIDIUM BROMIDE GRADIENTS

1. Measure the volume of the DNA solution. For every milliliter add exactly 1 g of solid cesium chloride. Mix gently until all of the salt is dissolved.

2. Add 0.8 ml of a solution of ethidium bromide (10 mg/ml in H_2O) for every 10 ml of cesium chloride solution. Mix well. The final density of the solution should be 1.55 g/ml ($\eta = 1.3860$), and the concentration of ethidium bromide should be approximately 600 μg/ml.

   Note. The furry, purple aggregates that float to the top of the solution are complexes formed between the ethidium bromide and bacterial proteins.

3. Transfer the cesium chloride solution (together with the protein aggregates) to a tube suitable for centrifugation in a Beckman Type-50 or Type-65 rotor. Fill the remainder of the tube with light paraffin oil.

4. Centrifuge at 45,000 rpm for 36 hours at 20°C.

5. Two bands of DNA should be visible in ordinary light. The upper band consists of linear bacterial DNA and nicked circular plasmid DNA; the lower band consists of closed circular plasmid DNA.

6. Remove the cap from the tube. Collect the lower band of DNA into a glass tube through a #21 hypodermic needle inserted into the side of the tube as described on page 82.

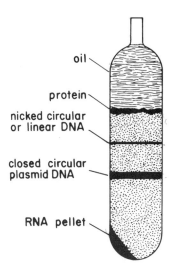

oil

protein

nicked circular or linear DNA

closed circular plasmid DNA

RNA pellet

Figure 3.3

7. Remove the ethidium bromide as follows:

   a. Add an equal volume of 1-butanol saturated with water *or* isoamyl alcohol.

   b. Mix the two phases by pipetting vigorously.

   c. Centrifuge at 1500*g* for 3 minutes at room temperature.

   d. Transfer the lower aqueous phase to a clean glass tube.

   e. Repeat the extraction 4–6 times until all the pink color disappears from the aqueous solution.

8. Dialyze the aqueous phase against several changes of TE (pH 8.0).

REMOVAL OF RNA FROM PREPARATIONS OF PLASMID DNA

For some purposes (e.g., for digestion with *Bal*31 or for using polynucleotide kinase to label the 5′ ends of restriction endonuclease fragments of plasmid DNA), it is essential to obtain DNA preparations that are free of contaminating RNA of lower molecular weight. Although the *weight* of such contaminants in plasmid DNA prepared by equilibrium centrifugation in cesium chloride–ethidium bromide gradients is small, the *number* of molecules can be relatively large and can contribute a significant proportion of the total 5′ ends in the restriction endonuclease digest.

RNA may be removed from plasmid preparations by either of the following two methods.

Centrifugation through 1 M NaCl[7]

1. Measure the volume of the DNA solution. Add 0.1 volume of 3 M sodium acetate (pH 5.2). Add 2 volumes of ethanol. Mix well and let stand at –20°C.

2. Recover the DNA by centrifugation. Discard the ethanol and dry the DNA pellet briefly in a vacuum desiccator.

3. Dissolve the DNA in TE (pH 8.0) at a concentration of at least 100 μg/ml.

4. Add DNase-free RNase (see page 451) to a final concentration of 10 μg/ml. Incubate at room temperature for 1 hour.

5. Prepare a Beckman SW50.1 centrifuge tube (or its equivalent) containing 4 ml of 1 M NaCl plus TE. Layer up to 1 ml of the RNase-treated plasmid DNA on top of the 1 M NaCl solution. If necessary, fill the tube with TE. Centrifuge for 6 hours at 40,000 rpm at 20°C in a Beckman SW50.1 rotor. The plasmid DNA sediments to the bottom of the tube while the ribo-oligonucleotides remain in the supernatant.

6. Discard the supernatant. Redissolve the pellet of plasmid DNA in the desired volume of TE.

[7] B. Seed (unpubl.).

Chromatography through Bio-Gel A-150[8]

1. Treat the preparation of plasmid DNA with RNase as described in steps 1 through 4 on the previous page.

2. Extract the solution once with an equal volume of equilibrated phenol.

3. Layer up to 1 ml of the aqueous phase on a column of Bio-Gel A-150 (1 cm × 10 cm) equilibrated in TE (pH 8.0) and 0.1% SDS.

4. Wash the DNA into the column, apply a reservoir of TE with 0.1% SDS, and immediately begin to collect 0.5-ml fractions.

5. When 15 fractions have been collected, clamp off the bottom of the column. Analyze 10 μl of each fraction by electrophoresis through an 0.7% agarose gel or by ethidium bromide fluorescence (see Appendix A) in order to locate the plasmid DNA.

6. Pool the fractions that contain plasmid DNA. Recover the DNA by precipitation with ethanol as described in steps 1–3.

Note

It is very difficult to remove all traces of plasmid DNA from the column of Bio-Gel A-150. To eliminate the possibility of contamination, use each column only once.

[8]Modification of a procedure developed by F. DeNoto and H. Goodman (unpubl.).

4

Enzymes Used in Molecular Cloning

Restriction Enzymes

Restriction endonucleases are enzymes, isolated chiefly from prokaryotes, that recognize specific sequences within double-stranded DNA. They can be classified into three groups. Type-I and type-III enzymes carry modification (methylation) and an ATP-requiring restriction (cleavage) activity in the same protein. Both types of enzymes recognize unmethylated recognition sequences in substrate DNA, but type-I enzymes cleave randomly, whereas type-III enzymes cut DNA at specific sites.

Type-II restriction/modification systems consist of a separate restriction endonuclease and modification methylase. A large number of type-II restriction enzymes have been isolated (see Roberts 1982), many of which are useful in molecular cloning. These enzymes cut DNA within or near to their particular recognition sequences, which typically are four to six nucleotides in length with a twofold axis of symmetry. The enzyme *Eco*RI, for example, recognizes the hexanucleotide sequence

$$5' \quad G\text{-}A\text{-}A\text{-}T\text{-}T\text{-}C \quad 3'$$
$$3' \quad C\text{-}T\text{-}T\text{-}A\text{-}A\text{-}G \quad 5'$$

Like many other restriction enzymes, *Eco*RI does not cleave exactly at the axis of dyad symmetry but at positions four nucleotides apart in the two DNA strands:

$$5' \ldots G \downarrow A\text{-}A\text{-}T\text{-}T\text{-}C \ldots 3'$$
$$3' \ldots C\text{-}T\text{-}T\text{-}A\text{-}A \uparrow G \ldots 5'$$

$$5' \ldots G \quad 3' \qquad 5' A\text{-}A\text{-}T\text{-}T\text{-}C \ldots 3'$$
$$3' \ldots C\text{-}T\text{-}T\text{-}A\text{-}A \quad 5' \qquad 3' \quad G \ldots 5'$$

This staggered breakage yields fragments of DNA with protruding, cohesive 5′ termini. Any one of these termini can form base pairs with any other. Thus, any DNA molecules containing such recognition sites can be joined to any others to form novel recombinant molecules.

Many restriction enzymes, like *Eco*RI, generate DNA fragments with protruding 5′ tails; others (e.g., *Pst*I) generate fragments with 3′ protruding, cohesive termini, whereas still others (e.g., *Bal*I) cleave at the axis of symmetry to produce blunt-ended fragments.

ISOSCHIZOMERS

In general, different restriction enzymes recognize different sequences (see Table 4.1). However, there are several examples of enzymes isolated from different sources that cleave within the same target sequences. These are known as isoschizomers. In addition, several enzymes have been discovered that recognize tetranucleotide sequences. In some cases these tetranucleotides occur within hexanucleotide target sequences of other enzymes. For example, *Mbo*I and *Sau*3A recognize the sequence

```
5'↓
  ...G-A-T-C...
  ...C-T-A-G...
3'          ↑ 5
```

whereas *Bam*HI recognizes

```
        ↓
5'  G   G-A-T-C-C   3'
  ...G'             ...
  ...C-C-T-A-G  G...
3'           ↑  5'
```

Assuming that restriction endonuclease sites are distributed randomly along DNA, the tetranucleotide target for *Mbo*I and *Sau*3A will occur on the average once every 4^4 (i.e., 256) nucleotides, whereas the hexanucleotide target for *Bam*HI will occur once in every 4^6 (i.e., 4096) nucleotides. DNA fragments generated by complete or partial digestion of eukaryotic DNA with *Mbo*I and *Sau*3A can therefore be cloned or subcloned in a vector DNA (such as bacteriophage λ or pBR322) that has been cleaved with *Bam*HI. Note that most *Mbo*I/*Bam*HI fusions do not regenerate a *Bam*HI site. Therefore, it is not usually possible to recover an *Mbo*I fragment inserted into a *Bam*HI vector by digesting the recombinant with *Bam*HI.

In a few cases, fragments generated by one restriction enzyme when ligated to fragments generated by a second enzyme give rise to hybrids that are recognized by neither of the parental enzymes. For example, when fragments generated by *Sal*I (G↓TCGAC) are ligated to fragments generated by *Xho*I (C↓TCGAG), the resulting hybrid target sites are cleaved by neither *Sal*I nor *Xho*I.

```
5'                            3'
  ...G            T-C-G-A-G...
  ...C-A-G-C-T  +         C...
3'                            5'
                 ↓
5'                  3'
  ...G-T-C-G-A-G...
  ...C-A-G-C-T-C...
3'                  5'
```

TABLE 4.1. RESTRICTION ENZYMES

Enzyme	Common isoschizomers	Salt[1]	Incubation temperature	Recognition sequence	Compatible cohesive ends[2]
AccI		med	37°C	GT↓($^{AG}_{CT}$)AC	AcyI[3], AsuII[3], ClaI[3] HpaII[3], Taq[3]
AcyI			37°C	G(^A_G)↓CG(^T_C)C	AccI[3], AsuII, ClaI, HpaII, TaqI
AluI		med	37°C	AG↓CT	blunt[4]
AosI			37°C	TGC↓GCA	blunt
ApyI	AtuI, EcoRII		37°C	CC↓(^A_T)GG	
AsuI			37°C	G↓GNCC	
AsuII			37°C	TT↓CGAA	AccI[3], AcyI, ClaI, HpaII, TaqI
AtuII			37°C	CC↓(^A_T)GG	
AvaI		med	37°C	G↓PyCGPuG	SalI[3], XhoI[3], XmaI[3]
AvaII		med	37°C	G↓G(^A_T)CC	Sau96I[3]
AvrII		low	37°C	CCTAGG	
BalI			37°C	TGG↓CCA	blunt
BamHI		med	37°C	G↓GATCC	BclI, BglII, MboI, Sau3A, XhoII
BbvI		low	37°C	GC(^T_A)GC	
BclI		med	60°C	T↓GATCA	BamHI, BglII, MboI, Sau3A, XhoII
BglI		med	37°C	GCCNNNN↓NGGC	
BglII		low	37°C	A↓GATCT	BamHI, BclI, MboI, Sau3A, XhoII
BpaI			37°C	GT↓(^C_A)(^G_T)AC	
BstEII		med	60°C	G↓GTNACC	
BstNI		low	60°C	CC↓(^A_T)GG	
ClaI			37°C	AT↓CGAT	AccI[3], AcyI, AsyII, HpaII, TaqI
DdeI		med	37°C	C↓TNAG	
DpnI	Sau3A	med	37°C	G^{Me}A↓TC	blunt
EcoRI		high	37°C	G↓AATTC	
EcoB			37°C	TGANNNNNNNNTGCT	
EcoK			37°C	AACNNNNNNGTGC	
EcoPI			37°C	AGACC	
EcoRI[1]			37°C	(^A_G)(^A_G)A↓T(^T_C)(^T_C)	blunt
EcoRI*			37°C	↓AATT	EcoRI
EcoRII	AtuI, ApyI	high	37°C	↓CC(^A_T)GG	
Fnu4HI		low	37°C	GC↓NGC	
FnuDII	ThaI	low	37°C	CG↓CG	blunt
HaeI		low	37°C	(^A_T)GG↓CC(^T_A)	blunt
HaeII		low	37°C	PuGCGC↓Py	
HaeIII		med	37°C	GG↓CC	blunt
HgaI		med	37°C	GACGCNNNNN↓ CTGCGNNNNNNNNNN[1]	
HgiAI		high	37°C	G(^T_A)GC(^T_A)↓C	
HhaI	CfoI	med	37°C	GCG↓C	
HincII		med	37°C	GTPy↓PuAC	blunt
HindII		med	37°C	GTPy↓PuAC	blunt
HindIII		med	37-55°C	A↓AGCTT	
HinfI		med	37°C	G↓ANTC	
HpaI		low	37°C	GTT↓AAC	blunt
HpaII		low	37°C	C↓CGG	AccI[3], AcyI, AsuII, ClaI, TaqI
HphI		low	37°C	GGTGANNNNNNNN↓ CCACTNNNNNNN[1]	
KpnI		low	37°C	GGTAC↓C	BamHI, BclI, BglII, XhoII
MboI	Sau3A	high	37°C	↓GATC	
MboII		low	37°C	GAAGANNNNNNNN↓ CTTCTNNNNNNN[1]	

TABLE 4.1.   RESTRICTION ENZYMES (continued)

Enzyme	Common isoschizomers	Salt[1]	Incubation temperature	Recognition sequence	Compatible cohesive ends[2]
*Mnl*I		high	37°C	CCTC	
*Msp*I		low	37°C	C↓CGG	
				C↓C^{Me}GG	
*Mst*I			37°C	TGCGCA	
*Pst*I		med	21–37°C	CTGCA↓G	
*Pvu*I		high	37°C	CGATCG	
*Pvu*II		med	37°C	CAG↓CTG	blunt
*Rsa*I		med	37°C	GT↓AC	blunt
*Sac*I	*Sst*I	low	37°C	GAGCT↓C	
*Sac*II		low	37°C	CCGC↓GG	
*Sac*III		high	37°C	ACGT	
*Sal*I		high	37°C	G↓TCGAC	*Ava*I[3], *Xho*I
*Sau*3A		med	37°C	↓GATC	*Bam*HI, *Bcl*I, *Bgl*II, *Mbo*I, *Xho*II
				G^{Me}ATC	
*Sau*96I		med	37°C	G↓GNCC	
*Sma*I	*Xma*I	(1)	37°C	CCC↓GGG	blunt
*Sph*I			37°C	GCATG↓C	
*Sst*I	*Sac*I	low	37°C	GAGCT↓C	
*Sst*II		low	37°C	CCGC↓GG	
*Sst*III		high	37°C	ACGT	
*Taq*I		low	65°C	T↓CGA	*Acc*I[3], *Acy*I, *Asu*II, *Cla*I, *Hpa*II
*Tha*I	*Fnu*DII	low	60°C	CG↓CG	blunt
*Xba*I		high	37°C	T↓CTAGA	
*Xho*I		high	37°C	C↓TCGAG	*Ava*I[3], *Sal*I
*Xho*II			37°C	(^A_G)↓GATC(^T_C)	*Bam*HI, *Bcl*I, *Bgl*II, *Mbo*I, *Sau*3A
*Xma*I	*Sma*I	low	37°C	C↓CCGGG	*Ava*I[3]
*Xma*III			37°C	C↓GGCCG	
*Xor*II	*Pvu*I, *Rsh*I	low	37°C	CGATC↓G	

[1]See Table 4.4 on page 104.

[2]This table shows termini produced by two different enzymes that can be ligated together. In some cases, however, the resulting hybrid site cannot be digested by either of the two enzymes. For example, *Bgl*II cleaves the sequence

    A↓G-A-T-C-T
    T-C-T-A-G↑A

whereas *Bam*HI cleaves the sequence

    G↓G-A-T-C-G
    C-C-T-A-G↑C

Both enzymes therefore yield fragments of DNA with identical protruding 5′ cohesive termini, which can be ligated together to form a hybrid site that is not recognized by either enzyme.

    A-G-A-T-C-C
    T-C-T-A-G-G

A similar situation occurs with *Sal*I and *Xho*I (see page 99).

[3]Enzymes that cleave degenerate sequences give rise to a population of DNA fragments with several different termini. Only some of the possible combinations of these termini will be ligatable.

[4]The blunt-ended fragments produced by these enzymes can be ligated to any other blunt-ended fragments.

METHYLATION

Most strains of *E. coli* contain two enzymes that methylate DNA—the *dam* methylase and the *dcm* methylase.

The dam *methylase.* This methylase introduces methyl groups at the N^6 position of adenine in the sequence $^{5'}$GATC$^{3'}$ (Hattman et al. 1978). The recognition sites of several restriction enzymes contain this sequence (*Pvu*I, *Bam*HI, *Bcl*I, *Bgl*II, *Xho*II, *Mbo*I, *Sau*3A) as do a proportion of the sites recognized by *Cla*I (1 site in 4), *Xba*I (1 site in 16), *Taq*I (1 site in 16), *Mbo*II (1 site in 16), and *Hph*I (1 site in 16). The effect of *dam* methylation on cleavage of DNA by these enzymes is summarized in Table 4.2.

The inhibition of digestion of prokaryotic DNA by *Mbo*I presents no practical problem because *Sau*3A recognizes exactly the same sequence as *Mbo*I but is unaffected by *dam* methylation. (Note that eukaryotic DNA is not methylated at the N^6 position of adenine and therefore either *Mbo*I or *Sau*3A can be used effectively.) However, when it is necessary to cleave prokaryotic DNA at every possible site with *Cla*I, *Xba*I, *Taq*I, *Mbo*II, or *Hph*I, or to cleave it at all with *Bcl*I, the DNA must be prepared from strains of *E. coli* that are *dam⁻* (Marinus 1973; Backman 1980; Roberts et al. 1980; McClelland 1981; J. Brooks, unpubl.).

TABLE 4.2. EFFECT OF *dam* METHYLATION ON CLEAVAGE OF DNA

Recognition sequence	Restriction enzyme	GATC	G^me ATC
G↓GATCC	*Bam*HI	+	+
T↓GATCA	*Bcl*I	+	–
A↓GATCT	*Bgl*II	+	+
↓GATC	*Mbo*I	+	–
↓GATC	*Sau*3A	+	+
CGAT↓CG	*Pvu*I	+	+
Pu↓GATCPy	*Xho*II	+	+
AT↓CGAT	*Cla*I	+	–[1]
T↓CTAGA	*Xba*I	+	–[2]
T↓CGA	*Taq*I	+	–[3]
GAAGA	*Mbo*II	+	–[4]
GGTGA	*Hph*I	+	–[5]
G^me ATC	*Dpn*I	–	+[6]

(+) Denotes cleavage; (–) denotes lack of cleavage.

[1] When *Cla*I recognition site is part of the sequence ATCG^me ATC.

[2] When *Xba*I recognition site is part of the sequence TCTAG^me ATC.

[3] When *Taq*I recognition site is part of the sequence TCG^me ATC.

[4] When *Mbo*II recognition site is part of the sequence GAAG^me ATCNNNNNN.

[5] When *Hph*I recognition site is part of the sequence GGTG^me ATC.

[6] Note that the enzyme *Dpn*I cleaves *dam* methylated DNA only.

The dcm *methylase.* This methylase introduces methyl groups at the C^5 position of cytosine in the sequences $^{5'}C^{me}CAGG^{3'}$ or $^{5'}C^{me}CTGG^{3'}$ (Marinus and Morris 1973; May and Hattman 1975). The chief enzyme affected by *dcm* methylation is *Eco*RII. For most purposes, this presents no practical problem because *Bst*NI recognizes exactly the same sequence as *Eco*RII (although it cuts the DNA at a different location within the sequence).

If it is not possible to substitute *Bst*NI for *Eco*RII, the DNA must be prepared from strains of *E. coli* that are *dcm*⁻ (Marinus 1973; Backman 1980; Roberts et al. 1980).

Mammalian DNA contains, in addition to the four normal bases, 5-methylcytosine residues: These are found chiefly at the 5′ side of G residues. Although only a proportion C_pG doublets are methylated, the pattern of methylation is highly specific (Bird and Southern 1978). Thus any given C_pG doublet is methylated in the majority of cells of a given population or in few of them. The effect of methylation on cleavage of mammalian DNA by commonly used restriction enzymes is shown in Table 4.3.

TABLE 4.3. EFFECT OF METHYLATION ON CLEAVAGE OF MAMMALIAN DNA

Restriction enzyme	Sequence cleaved	Sequence not cleaved
*Hha*I	GCGC	$G^{me}CGC$
*Hpa*II	CCGG	$C^{me}CGG$
*Msp*I	$CCGGC^{me}CGG$	$^{me}CCGG$
*Sal*I	GTCGAC	$GT^{me}CGAC$
*Taq*I	$TCGAT^{me}CGA$	—[1]
*Xho*I	CTCGAG	$CT^{me}CGAG$

Data from van der Ploeg and Flavell (1980).
[1] All occurrences of TCGA are cleaved, whether or not the cytosine is methylated.

DIGESTING DNA WITH RESTRICTION ENDONUCLEASES

Each restriction enzyme has a set of optimal reaction conditions, which are given on the information sheet supplied by the manufacturer. The major variables are the temperature of incubation and the composition of the buffer. Although the temperature requirements are fairly strict, the differences between buffers are often only slight. To avoid the labor involved in making up a separate buffer for every enzyme, it is convenient to divide the enzymes into three groups—those that work best at high ionic strength; those that prefer medium ionic strength; and those that have a preference for buffers of low ionic strength.

By following this scheme, only three stock buffers need be prepared (Table 4.4).

Usually all of the buffers in Table 4.4 are made up as 10× stock solutions, which may be stored at 4°C for periods of one to two weeks or at −20°C indefinitely.

Setting Up Digestions with Restriction Enzymes

Reactions typically contain 0.2-1 μg of DNA in a volume of 20 μl or less.

1. Mix water with the DNA solution in a sterile Eppendorf tube to give a volume of 18 μl.

2. Add 2.0 μl of the appropriate 10× digestion buffer. Mix by tapping tube.

3. Add 1 unit of restriction enzyme, and mix by tapping tube. (One unit of enzyme is usually defined as the amount required to digest 1 μg of DNA to completion in 1 hour in the recommended buffer and at the recommended temperature in a 20-μl reaction. In general, digestion for longer periods of time or with excess enzyme does not cause problems unless there is contamination with DNase or exonuclease. Such contamination is rare in commercial enzyme preparations.)

TABLE 4.4. BUFFERS FOR RESTRICTION ENDONUCLEASE DIGESTION

Buffer	NaCl	Tris · Cl (pH 7.5)	MgCl$_2$	Dithiothreitol
Low	0	10 mM	10 mM	1 mM
Medium	50 mM	10 mM	10 mM	1 mM
High	100 mM	50 mM	10 mM	1 mM

Because the enzyme *Sma*I will not work well in any of the above buffers, a separate buffer should be made up, consisting of:

20 mM KCl
10 mM Tris · Cl (pH 8.0)
10 mM MgCl$_2$
1 mM dithiothreitol

4. Incubate at the appropriate temperature for the required period of time.

5. Stop the reaction by the addition of 0.5 M EDTA (pH 7.5) to a final concentration of 10 mM.

   If the DNA is to be analyzed directly on a gel, add 6 μl of gel-loading dye I (see page 455), mix by vortexing briefly, and load the digest into the gel slot.

   If the restricted DNA is to be purified, extract once with phenol/chloroform, once with chloroform, and precipitate the DNA with ethanol (see page 461 for details).

Notes

Restriction enzymes are expensive! Costs can be kept to a minimum by following the advice given below.

   i. Many restriction enzymes are supplied by the manufacturer in concentrated form. Often 1 μl of many enzyme preparations is sufficient to digest 10 μg of DNA in an hour. To remove small quantities of enzyme from the container, touch the tip of a disposable, glass micropipette (0-5 μl) briefly to the surface of the fluid. In this way it is possible to remove as little as 0.1 μl of the enzyme preparation. Alternatively, a small piece of plastic tubing (1 cm long) can be attached to a 1-μl Hamilton syringe and used to transfer 0.1-μl volumes. The plastic tubing is discarded after each sample is pipetted.

   ii. Restriction enzymes are stable when they are stored at –20°C in buffer containing 50% glycerol. When carrying out restriction enzyme digestions, prepare the reactions to the point where all reagents except the enzyme have been mixed. Take the enzyme from the freezer and *immediately* put it into ice. *Use a fresh, sterile pipette every time you dispense enzyme.* Contamination of an enzyme with DNA or another enzyme can be costly and time-consuming. Work as quickly as possible, so that the enzyme is out of the freezer for as short a time as possible. Return the enzyme to the freezer *immediately* after use.

   iii. Keep reaction volumes to a minimum by reducing the amount of water in the reaction as much as possible. However, make sure that the restriction enzyme contributes less than $^1/_{10}$ volume of the final reaction mixture, otherwise the enzyme activity may be inhibited by glycerol.

   iv. Often the amount of enzyme can be reduced if the digestion time is increased. This can result in considerable savings when large quantities of DNA are cleaved. Small aliquots can be removed during the course of the reaction and analyzed on a minigel (page 163) to monitor the progress of the digestion.

v. When digesting many DNA samples with the same enzyme, calculate the total amount of enzyme that is needed. Remove the correct amount of enzyme solution from the container and mix it with the appropriate volume of water and 10× restriction buffer. Dispense aliquots of the enzyme/buffer mixture into the reaction mixtures.

vi. When DNA is to be cleaved with two or more restriction enzymes, the digestions can be carried out simultaneously if both enzymes work in the same buffer. Alternatively, the enzyme that works in the buffer of lower ionic strength should be used first. The appropriate amount of salt and the second enzyme(s) can then be added and the incubation continued.

vii. If the volume of the restriction enzyme reaction is too large to fit into the slot of a gel, the DNA may be concentrated by the following simple procedure: After the reaction has been stopped by the addition of EDTA, add $\frac{2}{3}$ volumes of 5 M ammonium acetate and 2 volumes of ethanol. Chill in a dry-ice/methanol bath for 5 minutes, then centrifuge for 5 minutes in an Eppendorf centrifuge. Discard the supernatant, which contains most of the protein. Dry the pellet briefly under vacuum. Dissolve the DNA in the appropriate volume of TE (pH 7.6) (see page 448).

Other Enzymes Used in Molecular Cloning

Many enzymes apart from restriction endonucleases are used routinely in molecular cloning. Their properties are given on the following pages. In some cases we have also provided protocols that describe the conditions required to carry out specific enzyme reactions. In other cases the reader is referred to places elsewhere in the manual where relevant protocols may be found. Finally, and for the sake of completeness, we have described the properties of a few enzymes that find occasional use in molecular cloning but that are not necessary to carry out any of the procedures described in this manual. The reader should obtain details of the practical use of these enzymes from the publications cited.

DNA Polymerase I

(E. coli)

The enzyme consists of a single polypeptide chain ($M_r = 109,000$) that carries three separate enzymatic activities (Kelley and Stump 1979).

Activity	Reaction	Template / primer or substrate

5' ⟶ 3' Polymerase

$$DNA_{OH} + ndNTP \xrightarrow{Mg^{++}} DNA\text{-}(pdN)_n + nPPi$$

single-stranded template; primer with 3' OH

For example:

```
5'
...pC -pC -pG_OH                                    + dATP
                                                      dCTP
...G_p-G_p-C_p-T_p-A_p-T_p-C_p-G_p-A_p...             dGTP
3'                                                    TTP
```

⟶ Mg^{++}

```
5'
...pC -pC -pG -pA -pT -pA -pG -pC -pT ...
...G_p-G_p-C_p-T_p-A_p-T_p-C_p-G_p-A_p...
```

5' ⟶ 3' Exonuclease

$$dsDNA \xrightarrow{Mg^{++}} \left\{ \begin{array}{l} 5' \\ {}_p N_{OH} \\ 5' \\ {}_p N_p N_{OH} \\ 5' \\ {}_p N_p N_p N_{OH} \end{array} \right. + ssDNA$$

degrades double-stranded DNA from a free 5' end

For example:

```
5'
  pC -pG -pC -pA -pT -pC -pT...
3' G_p-C_p-G_p-C_p-G_p-T_p-A_p-G_p-A_p...
```

⟶ Mg^{++}

```
      pC -pA -pT -pC -pT                  + pC
3' G_p-C_p-G_p-C_p-G_p-T_p-A_p-G_p-A_p...     pG
```

3' ⟶ 5' Exonuclease

$$\begin{array}{l} dsDNA \\ ssDNA \end{array} \xrightarrow{Mg^{++}} \begin{array}{l} 5' \\ {}_p N_{OH} \end{array}$$

degrades double-stranded or single-stranded DNA from a free 3'-OH end; activity on double-stranded DNA blocked by 5' ⟶ 3' polymerase activity

For example:

```
5'
...pC -pG -pC -pA -pT -pC -pT 3'
...G_p-C_p-G_p
3'                5'
```

⟶ Mg^{++}

```
5'                          pA
...pC -pG -pC_OH      +      pC
...G_p-C_p-G_p               pT
3'
```

Uses

1. Labeling of DNA by nick-translation.
2. The holoenzyme was originally used for second-strand synthesis in cDNA cloning (Efstratiadis et al. 1976) but is now superceded by reverse transcriptase or the Klenow fragment of DNA polymerase I.

ESCHERICHIA COLI **DNA POLYMERASE I**

Nick Translation of DNA

Escherichia coli DNA polymerase I adds nucleotide residues to the 3′-hydroxyl terminus that is created when one strand of a double-stranded DNA molecule is nicked. In addition, the enzyme, by virtue of its 5′ to 3′ exonucleolytic activity, can remove nucleotides from the 5′ side of the nick. The elimination of nucleotides from the 5′ side and the sequential addition of nucleotides to the 3′ side results in movement of the nick (nick translation) along the DNA (Kelly et al. 1970). By replacing the preexisting nucleotides with highly radioactive nucleotides, it is possible to prepare ^{32}P-labeled DNA with a specific activity well in excess of 10^8 cpm/μg (Maniatis et al. 1975; Rigby et al. 1977).

1. A typical reaction contains 1 μg of DNA in a volume of 50 μl. However, the reaction can be scaled down to volumes as small as 5 μl.

2. Although *E. coli* DNA polymerase I will work with concentrations of dNTPs as low as 2 μM, the enzyme synthesizes DNA much more efficiently when supplied with higher concentrations of substrates. For reasons of cost, nick-translation reactions usually contain minimal concentrations (2 μM) of labeled dNTPs and much greater concentrations (20 μM) of unlabeled dNTPs. A 50-μl reaction therefore contains 1 nmole of each unlabeled dNTP and 100 pmoles of each labeled dNTP. The specific activity of the final, nick-translated DNA depends in part on the ratio of labeled to unlabeled dNTPs in the reaction. When high specific activities ($>10^8$ cpm/μg DNA) are required (e.g., for screening of recombinant DNA libraries or for detecting single-copy sequences in Southern hybridizations of eukaryotic DNA), the nick translation should contain 200 pmoles of each of the four dNTPs labeled in the α position with ^{32}P (sp. act. $>$800 Ci/mmole).

    For most other purposes, it is adequate to use one dNTP labeled with α-^{32}P and three unlabeled dNTPs, or to dilute each [α-^{32}P]dNTP with an appropriate amount of the unlabeled dNTP.

3. Most commercial suppliers sell [α-^{32}P]dNTPs in a concentrated aqueous solution, which can be added directly to the nick-translation reaction. The specific activity ranges from 400 to 2000 Ci/mmole. If [α-^{32}P]dNTPs in an ethanol/water mixture are used, remove the ethanol and water as described on page 457.

4. Set up the nick-translation reaction as follows:

10× nick-translation buffer	5 μl
DNA	1 μg
unlabeled dNTPs (if needed)	1 nmole of each (1 μl of a 1 mM solution)
[α-^{32}P]dNTPs	100 pmoles
H$_2$O	to 44 μl

Chill the mixture to 0°C. Make a 10^4-fold dilution of a small quantity of a stock solution of DNase (1 mg/ml) in ice-cold, nick-translation buffer containing 50% glycerol. The diluted enzyme is stable when stored at −20°C in this buffer (see page 451).

5. Add 0.5 μl of diluted DNase I (0.1 μg/ml) to the reaction mixture. Mix by vortexing.

6. Add 5 units (as defined by Richardson et al. 1964) of *E. coli* DNA polymerase I. Mix.

7. Incubate at 16°C for 60 minutes.

   Note. If the reaction is carried out at higher temperature, a considerable amount of "snapback" DNA may be generated by DNA polymerase copying the newly synthesized strand.

9. Stop the reaction by adding 2 μl of 0.5 M EDTA.

10. Using the DE-81 binding or TCA precipitation assays described on page 473, determine the proportion of [α-^{32}P]dNTPs that have been incorporated into DNA.

11. Separate the nick-translated DNA from unincorporated dNTPs either by chromatography on or centrifugation through a small column of Sephadex G-50 (see pages 464–467).

Notes

i. The specific activity of the nick-translated DNA depends not only on the specific activity of the dNTPs, but also on the extent of nucleotide replacement of the template. This can be controlled by varying the amount of DNase I in the reaction. The aim is to establish conditions that will result in incorporation of about 30% of the [α-^{32}P]dNTPs into DNA.

TABLE 4.5.

Tube	DNase I (μl)	
	0.01 μg/ml	0.1 μg/ml
1	0	0
2	1	0
3	2	0
4	4	0
5	0	1

The size of DNA after nick translation also depends on the amount of DNase I added to the reaction and the amount of DNase contaminating the preparation of DNA polymerase I. Because the amount of contaminating nuclease differs widely between different batches of commercial DNA polymerase I and DNase I, each new preparation should be tested as follows:

Set up five nick-translation reactions. Add DNA polymerase I to all of the tubes and different quantities of DNase I as shown in Table 4.5.

Use the DE-81 binding or TCA precipitation assays (see page 473) to determine the proportion of [α-^{32}P]dNTPs that have been incorporated into DNA.

Analyze the size of the DNA by electrophoresis through an alkaline agarose gel (see page 171).

Choose a concentration of DNase I that results in incorporation of about 30% of the [α-^{32}P]dNTPs into DNA. The labeled DNA strands should be 400–800 nucleotides in length.

ii. Nick translation usually yields uniformly labeled DNA (Rigby et al. 1977), indicating that any bias in nicking by DNase I (Ehrlich et al. 1973) or in synthesis by DNA polymerase is too small to produce significant distortion in the pattern of labeling.

iii. The enzymes used in nick translation are sensitive to contaminants in agarose. Thus, DNA samples eluted from agarose gels should be carefully purified as described in Chapter 5 before they are used in nick-translation reactions.

iv. For some purposes (e.g., analysis of DNA · RNA hybrids by nuclease S1), it is desirable to obtain nick-translated DNA whose chain length is significantly longer than 400–800 nucleotides. In this case:

a. Carry out the nick-translation reaction using the amount of DNase I determined to be optimal as described above.

b. Add 2 μl of 0.5 M EDTA. Heat at 70°C for 5 minutes to inactivate DNA polymerase and DNase.

c. Add 150 μl of 10 mM MgCl$_2$ and 20 μl of 10× ligation buffer (see page 125).

d. Add 2 units of T4 ligase.

e. Incubate at room temperature for 2 hours.

f. Separate the nick-translated and ligated DNA from unincorporated dNTPs either by chromatography through a column of Sephadex G-50 or by spun-column chromatography (page 464–467).

g. Analyze the size of the DNA by electrophoresis through an alkaline agarose gel (see page 171).

h. If necessary, isolate DNA of the required size from a preparative alkaline agarose gel.

Stock Solutions

10× Nick-translation buffer

0.5 M Tris · Cl (pH 7.2)
0.1 M MgSO$_4$
1 mM dithiothreitol
500 μg/ml bovine serum albumin (BSA Pentax Fraction V)

Divide into small aliquots and store at −20°C.

DNase I. Prepare a stock solution containing 1 mg/ml of DNase I in 0.15 M NaCl and 50% glycerol. (Electrophoretically pure DNase I can be obtained from Worthington Biochemicals.) Divide into small aliquots and store at −20°C.

Deoxynucleotide triphosphates

dATP (1 mM)
dGTP (1 mM)
dCTP (1 mM)
TTP (1 mM)

Prepare stock solutions as described on page 449.

E. coli *DNA Polymerase I.* Most commercial suppliers sell the enzyme in buffer containing 50% glycerol. Usually 1 μl of the preparation contains 5 units of enzyme. (For a definition of a unit, see Richardson et al. 1964).

KLENOW FRAGMENT OF *E. COLI* DNA POLYMERASE I

Filling Recessed 3' Ends of Double-stranded DNA

The reaction conditions are identical to those used for nick translation of DNA (see page 109), except that:

The Klenow fragment of *E. coli* DNA polymerase is used instead of the holoenzyme.

No DNase is used.

Generally, only one of the four dNTPs is labeled.

Which dNTPs are added to the reaction depends on the sequence of the protruding 5' termini at the ends of the DNA; e.g., to fill in recessed 3' ends created by cleavage of DNA by *Eco*RI, only dATP and TTP need be present in the reaction:

$$p\text{-}C_p\text{-}T_p\text{-}T_p\text{-}A_p\text{-}A_p{}^{5'} \quad \xrightarrow[\text{TTP}]{\text{dATP}} \quad C_p\text{-}T_p\text{-}T_p\text{-}A_p\text{-}A_p{}^{5'}$$

$$p^G\!\rceil \qquad\qquad\qquad\qquad\qquad G\text{-}_pA\text{-}_pA\text{-}_pT\text{-}_pT\rceil$$
$$\quad OH_{3'} \qquad\qquad\qquad\qquad\qquad\qquad\qquad OH_{3'}$$

On the other hand, all four dNTPs are required to fill recessed ends created by *Hind*III:

$$p\text{-}T_p\text{-}T_p\text{-}C_p\text{-}G_p\text{-}A_p{}^{5'} \quad \xrightarrow[\substack{\text{dCTP}\\\text{TTP}}]{\substack{\text{dATP}\\\text{dGTP}}} \quad T_p\text{-}T_p\text{-}C_p\text{-}G_p\text{-}A_p{}^{5'}$$

$$p^A\!\rceil \qquad\qquad\qquad\qquad\qquad\qquad A\text{-}_pA\text{-}_pG\text{-}_pC\text{-}_pT\rceil$$
$$\quad OH_{3'} \qquad\qquad\qquad\qquad\qquad\qquad\qquad OH_{3'}$$

Finally, to repair the ends left after treatment of DNA with nuclease S1 or *Bal*31, all four dNTPs should be present during the reaction.

A typical reaction contains 1 μg of DNA in 20 μl. However, the reaction works well over a wide range of DNA concentrations (1–500 μg/ml).

1. Mix:

DNA	1 μg
10× nick-translation buffer	2 μl
unlabeled dNTPs (as needed)	2 nmoles of each (1 μl of a 2 mM solution)
$[\alpha\text{-}^{32}\text{P}]$dNTP	2 pmoles (sp. act. >400 Ci/mmole)
Klenow fragment of DNA polymerase	1 unit
H₂O	to 25 μl

2. Incubate at room temperature for 30 minutes.

3. Stop the reactions by adding 1 μl of 0.5 M EDTA. Extract once with phenol/chloroform (see page 458).

4. Separate the DNA from unincorporated dNTPs by chromatography on or centrifugation through small columns of Sephadex G-50 (see pages 464–467).

Large Fragment of DNA Polymerase I (Klenow fragment)

(E.coli)

The enzyme consists of a single polypeptide chain ($M_r = 76,000$) produced by cleavage of intact DNA polymerase I with subtilisin (Jacobsen et al. 1974). This peptide carries the $5' \longrightarrow 3'$ polymerase activity and the $3' \longrightarrow 5'$ exonuclease activity of intact DNA polymerase I but lacks the $5' \longrightarrow 3'$ exonuclease.

Activity	Reaction	Template/primer or substrate
$5' \longrightarrow 3'$ Polymerase	$DNA_{OH} + ndNTP \xrightarrow{Mg^{++}} DNA\text{-}(pdN)_n + {}_nPPi$	single-stranded template primer with 3' OH

For example:

$$5'$$
$$\ldots_pC -_pC -_pG$$
$$\ldots G_p - G_p - C_p - T_p - A_p - T_p - C_p - G_p - A_p$$
$$3'$$

+ dATP
   dCTP
   dGTP
   TTP

$$\Big\downarrow Mg^{++}$$

$$5'3'$$
$$\ldots_pC -_pC -_pG -_pA -_pT -_pA -_pG -_pC -_pT\ldots$$
$$\ldots G_p - G_p - C_p - T_p - A_p - T_p - C_p - G_p - A_p \ldots$$
$$3'5'$$

Activity	Reaction	Template/primer or substrate
$3' \longrightarrow 5'$ Exonuclease	$\begin{array}{c} ds\,DNA \\ ss\,DNA \end{array} \xrightarrow{Mg^{++}} 5\,pN_{OH}$	degrades double-stranded or single-stranded DNA from a free 3'-OH end; activity on double-stranded DNA blocked by $5' \longrightarrow 3'$ polymerase activity

Uses

1. Filling the 3' recessed termini created by digestion of DNA with restriction enzymes; for example:

$$5'\text{Klenow}5'$$
$$\ldots_pC -_pC_{OH}\xrightarrow[\text{dATP, TTP, Mg}^{++}]{\text{fragment of pol I}}\ldots_pC -_pC -_pT -_pT -_pA -_pA$$
$$\ldots G_p - G_p - A_p - A_p - T_p - T_p\ldots G_p - G_p - A_p - A_p - T_p - T_p$$
$$3'$$

2. Labeling the termini of DNA fragments (by using $[^{32}P]$ dNTPs in end-filling reactions).
3. Second-strand cDNA synthesis in cDNA cloning.
4. Sequencing DNA using the Sanger dideoxy system (Sanger et al. 1977).
5. At one time the $3' \longrightarrow 5'$ exonuclease activity of the Klenow fragment was used to digest protruding 3' termini created by some restriction enzymes. Lately, T4 DNA polymerase has become the enzyme of choice for this purpose because of its more active $3' \longrightarrow 5'$ exonuclease.

Rapid End-labeling of DNA[1]

The reaction conditions used are similar to those used for filling recessed 3′ ends of double-stranded DNA, except that:

Generally, only one $[\alpha\text{-}^{32}P]dNTP$ is used.

The reaction may be carried out immediately after digesting DNA with a restriction enzyme. There is no need to remove the restriction enzyme or to inactivate it, and there is no need to change buffers.

The $[\alpha\text{-}^{32}P]dNTP$ added to the reaction depends on the sequence of the protruding 5′ termini at the ends of DNA. For example, ends created by cleavage of DNA with *Eco*RI can be labeled with $[\alpha\text{-}^{32}P]dATP$.

Ends created by cleavage of DNA with *Bam*HI can be labeled with $[\alpha\text{-}^{32}P]dGTP$.

Blunt-ended DNA fragments are labeled by replacement of the nucleotide present at the 3′-hydroxyl terminus.

DNA termini with 3′ extensions are not labeled efficiently by the Klenow fragment of *E. coli* DNA polymerase. To label these molecules, use T4 DNA polymerase (see page 117).

1. Digest up to 1 μg of DNA with the desired restriction enzyme in 25 μl of the appropriate buffer.

2. Add 2.0 μCi of the appropriate $[\alpha\text{-}^{32}P]dNTP$.

3. Add 1.0 unit of the Klenow fragment of *E. coli* DNA polymerase, and incubate for 10 minutes at room temperature.

[1]Drouin (1980).

4. Either load the end-labeled DNA directly onto a gel or separate the labeled DNA from unincorporated dNTPs by chromatography on or centrifugation through small columns of Sephadex G-50 (see pages 464–467).

Notes

i. This procedure is the method of choice for generating labeled DNA fragments that can be used as size markers during gel electrophoresis. Because DNA fragments are labeled in proportion to their molar concentrations and not their sizes, both small and large fragments in a restriction enzyme digest become labeled to an equal extent. It is therefore possible to use autoradiography to locate bands of DNA that are too small to be visualized by staining with ethidium bromide.

ii. The procedure works well on relatively crude DNA preparations (e.g., minipreparations of plasmids; see pages 366ff).

iii. When the labeled DNA is to be used for sequencing by the Maxam-Gilbert technique or for mapping mRNA by the nuclease-S1 method, the amount of labeled dNTP added to the reaction should be raised to 250 pmoles. After the reaction has been allowed to proceed for 15 minutes at room temperature, 2 nmole of each of the four unlabeled dNTPs should be added (i.e., 1 μl of a 2 mM solution) and the incubation should be continued for a further 5 minutes at room temperature. This cold chase ensures that each recessed 3′ terminus will be completely filled and that all labeled DNA molecules will be exactly the same length.

iv. By choosing the appropriate, labeled dNTP, it is possible to label only one end of a duplex DNA molecule. If, for example, the DNA has been cleaved by *Hind*III at one end and *Bam*HI at the other, it can be labeled selectively by including either [α-^{32}P]dATP or [α-^{32}P]dGTP in the reaction.

T4 DNA POLYMERASE

The following protocols are adapted from O'Farrell (1981).

Rapid End-labeling of DNA

T4 DNA polymerase, like the Klenow fragment of *E. coli* DNA polymerase, can be used to label flush or recessed 3′ termini of DNA. However, T4 DNA polymerase carries a much more powerful 3′ → 5′ exonuclease activity than the Klenow fragment, and it is therefore the enzyme of choice for end-labeling DNA molecules with protruding 3′ tails.

In many cases, a single buffer (whose composition is given below) can be used both for cleavage of DNA with a restriction enzyme and for the subsequent end-labeling. However, not all restriction enzymes work in T4 polymerase buffer, and it is advisable to carry out preliminary pilot reactions with the particular batch of enzyme on hand. If the restriction enzyme will not work in the T4 DNA polymerase buffer, it is necessary to carry out the restriction endonuclease digestion and end-labeling in two separate steps. In this case, cleave the DNA in the appropriate restriction enzyme buffer, remove the restriction enzyme by extraction with phenol/chloroform, precipitate the DNA with ethanol, redissolve it in TE, and add the appropriate volume of 10× T4 DNA polymerase buffer.

10× T4 DNA polymerase buffer

0.33 M Tris-acetate (pH 7.9)
0.66 M potassium acetate
0.10 M magnesium acetate
5 mM dithiothreitol
1 mg/ml bovine serum albumin (BSA Pentax Fraction V)

The 10× stock should be divided into small aliquots and stored frozen at −20°C.

T4 DNA Polymerase

(T4-infected E.coli)

Like the Klenow fragment of E.coli DNA polymerase I, T4 DNA polymerase possesses a $5' \longrightarrow 3'$ polymerase activity and a $3' \longrightarrow 5'$ exonuclease activity. However, the exonuclease activity of T4 DNA polymerase is more than 200 times as active as that of DNA polymerase I.

Activity	Reaction	Template/primer or substrate
$5' \longrightarrow 3'$ Polymerase	$DNA_{OH} + ndNTP \xrightarrow{Mg^{++}} DNA(pdN)_n + nPPi$ For example: $5'$ $...C-_pC-_pG_{OH}$ $...G_p-G_p-C_p-T_p-A_p-T_p-C_p-G_p-A_p$ $3'$ $Mg^{++} \Big| \begin{matrix} dATP \\ dCTP \\ dGTP \\ TTP \end{matrix}$ $5'$ $...C-_pC-_pG-_pA-_pT-_pA-_pG-_pC-_pT$ $...G_p-G_p-C_p-T_p-A_p-T_p-C_p-G_p-A_p$ $3'$	single-stranded template primer with 3' OH
$3' \longrightarrow 5'$ Exonuclease	$ssDNA \xrightarrow{Mg^{++}} \begin{matrix} 5' \\ pN_{OH} \end{matrix}$ For example: $5'$ $..._pC-_pG-_pC-_pA-_pT-_pC-_pT \ 3'$ $...p-G_p-C_p-G_p$ $3'$ $\Big| Mg^{++}$ $5'$ $..._pC-_pG-_pC_{OH}$ $...p-G_p-C_p-G_p$ $3'$	considerably more active on single-stranded DNA than double-stranded DNA; activity on double-stranded DNA blocked by $5' \longrightarrow 3'$ polymerase activity

1. Mix:

DNA	0.5–2.0 μg
10× T4 polymerase buffer	2 μl
H₂O	to 19 μl

2. Add the desired restriction enzyme and incubate for the appropriate time.

3. Add directly to the reaction 1 μl of a 2 mM solution of three of the four dNTPs (e.g., 2 mM dGTP, 2 mM dATP, 2 mM TTP).

4. Add approximately 2 μCi of an aqueous solution of the fourth dNTP, labeled with α-^{32}P (in the example given above, this would be [α-^{32}P]dCTP).

Exchange reaction

If only one dNTP is present, the 3'——→5' exonuclease activity will degrade double-stranded DNA from the 3'-OH end until a base is exposed that is complementary to the dNTP. A continuous series of synthesis and exchange reactions then occurs at that position. For example:

$$5'$$
$$\cdots {}_pC - {}_pG - {}_pT - {}_pC - {}_pG - {}_pC_{OH}$$
$$\cdots p^{-}G_p - C_p - A_p - G_p - C_p - G_p$$
$$3'$$

$$\left[{}^{32}P\right]TTP \quad \bigg| \quad Mg^{++}$$

$$\downarrow$$

$$5'$$
$$\cdots {}_pC - {}_pG - {}_p$$
$$\cdots p^{-}G_p - C_p - A_p - G_p - C_p - G_p$$
$$3'$$

$$\downarrow$$

$$5'$$
$$\cdots {}_pC - {}_pG - {}_p{}^{*}T_{OH}$$
$$\cdots p^{-}G_p - C_p - A_p - G_p - C_p - G_p$$
$$3'$$

Uses

1. Labeling the termini of DNA fragments with protruding 5' ends (filling reaction).
2. Labeling the termini of blunt-ended DNA fragments or DNA fragments with protruding 3' termini (exchange reaction).
3. Labeling DNA fragments for use as hybridization probes by partial digestion of double-stranded DNA with 3'——→5' exonuclease, followed by a filling reaction with $\left[{}^{32}P\right]$ dNTPs (O'Farrell et al. 1980).

4. Sequencing DNA by the plus/minus method, see Sanger and Coulson (1975).

5. Add 1 μl (2.5 units) of T4 DNA polymerase.

6. Incubate for 5 minutes at 37°C.

7. Add 1 μl of a 1 mM solution of the unlabeled, fourth dNTP (in this case, dCTP). Continue incubation for about 10 minutes.

8. Stop the reaction by heating at 70°C for 5 minutes.

In the presence of dNTPs, the 3' exonuclease activity of the T4 DNA polymerase on double-stranded DNA is masked by the polymerization reaction. Thus, the product of the protocol described above will be a blunt-ended molecule with labeled nucleotides incorporated at or very near the termini of the DNA.

It is possible to obtain more extensive labeling by a replacement reaction, in which the 3′ exonuclease activity of the enzyme first digests duplex DNA to produce molecules with recessed 3′ termini. On subsequent addition of labeled dNTPs, the partially digested DNA molecules serve as primer · templates that are regenerated by the polymerase into intact, double-stranded DNA. Molecules labeled to high specific activity by this technique are used chiefly as hybridization probes. They have two advantages over probes prepared by nick-translation. First, they lack the artifactual hairpin structures that can be produced during nick translation. Second, they can easily be converted into strand-specific probes by cleavage with suitable restriction endonucleases (see Fig. 4.1).

However, this method, by contrast to nick translation, does not produce a uniform distribution of label along the length of the DNA. Furthermore, the 3′ exonuclease activity degrades single-stranded DNA much faster than double-stranded DNA, so that after a molecule has been digested to its midpoint, it will dissociate into two, half-length, single strands that will be rapidly degraded. It is therefore important to stop the exonuclease reaction before the enzyme reaches the center of the molecule. Consequently, the replacement synthesis method yields a population of molecules that are fully labeled at their ends but that contain progressively decreasing quantities of label toward their centers. Thus, the size of the smallest restriction fragment in a mixture of fragments dictates the maximum extent to which all the fragments can be labeled.

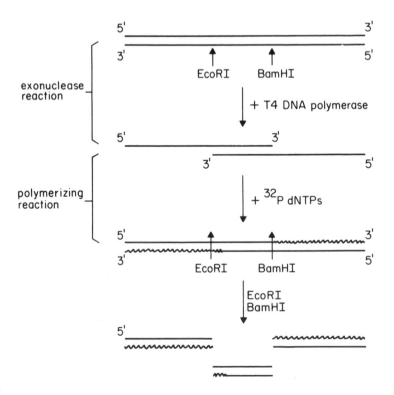

Figure 4.1

Replacement Synthesis

1. Mix:

DNA	0.2–0.5 μg
10 × T4 polymerase buffer	2 μl
H$_2$O	to 20 μl

2. Add the desired restriction enzyme and incubate for the desired time.

3. Add T4 DNA polymerase directly to the reaction. The amount of enzyme added and the ratio of enzyme to DNA in the reaction determine the rate of exonuclease digestion.

Ratio (units enzyme:μg DNA)	Number of Nucleotides Excised from Each 3′ End
0.62	10/min
1.25	20/min
1.75	30/min
2.5	40/min

Note. The rate of exonuclease digestion no longer increases linearly as a function of the enzyme-DNA ratio when the ratio exceeds 2.5.

4. Calculate the moles of nucleotide excised, using the following formula:

$$M = D \times \frac{N}{B}$$

where M = of nucleotide excised; D = amount of DNA in the reaction (expressed as moles of nucleotide); N = number of nucleotides excised from one end of DNA; B = number of base pairs in DNA. N.B.: 330 g = 1 mole of nucleotide in DNA.

5. Add directly to the reaction 1 μl of a 2 mM solution of three of the four dNTPs (e.g., 2 mM dGTP, 2 mM dATP, 2 mM TTP).

6. Add the fourth dNTP labeled with α-^{32}P (sp. act. $\geqslant$400 Ci/mmole) in an amount at least equivalent to the moles of nucleotide excised from the DNA by exonuclease. Incubate at 37°C for 1 hour.

7. Add 1 μl of a 2 mM solution of the fourth dNTP. Incubate at 37°C for a further 15 minutes.

8. Separate the labeled DNA from unincorporated dNTPs by chromatography on or centrifugation through small columns of Sephadex G-50 (see pages 464–467).

Note. The specific activity of the replaced sequences will be about 7×10^8 cpm/μg if a single [α-^{32}P]dNTP of specific activity of 400 Ci/mole is used.

LABELING THE 5′ ENDS OF DNA WITH T4 POLYNUCLEOTIDE KINASE

Forward Reaction—Using DNA Molecules with Protruding 5′ Termini as Templates[2]

1. Mix:

dephosphorylated DNA, 5′ ends	1–50 pmoles
10× kinase buffer I	10 μl
[γ-^{32}P]ATP (sp. act. = 3000 ci/mmole)	50 pmoles (150 μCi)
T4 polynucleotide kinase	10–20 units
H$_2$O	to 50 μl

   Incubate at 37°C for 30 minutes.

   10× Kinase buffer I

   0.5 M Tris·Cl (pH 7.6)
   0.1 M MgCl$_2$
   50 mM dithiothreitol
   1 mM spermidine
   1 mM EDTA

2. Add 2 μl of 0.5 M EDTA. Extract once with phenol/chloroform, and precipitate the DNA with ethanol.

3. Redissolve the DNA in 50 μl of TE (pH 7.9).

4. Separate the labeled DNA from unincorporated [γ-^{32}P]ATP by chromatography on or centrifugation through small columns of Sephadex G-50 (see pages 464–467).

Notes

   i. 1 mole of 5′ ends = 0.5 mole of DNA.
   For example:
   1 mole of linear pBR322 DNA = 3.2 × 10^6 g
   1 mole of 5′ ends of linear pBR322 DNA = 1.6 × 10^6 g
   1 pmole of 5′ ends of linear pBR322 DNA = 1.6 μg

   ii. Spermidine stimulates incorporation of [γ-^{32}P]ATP and inhibits a nuclease present in some preparations of polynucleotide kinase.

   iii. The ATP concentration in the reaction should be at least 1 μM.

   iv. The dephosphorylated DNA should be rigorously purified by gel electrophoresis or density gradient centrifugation in order to free it from low-molecular-weight nucleic acids. Although such contaminants may make

[2]Maxam and Gilbert (1980).

T4 Polynucleotide Kinase

(T4-infected E.coli)

The enzyme catalyzes the transfer of the δ-phosphate of ATP to a 5'-OH terminus in DNA or RNA (Richardson 1971).

Activity	Reaction		Substrate

Kinase
(forward reaction)

$$DNA^{5'}_{OH} \text{ or } RNA^{5'}_{OH} \xrightarrow[\substack{[\alpha^{-32}P]ATP \\ DTT}]{Mg^{++}} \quad [^{32}P]DNA \text{ or } [^{32}P]RNA^{5'} + ADP$$

Substrate: single or double-stranded DNA with 5'-OH terminus

For example:

$$A-P-P-P^{32} + {}^{L}C_p-G_p-C_p\cdots^{5'OH}$$

$$\updownarrow \quad DTT \mid Mg^{++}$$

$${}^{5'}_{32}p-C_p-G_p-C_p\cdots + A-P-P$$

Kinase
(exchange reaction)

$$A-P-P-P^{32} + \text{excess ADP} \quad + \quad {}^{5'}p-C_p-G_p-C_p\cdots$$

$$\updownarrow \quad DTT \mid Mg^{++}$$

$${}^{32}p-C_p-G_p-C\cdots + A-P-P + A-P-P-P$$

Substrate: single or double-stranded DNA with 5'-P terminus

The excess ADP drives the exchange reaction, causing polynucleotide kinase to transfer the terminal 5'-phosphate from DNA to ADP. The DNA is then rephosphorylated by transfer of the labeled γ-phosphate in the $[\alpha^{-32}P]ATP$ (Berkner and Folk 1977).

Uses

1. Labeling 5' termini in DNA for sequencing by the Maxam-Gilbert technique (Maxam and Gilbert 1977).

2. Phosphorylating synthetic linkers and other fragments of DNA lacking 5'-P termini prior to ligation.

up only a small fraction of the weight of the nucleic acid in the preparation, they provide a much larger proportion of the 5' ends. Unless steps are taken to remove them, contaminating low-molecular-weight DNAs and RNAs can be the predominant species of nucleic acid that are labeled in polynucleotide kinase reactions.

v. Ammonium ions are strong inhibitors of polynucleotide kinase. Therefore DNA should not be dissolved in or precipitated from buffers containing ammonium salts prior to kinasing.

Forward Reaction—Using DNA Molecules with Blunt Ends or Recessed 5′ Termini as Templates

DNA molecules with blunt ends or recessed 5′ ends are labeled less efficiently with polynucleotide kinase than molecules with protruding 5′ ends. The efficiency of labeling may be improved as follows:

1. Mix:

   dephosphorylated DNA, 5′ ends   1–50 pmoles

   a solution of $\begin{cases} 0.2 \text{ M Tris} \cdot \text{Cl (pH 9.5)} \\ 10 \text{ mM spermidine} \\ 1 \text{ mM EDTA} \end{cases}$ 4 μl

   H_2O to 40 μl

   Heat to 70°C and chill quickly on ice.

2. Add 5 μl of 10× blunt-end kinase buffer.

   10× Blunt-end kinase buffer
   0.5 M Tris·Cl (pH 9.5)
   0.1 M $MgCl_2$
   50 mM dithiothreitol
   50% glycerol

3. Add at least 50 pmoles of [γ-^{32}P]ATP (sp. act. = 3000 Ci/mmole) in a volume of 5 μl.

4. Add 20 units of T4 polynucleotide kinase.

5. Mix and incubate at 37°C for 30 minutes.

6. Add 2 μl of 0.5 M EDTA.

7. Extract once with phenol/choloroform.

8. Add 5 μl of 3 M sodium acetate (pH 5.2) to the aqueous phase and precipitate the DNA with ethanol.

9. Redissolve the DNA in 50 μl of TE (pH 7.6).

10. Separate the labeled DNA from unincorporated [γ-^{32}P]ATP by chromatography on or centrifugation through small columns of Sephadex G-50 (see pages 464–467).

Forward Reaction—Using Synthetic Linkers as Templates[3]

1. Dissolve 1 OD_{260} of linkers (as supplied by the manufacturer) in 100 μl of TE (pH 7.6).

2. Mix:

linkers (0.5 mg/ml)	4 μl ($\sim 2 \mu$g)
[γ-^{32}P]ATP (sp. act. >3000 Ci/mmole)	6.6 pmoles (20 μCi)
10× linker-kinase buffer	1 μl
H_2O	to 9 μl

   Add 1 μl (10 units) polynucleotide kinase.

   10× Linker-kinase buffer
   0.7 M Tris·Cl (pH 7.6)
   0.1 M $MgCl_2$
   50 mM dithiothreitol

3. Incubate at 37°C for 15 minutes.

4. Add:

10× linker-kinase buffer	1 μl
10 mM ATP	1 μl
H_2O	7 μl
polynucleotide kinase	1 μl (10 units)

   Incubate for a further 30 minutes at 37°C. Store the kinased linkers at -20°C.

5. Check that the kinased linkers can be ligated according to the following steps.

   a. Mix:

kinased linkers	5 μl
10× ligation buffer	1 μl
H_2O	3 μl
T4 DNA ligase	1 μl

   Incubate at 4°C for 4 hours.

[3]Maniatis et al. (1978).

10× Ligation buffer

    0.66 M Tris · Cl (pH 7.6)
    50 mM $MgCl_2$
    50 mM dithiothreitol
    10 mM ATP

b. Heat to 65°C for 15 minutes (to inactivate ligase).

c. Remove 5 μl of the kinased, ligated linkers to a fresh tube. Add:

H_2O	10 μl
10× restriction buffer	2.5 μl
restriction enzyme	10 units

Incubate at 37°C for 1 hour.

d. Analyze unligated, kinased linkers (2 μl), ligated, kinased linkers (5 μl), and ligated, kinased, restricted linkers (5 μl) by electrophoresis through a 10% acrylamide gel, cast and run in 0.5× TBE. Run the gel until the bromophenol blue has migrated half the length of the gel.

e. Cover the gel with Saran Wrap and expose for autoradiography at −70°C. The autoradiograph of ligated, kinased linkers should look like Figure 4.2.

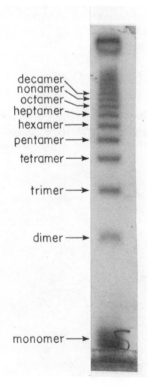

decamer
nonamer
octamer
heptamer
hexamer
pentamer
tetramer
trimer
dimer
monomer

Figure 4.2

Exchange Reaction—DNA Molecules with Protruding 5′ Phosphoryl Termini as Templates

This procedure (Berkner and Folk 1977) is faster than the forward reaction because it does not require that the DNA be dephosphorylated. In our hands, however, it is less efficient.

1. Mix:

DNA with 5′ terminal phosphates	1–50 pmoles of 5′ ends
10× exchange-reaction buffer	5 μl
5 mM ADP	3 μl
[γ-^{32}P]ATP (sp. act. = 3000 Ci/mmole)	100 pmoles (i.e., 30 μl of a 10 mCi/ml solution)
H$_2$O	to 50 μl
T4 polynucleotide kinase	1 μl (20 units)

10× Exchange-reaction buffer
0.5 M imidazole · Cl (pH 6.6)
0.1 M MgCl$_2$
50 mM dithiothreitol
1 mM spermidine
1 mM EDTA

2. Incubate at 37°C for 30 minutes.

3. Add 2 μl of 0.5 M EDTA.

4. Extract once with phenol/chloroform.

5. Add 5 μl of 3 M sodium acetate (pH 5.2) to the aqueous phase and precipitate the DNA with ethanol.

6. Redissolve the DNA in 50 μl of TE (pH 7.6).

7. Separate the labeled DNA from unincorporated [γ-^{32}P]ATP by chromatography on or centrifugation through small columns of Sephadex G-50 (see pages 464–467).

RNA-dependent DNA Polymerase (reverse transcriptase)

(avian myeloblastosis virus)

The enzyme consists of two polypeptides (one of which carries both a $5' \rightarrow 3'$ polymerase activity and a processive 5' and 3' riboexonuclease that is specific for RNA·DNA hybrids (Verma 1977).

Activity	Reaction	Template/primer or substrate
$5' \rightarrow 3'$ DNA polymerase	DNA_{OH} or $+$ ndNTP RNA_{OH} $\downarrow Mg^{++}$ $DNA\text{-}(pdN)_n + nPPi$ or $RNA\text{-}(pdN)_n + nPPi$	RNA or DNA template RNA or DNA primer with $3'_{OH}$

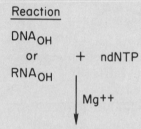

$$\begin{array}{l} 5' \qquad\qquad\qquad\quad OH \\ \,_{,''p}T-_pT-_pT-_pT-_pT\rfloor \\ ...A_p-A_p-A_p-A_p-A_p-U_p-C_p-U_p-G_p-U_p-C_p-C_p-U_p-A_p \\ 3' \qquad\qquad\qquad\qquad\qquad\qquad\qquad\qquad\qquad 5' \end{array}$$

$$\left. \begin{array}{l} dATP \\ dCTP \\ dGTP \\ TTP \end{array} \right| Mg^{++} \;\downarrow$$

$$\begin{array}{l} 5' \qquad\qquad\qquad\qquad\qquad\qquad\qquad\qquad OH \\ _pT-_pT-_pT-_pT-_pT-_pA-_pG-_pA-_pC-_pA-_pG-_pG-_pA-_pT\rfloor \\ A_p-A_p-A_p-A_p-A_p-U_p-C_p-U_p-G_p-U_p-C_p-C_p-U_p-A_p \\ 3' \qquad\qquad\qquad\qquad\qquad\qquad\qquad\qquad\qquad 5' \end{array}$$

$5' \rightarrow 3'$ Exoribonuclease $3' \rightarrow 5'$ Exoribonuclease (RNase H)	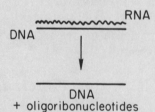	specifically degrades RNA in a DNA·RNA hybrid by a processive mechanism

Uses

1. Synthesis of cDNA for cloning (both first and second strands) or for hybridization probes.

2. Labeling the termini of DNA fragments with protruding 5' ends (filling reaction).

RNA-DEPENDENT DNA POLYMERASE

RNA-dependent DNA polymerase (reverse transcriptase) is used chiefly to transcribe mRNA into double-stranded DNA, which can be inserted into prokaryotic vectors. The details of this procedure are given in Chapter 7. Reverse transcriptase can also be used with either single-stranded DNA or RNA templates to make probes for use in hybridization experiments. The primers for these reactions can be oligo(dT) (see Chapter 7) or a collection of a large number of randomly generated oligodeoxynucleotides (Taylor et al. 1976). The diversity of these oligodeoxynucleotides is so large as to guarantee that some of them will be complementary to sequences in the template nucleic acid. After annealing to the template, these oligonucleotides serve as primers for reverse transcriptase. Because different oligonucleotides bind to different sequences in the template, all parts of the template are represented in the resulting DNA at equal frequency (at least in theory). Furthermore, random oligonucleotides can be used as primers on any single-stranded DNA or RNA template. By contrast, oligo(dT) binds only to poly(A), which primes the synthesis of DNA that is heavily biassed toward sequences at the 3' end of an mRNA template.

Preparation of Oligodeoxynucleotide Primers

1. Dissolve approximately l g of commercial calf thymus DNA in 30 ml of 20 mM Tris·Cl (pH 7.4) and 10 mM $MgCl_2$.

2. Add 2 mg of powdered pancreatic DNase and mix by vortexing. The viscosity of the solution should decrease rapidly.

3. Incubate at 37°C for 30 minutes.

4. Add 3 ml of 10% SDS and 1.5 ml of pronase (20 mg/ml).

5. Incubate at 37°C for 45 minutes.

6. Remove proteins by extracting twice with phenol/chloroform.

7. Transfer the aqueous phase to a Pyrex glass tube. Denature the DNA by heating for 15 minutes in a boiling-water bath. Cool quickly by immersing the tube in an ice-water bath.

8. Add 0.6 ml of 5 M NaCl (final concentration of NaCl = 0.1 M) and load the solution onto a column (20-ml bed volume) of DEAE-cellulose (Whatman DE-52) equilibrated with a solution of 5 mM Tris·Cl (pH 7.4), 1 mM EDTA, and 0.1 M NaCl.

9. Wash the column with approximately 300 ml of the same buffer. The OD_{260} of the final effluent fractions should now be less than 0.05.

10. Recover the primers by washing the column with a solution of 5 mM Tris·Cl (pH 7.4), 1 mM EDTA, and 0.3 M NaCl until all the material absorbing at 260 nm has eluted. This usually requires approximately 150 ml of elution buffer.

11. Concentrate the primers by precipitation with ethanol for several hours at −20°C.

12. Recover the primers by centrifugation (2000*g* for 20 minutes at 0°C). After drying the pellet under vacuum, dissolve the primers in 1 ml of H_2O. Measure the OD_{260} of a 1:1000 dilution of the preparation. The yield of primers varies between 5% and 30% from preparation to preparation. Adjust the concentration of primers to 50 mg/ml, divide the preparation into small aliquots, and store frozen at −20°C.

13. Test the primers in a reaction with reverse transcriptase in the presence and absence of template (see page 131). The number of counts incorporated into DNA in the absence of template should be less than 1% of the amount incorporated in the presence of template.

Preparation of Hybridization Probes Using Reverse Transcriptase and Random Primers

This is the method of choice for synthesizing ^{32}P-labeled probes of high specific activity from single-stranded DNA or from RNA templates.

1. Mix 1.0 μg of template (linear double-stranded or single-stranded DNA or RNA) with 20 μl of water. Heat to 100°C in a boiling-water bath for 5 minutes. Chill quickly in ice water.

2. Add:

calf thymus or salmon sperm primer (50 mg/ml)	10 μl
5× random-primer buffer	20 μl
2 mM solution of each unlabeled dNTP	2 μl
[α-^{32}P]dNTP (sp. act. > 400 Ci/mM)	250 pmoles (100 μCi)
reverse transcriptase	200 units
H_2O	to 100 μl

5× Random-primer buffer

0.25 M Tris · Cl (pH 8.1)
10 mM dithiothreitol
25 mM $MgCl_2$
0.2 M KCl

3. Mix and incubate at 37°C for 1 hour.

4. Add 2 μl of 0.5 M EDTA. Separate the labeled DNA from unincorporated dNTPs either by chromatography through a column of Sephadex G-50 or by spun-column chromatography (see pages 464–467). Approximately 30% of the [α-^{32}P]dNTP should have been incorporated into DNA.

Notes

i. RNA templates can be removed at the end of the reaction as follows: After addition of EDTA (step 4), add 12 μl of 3 M NaOH and incubate for 12 hours at 37°C. The alkaline solution can then be applied directly to Sephadex G-50 equilibrated in TE (pH 7.6).

ii. Hybridization probes made by reverse transcriptase from DNA are single-stranded copies of the templates:

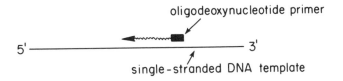

5' ————————————————————————— 3'
single-stranded DNA template
oligodeoxynucleotide primer

Those made from RNA, however, consist of both single-and double-stranded molecules that are synthesized by two different mechanisms:

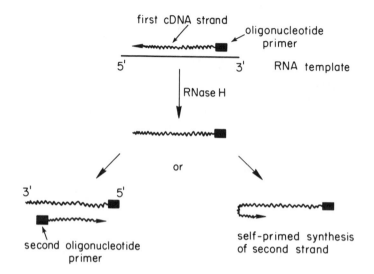

Self-primed synthesis is much the less efficient of the two reactions, and self-complementary hairpin molecules usually constitute less than 2–3% of the final yield of ^{32}P-labeled DNA. If necessary, the synthesis of second-strand (both self- and exogenously primed) can be inhibited by including actinomycin D in the reaction mixture at a final concentration of 1 mg/ml. In the presence of actinomycin D, the incorporation of [α-^{32}P]dNTP into DNA is reduced by approximately 50%.

DEPHOSPHORYLATION OF DNA

The terminal 5' phosphates can be removed from DNA by treatment either with bacterial alkaline phosphatase (BAP) or with calf intestinal alkaline phosphatase (CIP) (Chaconas and van de Sande 1980). The latter enzyme has the considerable advantage that it can be completely inactivated by heating to 68°C in SDS. BAP, on the other hand, is heat-resistant (in fact, reactions with BAP are usually carried out at 68°C to suppress the activity of an exonuclease that often contaminates preparations of the enzyme). Either multiple extractions with phenol/chloroform or purification of the DNA fragment by gel electrophoresis are required to remove all traces of BAP activity. For most purposes, therefore, CIP is the preferred enzyme.

1. Digest the DNA to completion with the restriction enzyme of choice. Extract once with phenol/chloroform and precipitate the DNA with ethanol.

2. Dissolve the DNA in a minimum volume of 10 mM Tris · Cl (pH 8.0). Add:

   10× CIP buffer   5 μl
   H$_2$O   to 48 μl
   CIP (see note below).

   0.01 units of CIP are needed to remove the terminal phosphates from 1 pmole of 5' ends of DNA (1 pmole of 5' ends of a 4-kb linear DNA is 1.6 μg).

   10× CIP buffer
   0.5 M Tris · Cl (pH 9.0)
   10 mM MgCl$_2$
   1 mM ZnCl$_2$
   10 mM spermidine

   To dephosphorylate protruding 5' termini, incubate at 37°C for 30 minutes, add a second aliquot of CIP, and continue the incubation for a further 30 minutes.
   To dephosphorylate DNA with blunt ends or recessed 5' termini, incubate for 15 minutes at 37°C and 15 minutes at 56°C. Then add a second aliquot of CIP and repeat the incubations at both temperatures.

3. Add 40 μl of H$_2$O, 10 μl of 10× STE, and 5 μl of 10% SDS. Heat to 68°C for 15 minutes.

   10× STE (TNE)
   100 mM Tris · Cl (pH 8.0)
   1 M NaCl
   10 mM EDTA

Alkaline Phosphatase

(bacteria [BAP] ; calf intestine [CAP])

The enzyme catalyzes the removal of 5'-phosphate residues from DNA and RNA, ribo- and deoxyribonucleoside triphosphates.

Activity	Reaction			Substrates
Phosphatase	$5'_{pDNA}$		$5'_{OH}$-DNA	single- or double-stranded DNA and RNA ;
	or	$\longrightarrow$		
	$5'_{pRNA}$		$5'_{OH}$-RNA	ribo- and deoxy-ribotriphosphates

Uses

1. To remove 5' phosphates from DNA or RNA prior to labeling the 5' end with ^{32}P.

2. To remove 5' phosphates from fragments of DNA in order to prevent self-ligation.

4. Extract twice with phenol/chloroform and twice with chloroform.

5. Pass the aqueous phase through a spun column of Sephadex G-50 equilibrated in TE (465–467).

6. Precipitate the DNA with ethanol. It is now ready for ligation or kinasing.

Notes

i. CIP is usually supplied as a suspension in ammonium sulfate. Unless the ammonium sulfate is removed at the end of the reaction by passing the reaction mixture through Sephadex G-50, it will be precipitated by the ethanol. Multiple washings of the pellet with 70% ethanol are required to get rid of the ammonium sulfate. This is important because ammonium ions inhibit the activity of polynucleotide kinase.

Prepare CIP for use as follows:

1. Centrifuge the ammonium sulfate suspension for 1 minute at 4°C in an Eppendorf centrifuge.

2. Discard the supernatant. Redissolve the pellet in a minimum amount of water. The enzyme is stable for at least a month when stored at 4°C in this form.

ii. If heat inactivation in the presence of SDS proves to be insufficient, add trinitriloacetic acid to 10 mM to chelate the divalent zinc ions. CIP is more thermolabile in the absence of zinc ions.

NUCLEASE *Bal*31

*Bal*31 degrades both the 3' and 5' strands of DNA in a progressive manner. Under suitable conditions, a linear DNA molecule can be digested from both ends in a controlled fashion.

The rate at which *Bal*31 removes nucleotides from the 5' and 3' termini of duplex DNA is given by the formula

$$\frac{\mathrm{d}M_t}{\mathrm{d}t} = -2\ V_m M_n / [K_m + (S)_o]$$

where:

M_t = molecular weight of linear duplex DNA after t minutes of digestion;
V_m = maximum reaction velocity
    (expressed as moles of nucleotide removed/liter/min);
M_n = average molecular weight of sodium mononucleotide
    (taken as 330 daltons);
K_m = Michaelis-Menten constant
    (expressed as moles of duplex ends DNA/liter)
$(S)_o$ = moles of duplex DNA ends/liter at $t = 0$ minutes
    (remains constant until a significant portion of the molecules have undergone complete degradation).

At an enzyme concentration of 40 units/ml,

$V_m = 2.4 \times 10^{-5}$ moles of nucleotide removed/liter/min.
$K_m = 4.9 \times 10^{-9}$ moles of duplex termini/liter.

Assuming that the value of V_m alters in proportion to enzyme dilution, then the rate of degradation of a 2-kb fragment of DNA in a 100-μl reaction containing 10 μg of DNA and 0.5 units of enzyme will be

$$= \frac{-2(\tfrac{5}{40})(2.4 \times 10^{-5})(330)}{(4.9 \times 10^{-9}) + (1.5 \times 10^{-7})}$$

$$= \frac{-\tfrac{1}{4} \times 2.4 \times 10^{-5} \times 330}{1.5049 \times 10^{-7}}$$

$$= -1.31 \times 10^4 \text{ daltons/min/DNA molecule}$$

$$= -19 \text{ bp/min/DNA molecule}$$

$$= -10 \text{ bp/min/end of dsDNA}$$

In most cases (i.e., when digesting DNAs less than 10 kb in length and when the concentration of DNA in the reaction exceeds 20 μg/ml), the term for K_m in the denominator of the equation becomes insignificant.

Nuclease Bal3l

(Alteromonas espejiana Bal3l)

The enzyme carries two activities (Lau and Gray 1979) : (1) a highly specific, single-stranded endodeoxyribonuclease and exonuclease that catalyzes the removal of small oligonucleotides or mononucleotides from both 5' and 3' termini of double-stranded DNA (both strands of DNA are degraded at approximately the same rate); and (2) a single-strand-specific endonuclease similar to nuclease S1.

Activity	Reaction	Substrate

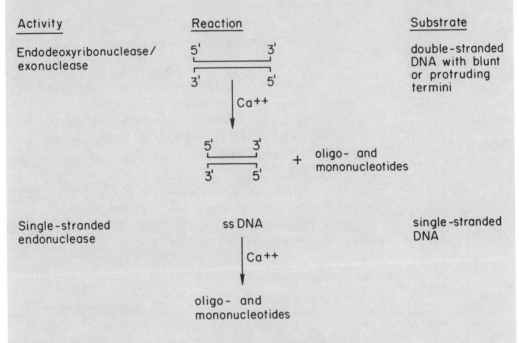

| Endodeoxyribonuclease/ exonuclease | | double-stranded DNA with blunt or protruding termini |
| Single-stranded endonuclease | ss DNA | single-stranded DNA |

Uses

1. Removing nucleotides from the termini of double-stranded DNA. The resulting molecules are shortened and can be joined to synthetic linkers or to vectors by T4 polynucleotide ligase.
2. Mapping restriction sites in DNA (Legerski et al. 1978).

The ability to control the rate of digestion by *Bal*31 and the fact that the enzyme is absolutely dependent on calcium and therefore can be completely inactivated with EGTA have allowed the development of a very simple method to map restriction sites in small fragments of DNA (Legerski et al. 1978) (see below).

A reaction with *Bal*31 is set up and samples are withdrawn at different times into EGTA. After digestion of these samples with the restriction enzyme of interest, restriction fragments can be seen to disappear in a defined order. By using a DNA consisting of vector sequences at one end (for which the restriction map is known) and unmapped sequences at the other, it is possible to distinguish fragments from the two ends and to deduce the order of the fragments in the unmapped DNA.

*Bal*31-treated DNA can also be used for subsequent cloning. After repair with the Klenow fragment of *E. coli* DNA polymerase, synthetic linkers are added to the DNA, which is then inserted into a suitable plasmid vector. In this way, it is possible to generate a set of deletions from a defined endpoint in DNA.

Mapping Restriction Sites in DNA with *Bal*31[4]

1. Using the equation given on page 135, calculate the time required to remove the desired number of nucleotides from your DNA.

2. Precipitate the DNA with ethanol. Dissolve the DNA in a solution of BSA (500 μg/ml).

3. Add an equal quantity of 2× *Bal*31 buffer.

   2× Bal31 buffer
   24 mM $CaCl_2$
   24 mM $MgCl_2$
   0.4 M NaCl
   40 mM Tris·Cl (pH 8.0)
   2 mM EDTA

4. Incubate at 30°C for 3 minutes.

5. Add the appropriate amount of *Bal*31.

6. At appropriate times, remove samples from the reaction. Add EGTA (from a 0.2 M solution, pH 8.0) to a final concentration of 20 mM. Store samples on ice.

7. At this stage two alternatives are available:

   a. Aliquots of the samples may be diluted threefold with water in order to lower the concentration of NaCl from 0.2 M to 66 mM. After addition of $^1/_{10}$ volume of restriction enzyme buffer (prepared without NaCl), the restriction enzyme of interest is added and digestion is carried out for the appropriate time. The aliquots are then analyzed by gel electrophoresis.

   Note. EGTA chelates divalent calcium ions specifically and therefore inhibits the activity of *Bal*31 without seriously affecting the digestion of DNA by restriction endonucleases.

[4]Legerski et al. (1978).

b. The samples of *Bal*31-digested DNA may be purified by extraction with phenol/chloroform and precipitated with ethanol. After washing with 70% ethanol, the DNA pellets are dissolved in the appropriate buffer and digested with a restriction enzyme. This procedure is used when the restriction enzyme of interest is inhibited by 66 mM NaCl.

Cloning Fragments of DNA That Have Been Digested With *Bal*31

1. Digest the DNA with *Bal*31 for the appropriate time. Wherever possible, use digestion with restriction enzymes to check that the actual rate of degradation by *Bal*31 matches the predicted rate (see above).

   Notes. It is often useful to include a control reaction with a DNA for which the complete restriction map is known. The time at which particular fragments disappear from the restriction digest should be consistent with the predicted rate of digestion with *Bal*31.
   *Bal*31 is active at temperatures as low as 16°C. The slower rates of digestion that occur under these conditions may be useful when only a few nucleotides are to be removed from the ends of DNA.

2. Stop the reaction with EGTA and purify the DNA by extraction with phenol/chloroform and precipitation with ethanol.

3. Because removal of the 3′ and 5′ termini of duplex DNA is not synchronous, only a fraction of the DNA molecules in the reaction mixture at any one time will have blunt ends that are suitable for ligation. DNA treated with *Bal*31 should therefore be repaired in an end-filling reaction using the Klenow fragment of *E. coli* DNA polymerase I (see page 113). If the *Bal*31-digested DNA is unlabeled, the repair reaction should be carried out with one ^{32}P-labeled and three unlabeled dNTPs in order to facilitate tracing the DNA through the subsequent steps in the procedure (particularly step 6).

4. Attach phosphorylated synthetic linkers (see page 396) to the repaired DNA.

   Note. There is no need to purify the DNA after the Klenow reaction before setting up the ligation reaction.

5. Cleave the DNA with the appropriate restriction enzyme(s) to digest the linkers and, if desired, any restriction sites.

6. Separate the DNA from the fragments of linkers by chromatography on Sepharose CL-4B (see pages 464–465).

7. Ligate the DNA to a plasmid vector cleaved with a suitable restriction enzyme(s) (see page 391).

8. Transform *E. coli* (see Chapter 8).

Note

*Bal*31 should not be frozen. Store at 4°C.

Nuclease S1

(Aspergillus oryzae)

The enzyme degrades single-stranded DNA (Vogt 1973) to yield 5' phosphoryl mono- or oligonucleotides. Double-stranded DNA, double-stranded RNA, and DNA·RNA hybrids are relatively resistant to the enzyme. However, duplex nucleic acids are digested completely by S1 if they are exposed to very large amounts of the enzyme.

Activity	Reaction	Substrate
Single-stranded-specific nuclease		single-stranded DNA or RNA; more active on DNA than RNA

ssDNA or ssRNA

$$\downarrow \text{pH}4.5 \quad Zn^{++}$$

$5'pdN$ or $5'prN$

Moderate amounts of the enzyme will cleave duplex nucleic acids at nicks or small gaps (Kroeker and Kawalski 1978).
For example:

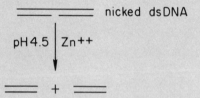

nicked dsDNA

$$\downarrow \text{pH}4.5 \quad Zn^{++}$$

═══ + ═══

Uses

1. Analyzing the structure of DNA·RNA hybrids (Berk and Sharp 1978). For a detailed description of this technique, see Favaloro et al. (1980).

2. Removing single-stranded tails from DNA fragments to produce blunt ends.

3. Opening the hairpin loop generated during synthesis of ds-cDNA.

Mung-bean Nuclease

(mung-bean sprouts)

The enzyme degrades single-stranded DNA to mono- or oligonucleotides with phosphate groups at their 5' ends (Laskowski 1980). Double-stranded DNA, double-stranded RNA, and DNA·RNA hybrids are relatively resistant to the enzyme. However, duplex nucleic acids are digested completely by mung-bean nuclease if they are exposed to very large amounts of the enzyme (Kroeker and Kowalski 1976).

Although mung-bean nuclease and nuclease S1 are similar to one another in their physical and catalytic properties, there are indications that mung-bean nuclease may be less severe in its action from nuclease S1. For example, nuclease S1 has been shown to cleave the DNA strand opposite a nick in a duplex, whereas mung-bean nuclease will only attack the nick after it has been enlarged to a gap several nucleotides in length (Kroeker and Kowalski 1976).

Uses

1. Converting protruding termini of DNA to blunt ends.

Ribonucleases

1. Ribonuclease A (bovine pancreas): The enzyme is an endoribonuclease that specifically attacks pyrimidine nucleotides at the 3'-phosphate group and cleaves the 5'-phosphate linkage to the adjacent nucleotide. The end products are pyrimidine 3' phosphates and oligonucleotides with terminal pyrimidine 3' phosphates (Davidson 1972). For example:

$$\overset{5'}{_p}A-_pG-_pG-_pC-_p\!\uparrow\!C-_p\!\uparrow\!G-_pA-_pA-_pG-_pU-_p\!\uparrow\!G-_pC-_p\!\uparrow\!A-_pG-_pG\overset{3'}{\uparrow}$$

$$\downarrow \text{RNase A}$$

$$\overset{5'}{_p}A-_pG-_pG-_pC-_p \;+\; C-_p \;+\; G-_pA-_pA-_pG-_pU-_p \;+\; G-_pC-_p \;+\; A-_pG-_pG^{3'}$$

2. Ribonuclease T1 (Aspergillus oryzae): The enzyme is an endoribonuclease that specifically attacks the 3'-phosphate groups of guanosine nucleotides and cleaves the 5'-phosphate linkage to the adjacent nucleotide. The end products are guanosine 3' phosphates and oligonucleotides with guanosine 3'-phosphate terminal groups (Davidson 1972). For example:

$$\overset{5'}{_p}A-_pG-_p\!\uparrow\!G-_p\!\uparrow\!C-_pC-_pG-_p\!\uparrow\!A-_pA-_pG-_p\!\uparrow\!U-_pG-_p\!\uparrow\!C-_pA-_pG-_p\!\uparrow\!C\overset{3'}{}$$

$$\downarrow \text{RNase T1}$$

$$\overset{5'}{_p}A-_pG-_p \;+\; G-_p \;+\; C-_pC-_pG-_p\!\cdot\!+\; A-_pA-_pG-_p \;+\; U-_pG-_p \;+\; C-_pA-_pG-_p +C^{3'}$$

Deoxyribonuclease I (DNase I)

(bovine pancreas)

The enzyme is an endonuclease that hydrolyzes double-stranded or single-stranded DNA to a complex mixture of mono- to oligonucleotides with 5'-phosphate termini.

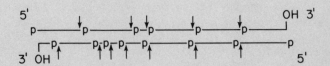

In the presence of Mg^{++}, DNase I attacks each strand of DNA independently and the sites of cleavage are distributed in a statistically random fashion.

In the presence of Mn^{++}, DNase I cleaves both strands of DNA at approximately the same site (Melgar and Goldthwaite 1968).

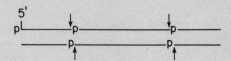

to yield fragments of DNA that are blunt-ended or have protruding termini only one or two nucleotides in length.

Exonuclease VII

(E. coli)

The enzyme is a processive exonuclease that releases small oligonucleotides from the 3' and 5' ends of single-stranded DNA (Chase and Richardson 1964). Exonuclease VII does not need Mg^{++} and works in the presence of EDTA.

Activity	Reaction		Substrate
Exonuclease	5'⌐————3'⌐	5'— 3'	single-stranded DNA
	⌐3'————5'⌐	3'— 5'	
	or	→	
	5'⌐————3'	5'— 3'	
	3'— 5'	3'— 5'	

Uses

1. Recovering cDNA inserted into plasmid vectors by dA·dT tailing (Goff and Berg 1978).
2. Mapping the position of introns and exons in DNA (Berk and Sharp 1978).

Exonuclease III
(E.coli)

The enzyme catalyzes the stepwise $3' \longrightarrow 5'$ removal of 5' mononucleotides from double-stranded DNA carrying a 3'-OH end (Weiss 1976). The enzyme also carries three other activities — an endonuclease specific for apurinic DNA, an RNase H activity (Rogers and Weiss 1980) and a 3' phosphatase activity.

Activity	Reaction	Substrate

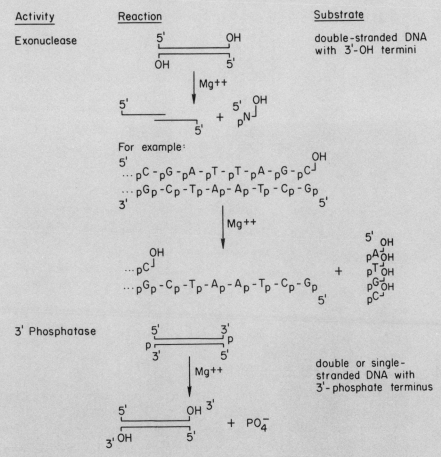

Exonuclease — double-stranded DNA with 3'-OH termini

3' Phosphatase — double or single-stranded DNA with 3'-phosphate terminus

Uses

1. Preparing linear DNA as a template for DNA polymerase (e.g. in sequencing by the dideoxy technique) (Sanger et al. 1977).

2. Preparing strand-specific probes; for example:

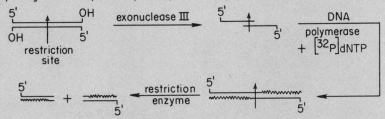

3. Deleting sequences from the termini of DNA fragments; for example:

λ Exonuclease

(λ-infected E.coli)

The enzyme catalyzes the stepwise release of 5' mononucleotides from double-stranded DNA with a terminal 5' phosphate (Little et al. 1967).

Activity	Reaction	Substrate

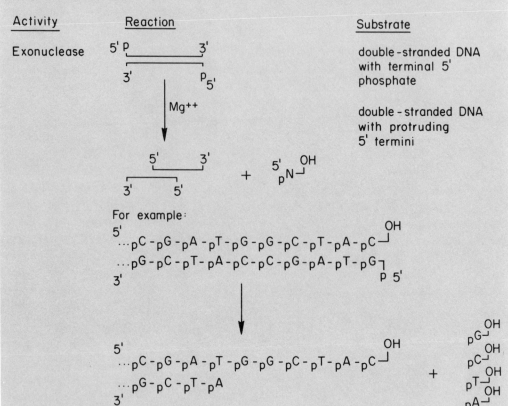

Exonuclease — double-stranded DNA with terminal 5' phosphate

double-stranded DNA with protruding 5' termini

Uses

1. Preparing DNA for sequencing by the dideoxy technique (Sanger et al. 1977).
2. Removing protruding 5' termini from double-stranded DNA prior to tailing with terminal transferase.

Poly (A) Polymerase

(E.coli)

The enzyme polymerizes AMP (derived from ATP) onto a free 3'-OH terminus of RNA (Sippel 1973).

Activity	Reaction	Substrate
Poly (A) polymerase		single-stranded RNA with free 3'-OH terminus

$$5' \underline{\quad RNA \quad}$$
$$OH \; 3'$$

$$ATP \;\Big|\; \begin{matrix} Mg^{++} \\ Mn^{++} \end{matrix}$$

$$5' \underline{\quad RNA \quad} A_p A_p (A_p)_n A \atop OH$$

n can be as large as 10^4.

Uses

1. Preparing poly(A)$^-$ RNA for cloning.
   For example:

   $$poly(A)^- RNA \xrightarrow[\text{polymerase}]{poly(A)} 5' \underline{\quad RNA \quad} (A)_n \xrightarrow[\substack{\text{primed} \\ \text{cDNA} \\ \text{synthesis}}]{\text{oligo [dT]}}$$

   For a detailed descripton, see Gething et al. 1980

2. Labeling the 3' terminus of RNA with $[\alpha\text{-}^{32}P]$ ATP to generate hybridization probes.

T4 DNA Ligase

(T4-infected E.coli)

The enzyme, a single polypeptide ($M_r = 68,000$), catalyzes the formation of a phosphodiester bond between adjacent 3'-OH and 5'-P termini in DNA (Weiss et al. 1968).

Activity	Reaction	Substrates
Ligation of cohesive ends	For example:	(a) double-stranded DNA molecules with complementary cohesive termini that base pair with one another and bring together 3'-OH and 5'-P termini.

Reaction (Ligation of cohesive ends):

5'
$\ldots _pA-_pC-_pG \rfloor_pA-_pA-_pT-_pT-_pC-_pG-_pT \ldots$ 3'   (OH)

$\ldots T_p-G_p-C_p-T_p-T_p-A_p-A_p \lceil G_p-C_p-A_p \ldots$
3'   (OH)   5'

ATP | Mg++ ↓

$\ldots _pA-_pC-_pG-_pA-_pA-_pT-_pT-_pC-_pG-_pT \ldots$
$T_p-G_p-C_p-T_p-T_p-A_p-A_p-G_p-C_p-A_p$

| Ligation of blunt ends | For example: | high concentrations of blunt-ended, double-stranded DNA with 5'-P and 3'-OH termini. |

(b) "nicked" DNA

Reaction (Ligation of blunt ends):

5'   OH   3'
$\ldots _pC-_pG-_pA \rfloor_pC-_pG-_pT-_pA \ldots$

$\ldots G_p-C_p-T_p \lceil G_p-C_p-A_p-T_p \ldots$
3'   OH   5'

ATP | Mg++ ↓

$\ldots _pC-_pG-_pA-_pC-_pG-_pT-_pA \ldots$
$\ldots G_p-C_p-T_p-G_p-C_p-A_p-T_p \ldots$

Uses

1. Joining together DNA molecules with compatible cohesive termini.

2. Joining blunt-ended, double-stranded DNA molecules to one another or to synthetic linkers. The activity of T4 ligase on blunt-ended DNA molecules can be stimulated approximately 20-fold by the addition of T4 RNA ligase (Sugino et al. 1977).

T4 RNA Ligase

(T4-infected E.coli)

The enzyme catalyzes the covalent joining of 5'-phosphoryl, single-stranded DNA or RNA to 3'-hydroxyl, single-stranded DNA or RNA (Sugino et al. 1977).

Activity	Reaction	Substrates
Ligase	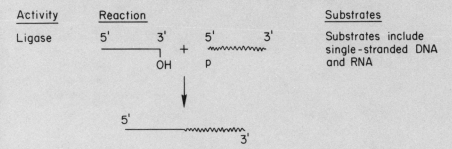	Substrates include single-stranded DNA and RNA

Uses

T4 RNA ligase increases the efficiency of blunt-end ligation of double-stranded DNA catalyzed by T4 DNA ligase (Sugino et al. 1977).

EcoRI Methylase

(E.coli)

The enzyme catalyzes the transfer of methyl groups from S-adenosyl-methionine (SAM) to the adenines marked ∗ in the EcoRI recognition sequence, $\overset{*}{\text{GAATTC}} \atop \text{CTTAAG}$. The modification of adenine to 6-methylaminopurine protects the $\text{DNA}\overset{}{}$ from cleavage by EcoRI (Greene et al. 1975).

Activity	Reaction	Substrate
Methylase	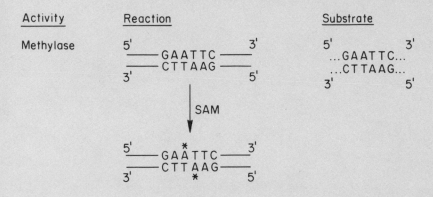	5' 3' ...GAATTC... ...CTTAAG... 3' 5'

Uses

Protecting the "natural" EcoRI sites within DNA that is to be cloned by addition of synthetic linkers. For example, during construction of libraries of eukaryotic DNA, EcoRI methylase has been used to modify EcoRI sites within the sheared fragments of DNA. Synthetic EcoRI linkers were then added to the ends of the modified DNA. The linkers were cleaved with EcoRI, and the DNA fragments were inserted into the arms of bacteriophage λDNA (Maniatis et al. 1978).

Terminal Deoxynucleotidyl Transferase (terminal transferase)
(calf thymus)

The enzyme catalyzes the addition of deoxynucleotides to the 3'-OH end of DNA molecules (Bollum 1974).

Activity	Reaction	Template/primer or substrate
Terminal transferase	ss DNA$_{OH}$ + ndNTP $\downarrow$ Mg++ DNA-(pdN)$_n$ + nPPi	single-stranded DNA with a 3'-OH terminus (or double-stranded DNA with a protruding 3'-OH terminus).
Terminal transferase	ds DNA$_{OH}$ + ndNTP $\downarrow$ Co++ DNA-(pdN)$_n$ + nPPi	blunt-ended, double-stranded DNA or DNA with a recessed 3'-OH terminus serves as template if Co++ is supplied as cofactor (Roychoudhury et al. 1976).

Uses

1. Adding complementary homopolymer tails to vector and cDNA (see chapter 7).
2. Labeling the 3' ends of DNA fragments with a ^{32}P-labeled-3'-deoxynucleoside (Tu and Cohen 1980) or a 3'-ribonucleoside (Wu et al. 1976). For labeling with ribonucleosides, $[\alpha-^{32}P]$ rNTP is used, followed by treatment with alkali.

5

Gel Electrophoresis

AGAROSE GEL ELECTROPHORESIS

The standard method used to separate, identify, and purify DNA fragments is electrophoresis through agarose gels. The technique is simple, rapid to perform, and capable of resolving mixtures of DNA fragments that cannot be separated adequately by other sizing procedures, such as density gradient centrifugation. Furthermore, the location of DNA within the gel can be determined directly: Bands of DNA in the gel are stained with low concentrations of the fluorescent, intercalating dye ethidium bromide; as little as 1 ng of DNA can then be detected by direct examination of the gel in ultraviolet light (Sharp et al. 1973).

The electrophoretic migration rate of DNA through agarose gels is dependent upon four main parameters, which are discussed below.

The molecular size of the DNA. Molecules of linear, duplex DNA, which are believed to migrate in an end-on position (Fisher and Dingman 1971; Aaij and Borst 1972), travel through gel matrices at rates that are inversely proportional to the $\log_{10}$ of their molecular weights (Helling et al. 1974) (see Fig. 5.1).

The agarose concentration. A DNA fragment of a given size migrates at different rates through gels containing different concentrations of agarose. There is a linear relationship between the logarithm of the electrophoretic mobility of DNA (μ) and gel concentration (τ), which is described by the equation:

$$\log \mu = \log \mu_0 - K_r \tau$$

where μ_0 is the free electrophoretic mobility and K_r is the retardation coefficient, a constant that is related to the properties of the gel and the size and shape of the migrating molecules. Thus, by using gels of different concentrations, it is possible to resolve a wide size-range of DNA fragments.

Amount of Agarose in Gel (%)	Efficient Range of Separation of Linear DNA Molecules (kb)
0.3	60–5
0.6	20–1
0.7	10–0.8
0.9	7–0.5
1.2	6–0.4
1.5	4–0.2
2.0	3–0.1

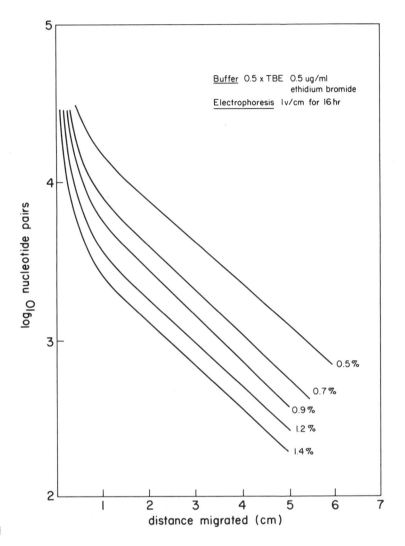

Figure 5.1

The conformation of the DNA. Closed circular (form-I), nicked circular (form-II), and linear (form-III) DNA of the same molecular weight migrate through agarose gels at different rates (Thorne 1966, 1967). The relative mobilities of the three forms are dependent primarily on the agarose concentration in the gel but are also influenced by the strength of the applied current, the ionic strength of the buffer, and the density of superhelical twists in the form-I DNA (Johnson and Grossman 1977). In some conditions, form-I DNA migrates faster than form-III; in other conditions, the order is reversed. An unambiguous method for identifying the different conformational forms of DNA is to carry out electrophoresis in the presence of increasing quantities of ethidium bromide. As the concentration of ethidium bromide increases, more of the dye becomes bound to DNA. The negative superhelical turns in form-I molecules are progressively removed, and their rate of migration is slowed. At the critical free-dye concentration, where no superhelical turns remain, the rate of migration of form-I DNA reaches its minimum value. As still more ethidium bromide is added, positive superhelical turns are generated, and the mobility of form-I DNA increases rapidly. Simultaneously, the mobilities of form-II and form-III DNA are differentially decreased as a result of charge neutralization and of the greater stiffness imparted to the DNA by the ethidium bromide. For most preparations of form-I DNA, the critical concentration of free ethidium bromide is in the range 0.1–0.5 μg/ml.

The applied current. At low voltages, the rate of migration of linear DNA fragments is proportional to the voltage applied. However, as the electric field strength is raised, the mobility of high-molecular-weight fragments of DNA is increased differentially. Thus, the effective range of separation of agarose gels decreases as the voltage is increased. To obtain maximum resolution of DNA fragments, gels should be run at no more than 5 V/cm.

Base composition and temperature. The electrophoretic behavior of DNA in agarose gels (by contrast to polyacrylamide gels [Allett et al. 1973]) is not significantly affected either by the base composition of the DNA (Thomas and Davis 1975) or the temperature at which the gel is run. Thus, in agarose gels the relative electrophoretic mobilities of DNA fragments of different sizes do not change between 4°C and 30°C. In general, agarose gels are run at room temperature. However, gels containing less than 0.5% agarose are rather flimsy, and it is best to run them at 4°C, where they gain some rigidity.

GEL ELECTROPHORESIS TANKS

Over the 15 years since agarose gel electrophoresis was introduced, many different designs of apparatus have been used. Currently, however, almost all agarose gel electrophoresis is carried out with horizontal slab gels. There are at least four advantages of this system:

Low agarose concentrations can be used because the entire gel is supported from beneath.

Gels can be cast in a wide variety of sizes.

The gels are very simple to load, pour, and handle.

The apparatus is durable and inexpensive to construct.

A number of well-designed horizontal gel tanks are available (see Fig. 5.2). Most of them are modifications of a design by W. Schaffner in which the gel is poured on a removable glass plate. The plate is installed on a platform so that the gel is submerged just beneath the surface of the electrophoresis buffer. The resistance to the passage of electrical current of the gel is almost the same as that of the buffer, so a considerable fraction of the applied current passes along the length of the gel.

Davis system - gel unit - 14.5 cm long gel

Material: ¼" clear acrylic plastic

⅛" clear acrylic ultraviolet translucent (u.v.t.) plastic

Parts List

letter	size	no. req.
A	⅛ x 5¾ x 5¾	1 - u.v.t. material
B	¼ x 5¾ x 1⅜	2
C	¼ x 5¾ x 3	2
D	¼ x 5¾ x 2¾	2
E	¼ x 2¾ x 10¾	2

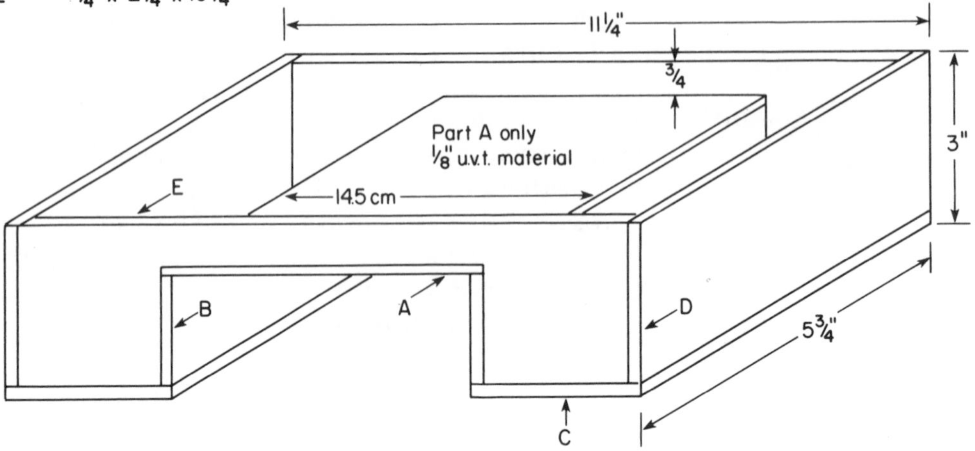

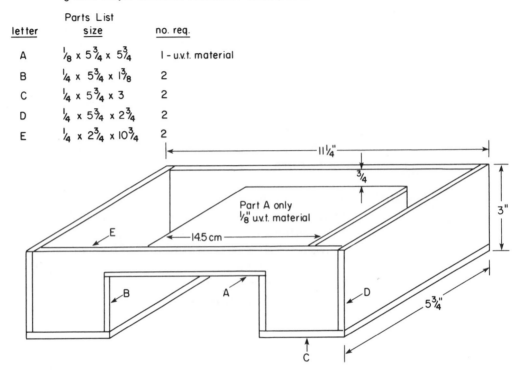

Davis system - gel unit - 14.5 cm long model

Material: ⅛" clear acrylic plastic

Parts List

letter	size	no. req.
F	⅛ x ¼ x 2	4
G	⅛ x 2 x 5¼	2 - note 5¼" - fit to suit!
H	.040 x .002 - platinum wire ribbon	

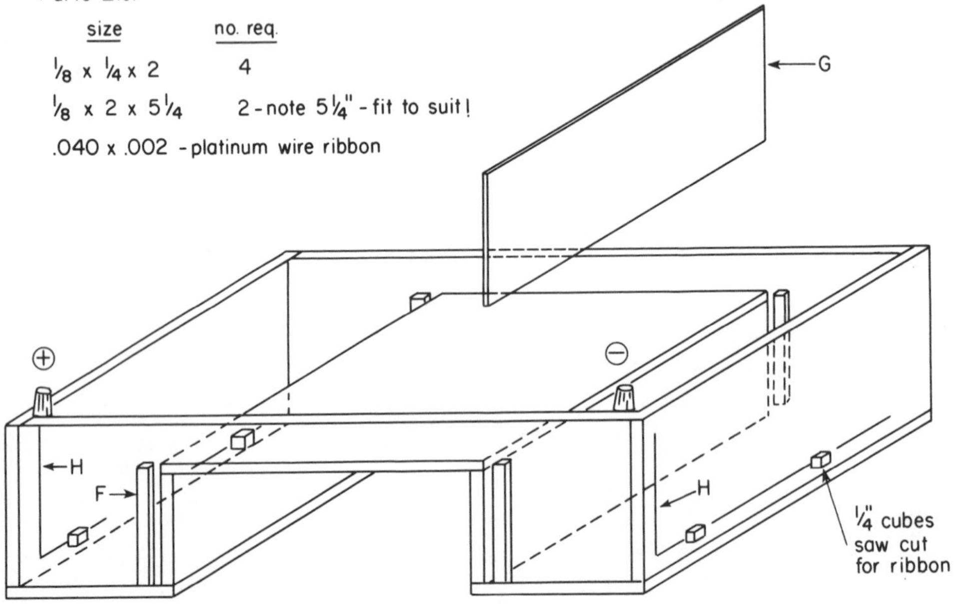

Figure 5.2

This gel system was developed in the laboratory of R.W. Davis.

Davis system - gel unit: 14.5cm long gel model
Part I: comb holder
Material: 1/4" clear acrylic plastic
Size: 1/4 x 7/8 x 5 3/4

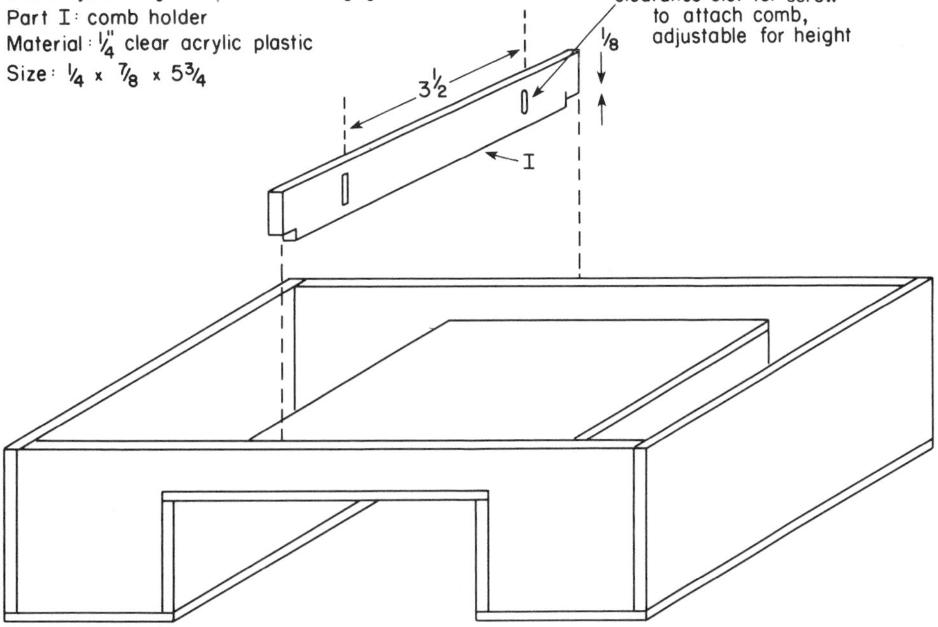
clearance slot for screw
to attach comb,
adjustable for height
1/8
3 1/2
I

Davis system - gel unit: 20 cm long model
Material: 1/4" clear acrylic plastic
1/8" clear acrylic u.v.t. plastic

Parts List

letter	size	no. req.
A	1/8 x 5 3/4 x 7 7/8	1 - u.v.t.
B	1/4 x 5 3/4 x 1 7/8	2
C	1/4 x 5 3/4 x 3	2
D	1/4 x 5 3/4 x 2 3/4	2
E	1/4 x 2 3/4 x 12 7/8	2

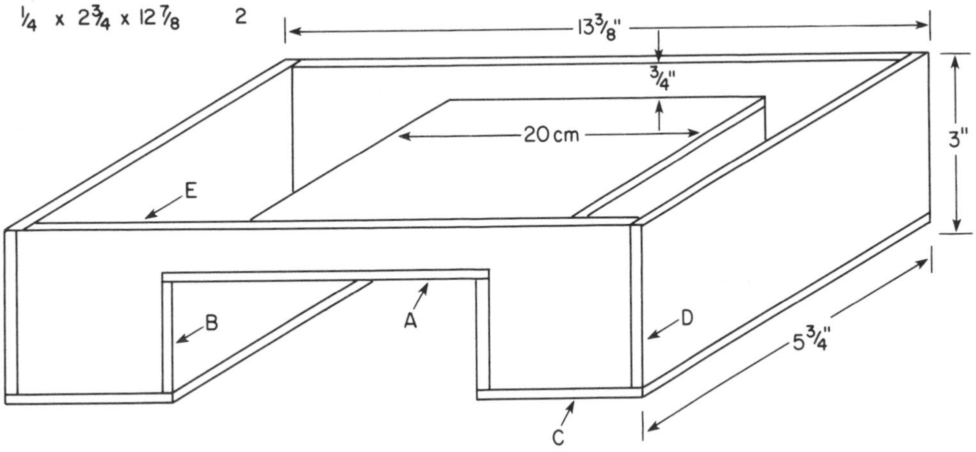
13 3/8"
3/4"
20 cm
3"
E
B
A
D
C
5 3/4"

Figure 5.2 (continued)

BUFFERS

Several different electrophoresis buffers are available containing Tris-acetate, -borate, or -phosphate at about 50 mM and pH 7.5–7.8 (see Table 5.1). These are usually made up as concentrated solutions and stored at room temperature.

For historical reasons, Tris-acetate is the most commonly used buffer. However, its buffering capacity is rather low, and it tends to become exhausted during extended electrophoresis (the anode becomes alkaline, the cathode acidic). Recirculation of buffer between the two reservoirs is advisable. Both Tris-phosphate and Tris-borate give equally good resolution of DNA fragments and have significantly higher buffering capacity. Recirculation of buffer is therefore unnecessary. Tris-phosphate has the additional, though marginal, advantage that gels made with it can be dissolved in chaotropic agents such as sodium perchlorate or potassium iodide. This forms the basis of a method (now little used) to recover DNA fragments from gels.

TABLE 5.1. COMMONLY USED BUFFERS

Buffer	Working solution	Concentrated stock solution (per liter)
Tris-acetate (TAE)	0.04 M Tris-acetate 0.002 M EDTA	50 ×: 242 g Tris base 57.1 ml glacial acetic acid 100 ml 0.5 M EDTA (pH 8.0)
Tris-phosphate (TPE)	0.08 M Tris-phosphate 0.008 M EDTA	10 ×: 108 g Tris base 15.5 ml of 85% phosphoric acid (1.679 μg/ml) 40 ml 0.5 M EDTA (pH 8.0)
Tris-borate (TBE)	0.089 M Tris-borate 0.089 M boric acid 0.002 M EDTA	5 ×: 54 g Tris base 27.5 g boric acid 20 ml 0.5 M EDTA (pH 8.0)

PREPARATION OF AGAROSE GELS

Many different grades of agarose are available, and even within one grade there is considerable variation from batch to batch. In our experience, the best general-purpose agarose is type-II, low-endo-osmotic agarose. It melts easily to give transparent solutions, and the resulting gels are resilient, even at low concentrations. However, type-II agarose is contaminated with sulfated polysaccharides, which inhibit enzymes such as ligases, polymerases, and restriction endonucleases. Therefore, DNA fragments eluted from such gels have to be extensively purified before they can be used as templates or substrates for these enzymes.

Agarose gels are prepared as discussed on the following pages and shown in Figures 5.3 and 5.4.

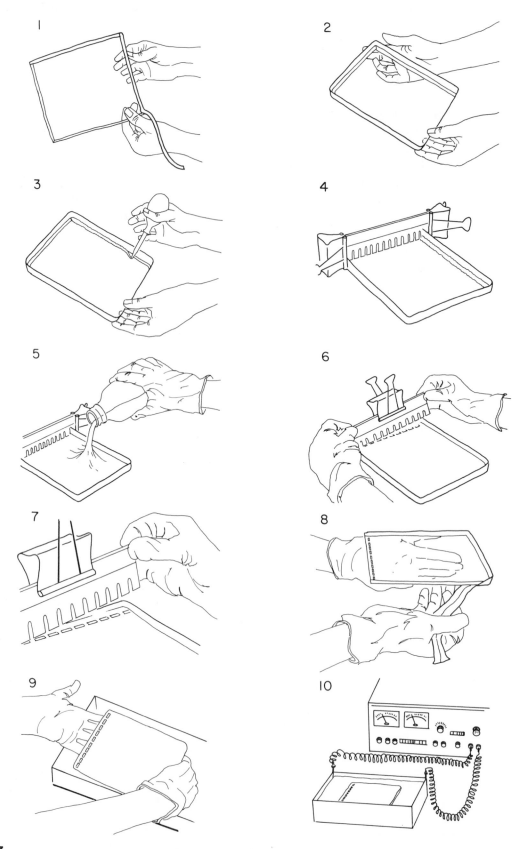

Figure 5.3

This method of pouring gels originated in the laboratory of W. Schaffner.

1. Add the correct amount of powdered agarose to a measured quantity of electrophoresis buffer.

2. Heat the slurry in a boiling-water bath or in a microwave oven until the agarose dissolves.

3. Cool the solution to 50°C, and add ethidium bromide (from a stock solution of 10 mg/ml in water, stored at 4°C in a light-proof bottle) to a final concentration of 0.5 μg/ml.

4. Seal the edges of a clean, dry, glass plate with autoclave tape so as to form a mold (see Fig. 5.3).

5. Using a pasteur pipette, seal the edges of the mold with a small quantity of the agarose solution.

6. When the seal is set, pour the rest of the warm agarose solution into the mold and immediately clamp the comb, the teeth of which will form the sample wells, into position near one end of the gel. Check to see that there is 0.5–1.0 mm of agarose between the bottom of the teeth and the base of the gel, so that the sample wells are completely sealed (Fig. 5.4).

7. After the gel is completely set (30–45 minutes at room temperature), carefully remove the comb and autoclave tape and mount the gel in the electrophoresis tank.

8. Add just enough electrophoresis buffer (containing 0.5 μg/ml of ethidium bromide, if desired; see Table 5.1) to cover the gel to a depth of about 1 mm.

9. Samples are mixed with loading buffer and are loaded into the slots of the submerged gel, which contains 5–10% glycerol, 7% sucrose or 2.5% Ficoll, and one or more tracking dyes (0.025% bromophenol blue or xylene cyanol). Usually loading buffer (see Table 5.2) is made up as a 6-fold to 10-fold concentrated solution, which is mixed with the sample and then slowly applied to the gel using a disposable micropipette, an automatic micropipettor (Eppendorf or Gilson), or, if you have a steady hand, a pasteur pipette.

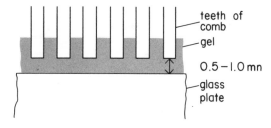

Figure 5.4

The maximum amount of DNA that can be applied to a slot depends on the number of fragments in the sample and their sizes. The minimum amount of DNA that can be detected by photography of ethidium-bromide-stained gels (see below) is about 2 ng in a 0.5-cm-wide band (the usual width of a slot). If there is more than 200 ng of DNA in a band of this width, the slot will be overloaded, resulting in trailing and smearing—a problem that becomes more severe with increasing size of DNA.

When simple populations of DNA molecules are to be analyzed, 0.2–0.5 μg of DNA should be loaded per 0.5-cm slot. When, however, the samples consist of a very large number of DNA fragments of different sizes (e.g., restriction enzyme digests of mammalian DNA), it is possible to load 5–10 μg per slot without significant loss of resolution.

TABLE 5.2. GEL-LOADING BUFFERS

Buffer type	6× buffer	Storage temperature
I	0.25% bromophenol blue 0.25% xylene cyanol 40% (w/v) sucrose in H_2O	4°C
II	0.25% bromophenol blue 0.25% xylene cyanol 15% (Ficoll type 400) in H_2O	room temp.
III	0.25% bromophenol blue 0.25% xylene cyanol 30% glycerol in H_2O	4°C
IV	0.25% bromophenol blue 40% (w/v) sucrose in H_2O	4°C

STAINING DNA IN AGAROSE GELS

The most convenient method of visualizing DNA in agarose gels is by use of the fluorescent dye ethidium bromide (Sharp et al. 1973). This substance contains a planar group that intercalates between the stacked bases of DNA. The fixed position of this group and its close proximity to the bases causes dye bound to DNA to display an increased fluorescent yield compared to dye in free solution. UV-irradiation absorbed by the DNA at 260 nm and transmitted to the dye, or irradiation absorbed at 300 nm and 360 nm by the bound dye itself, is emitted at 590 nm in the red-orange region of the visible spectrum.

Ethidium bromide can be used to detect both single- and double-stranded nucleic acids (both DNA and RNA). However, the affinity of the dye for single-stranded nucleic acid is relatively low and the fluorescent yield is poor.

Usually ethidium bromide (0.5 μg/ml) is incorporated both into the gel and the running buffer. Although the electrophoretic mobility of linear duplex DNA is reduced in the presence of the dye by approximately 15%, the ability to examine the gel directly under UV-illumination during or at the end of the run is a distinct advantage. If you prefer, however, you may run the gel in the absence of ethidium bromide and stain the DNA after electrophoresis is complete. The gel is immersed in electrophoresis buffer or water containing ethidium bromide (0.5 μg/ml) for 45 minutes at room temperature. Destaining is not usually required. However, detection of very small amounts (<10 ng) of DNA is made easier if the background fluorescence caused by unbound ethidium bromide is reduced by soaking the stained gel in 1 mM MgSO$_4$ for 1 hour at room temperature.

Caution

Ethidium bromide is a powerful mutagen. Always wear gloves while handling gels or solutions containing the dye.

PHOTOGRAPHY

Photographs of gels may be made using transmitted or incident UV light (Fig. 5.5).

The most sensitive film is Polaroid Type 57 or 667 (ASA 3000). With an efficient UV source ($>2500 \ \mu W/cm^2$) and a good lens, an exposure of a few seconds is sufficient to obtain images of bands containing as little as 10 ng of DNA. With long exposure and a strong UV source, as little as 1 ng of DNA can be detected.

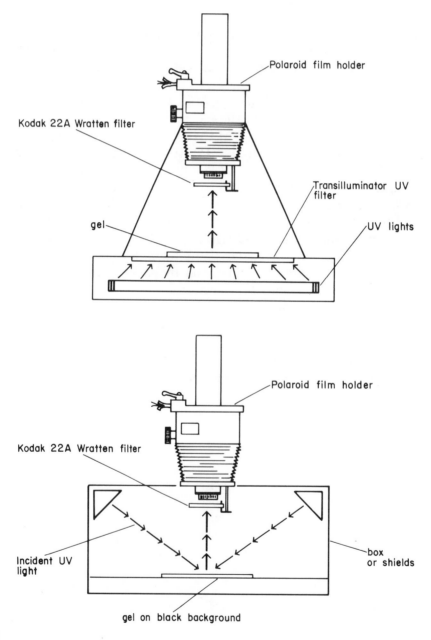

Figure 5.5

MINIGELS

Recently, methods have been developed in which small quantities of DNA can be analyzed very rapidly by agarose gel electrophoresis (the use of minigels was developed in the laboratory of D. Hogness). Several kinds of apparatus are manufactured commercially, but most of these are no more than miniaturized versions of the full-sized apparatus described earlier. Because of the rather delicate milling involved in shaping the tiny combs and other parts, miniature gel tanks are quite expensive. However, an inexpensive version can be constructed easily in the laboratory.

A convenient electrophoresis chamber for minigels is a small (2 $\frac{1}{2}$-inch × 6 $\frac{1}{2}$-inch) polystyrene snap-lock box. Four inches of platinum wire are required for each of the two electrodes, which are glued in place with silicone rubber cement. A platform composed of four 2-inch × 3-inch microscope slides glued together in a stack is cemented in the center of the box (see Fig. 5.6).

The gels themselves are poured on 2-inch × 3-inch lantern slides, adding 10–12 ml of melted agarose solution (0.5–2.0%) containing ethidium bromide (0.5 μg/ml) from a disposable 10-ml pipette. The agarose solution flows better when the pipette tip is broken off. Each gel can accommodate 8 slots, 2.5-mm long and 1-mm wide, which are formed by a miniature lucite comb of standard design. Several gels may be prepared simultaneously by mounting a long gel comb across slides placed side by side on a sheet of parafilm.

Each gel slot holds 3–5 μl of fluid depending on the thickness of the gel. Usually, 10–20 ng of DNA is applied to a slot in 2–3 μl of loading buffer. The gel is usually run for 30 minutes at high voltage (15 V/cm or even greater). During this time the bromophenol blue migrates almost the full length of the gel. The gel is then photographed as described previously. Minigels are particularly useful when a rapid answer is required before the next step in a cloning protocol can be undertaken.

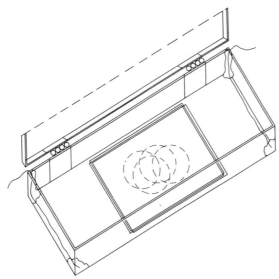

Figure 5.6

RECOVERY OF DNA FROM AGAROSE GELS

Many methods have been developed (Southern 1975; reviewed by Wu et al. 1976; Smith 1980) to recover DNA from agarose gels, and none of them is entirely satisfactory. There are two major problems. First, most grades of agarose are contaminated by sulfated polysaccharides, which are extracted from the gel together with the DNA; these substances are potent inhibitors of many of the enzymes (restriction endonucleases, ligases, kinases, polymerases) that are commonly used in subsequent cloning steps. Second, the efficiency with which DNA is extracted from agarose gels is a function of its molecular weight. DNA fragments less than 1 kb in length can be recovered in virtually quantitative yield. As the molecular weight of the DNA increases, the yield drops steadily so that fragments greater than 20 kb in size are rarely recovered in better than 20% yield.

The following methods have worked well in our laboratory.

Electroelution into Dialysis Bags

1. Run the gel and localize the band of interest using a long-wave-length (300–360 nm) UV lamp to minimize damage to the DNA.

2. Using a sharp scalpel, cut out a slice of agarose containing the band.

3. Photograph the gel after cutting out the band, so you will have a record of which band was eluted.

4. Fill a dialysis bag to overflowing (see page 456 for preparation of dialysis tubing) with 0.5× TBE. Holding the neck of the bag, pick up the gel slice with forceps and place it in the fluid-filled bag.

5. Allow the gel slice to sink to the bottom of the bag. Remove most of the buffer, leaving just enough fluid to keep the gel slice in constant contact with the electrophoresis buffer. Then tie the bag just above the gel slice; avoid trapping air bubbles (see Fig. 5.7).

6. Immerse the bag in a shallow layer of 0.5× TBE in an electrophoresis tank. Pass electric current through the bag (usually 100 V for 2–3 hours). During this time, the DNA is electroeluted out of the gel and onto the inner wall of the dialysis bag.

7. Reverse the polarity of the current for 2 minutes to release the DNA from the wall of the dialysis bag.

8. Open the dialysis bag and carefully recover all the buffer surrounding the gel slice. Using a pasteur pipette, wash out the bag with a small quantity of 0.5× TBE.

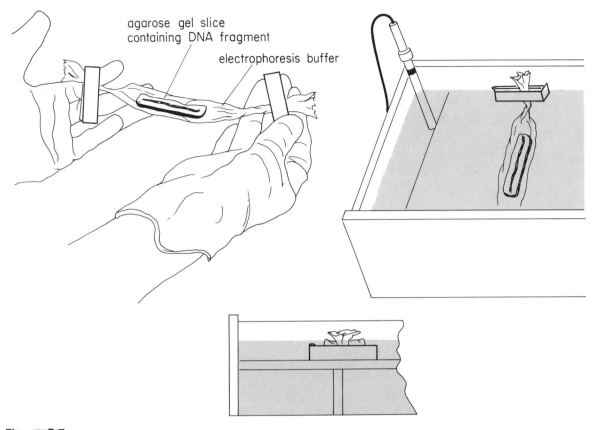

agarose gel slice
containing DNA fragment

electrophoresis buffer

Figure 5.7

This method of dialysis was originally described by McDonnell et al. (1977).

9. Stain the gel slice with ethidium bromide (for 30 minutes in 0.5× TBE containing 0.5 μg/ml of ethidium bromide). Examine under UV light to check that all the DNA is eluted.

10. Purify the DNA by one of the two following methods (page 166).

Passage through DEAE-Sephacel

1. Equilibrate DEAE-Sephacel in

   10 mM Tris·Cl (pH 7.6)
   1 mM EDTA
   60 mM NaCl

2. Pack 0.6 ml (sufficient to bind 20 μg of DNA) of the DEAE-Sephacel slurry into a small column. Dispocolumns made by Bio-Rad are most convenient for this purpose.

3. Wash the column with

TE (pH 7.6) containing 0.6 M NaCl	3 ml
TE alone	3 ml
TE containing 0.1 M NaCl	3 ml

4. Load the DNA in gel buffer directly onto the column; collect the flow-through and reapply it to the column.

5. Wash the column twice with 1.5 ml of TE containing 0.3 M NaCl.

6. Elute the DNA with three 0.5-ml washes of TE containing 0.6 M NaCl.

7. Extract the pooled eluate once with phenol and once with chloroform.

8. Recover the DNA by precipitation with ethanol.

Direct Extraction

This method works for most, but not all, lots of agarose.

1. Pass the DNA recovered from the gel in 0.5× TBE over a column of packed siliconized glass wool in a pasteur pipette to remove pieces of gel.

2. Extract the eluate twice with phenol, once with phenol/chloroform, and once with chloroform.

3. Recover the DNA by ethanol precipitation.

4. Resuspend the precipitate in 200 μl of H_2O, add 25 μl of 3 M sodium acetate (pH 5.2), and precipitate the DNA again with ethanol.

5. Rinse the pellet once with 70% ethanol.

6. Dry the pellet and resuspend in an appropriate volume of TE (pH 7.6).

Electroelution into Troughs

DNA can also be electroeluted from horizontal slab gels into troughs cut into the gel, a method developed in the laboratory of D. Hogness.

1. Run the gel and localize the band of interest using a long-wave-length UV light (as in procedure above).

2. Using a sharp scalpel or razor blade, cut a trough directly in front of the leading edge of the band and about 2 mm wider than the band on each side.

3. Fill the trough with electrophoresis buffer and resume electrophoresis.

4. Every 2 or 3 minutes recover the fluid from the trough. Refill the trough with fresh buffer and continue electrophoresis until all the DNA in the band has moved from the gel and has been recovered in the fluid taken from the trough.

5. The DNA is then purified by chromatography on DEAE-Sephacel or by direct extraction as described earlier (see page 166).

Although this method of recovering DNA from agarose gels can be efficient in terms of yield, it is time-consuming and requires constant attention during the period of electroelution. These problems are avoided in the following modification of the method.

Electrophoresis onto a Dialysis Membrane[1]

1. Run the gel and localize the band of interest using a long-wave-length UV light.

2. Using a sharp scalpel or razor blade, make an incision in the gel directly in front of the leading edge of the band and about 2 mm wider than the band on each side.

3. Cut a piece of Whatman 3MM paper and a piece of single dialysis membrane the width of the slot and slightly deeper than the gel. Soak the paper and the membrane for 5 minutes in electrophoresis buffer.

4. Using wide-edged forceps, hold apart the walls of the incision and insert the 3MM paper backed by the dialysis membrane, with the 3MM paper nearest the DNA band.

5. Remove the forceps and allow the incision to close, being careful that no air bubbles are trapped.

6. Continue electrophoresis until the band of DNA has migrated into the paper.

7. When all the DNA has left the gel and is trapped on the paper, turn off the current. Pierce a hole in the bottom of a 400-μl microfuge tube with a red-hot needle and place it inside a 1.5-ml Eppendorf tube. Remove the caps of both tubes.

8. Withdraw the paper and dialysis tubing from the gel and place them inside the 400-μl microfuge tube. Centrifuge for 15 seconds. Recover the eluate from the 1.5-ml tube.

9. Add 100 μl of elution buffer (0.2 M NaCl, 50 mM Tris [pH 7.6], 1 mM EDTA, and 0.1% SDS) to the 400-μl tube containing the paper and dialysis membrane.

10. Centrifuge as above and recover the eluate.

11. Repeat steps 9 and 10 twice more.

12. Extract the pooled eluates once with phenol. Recover the aqueous phase and extract once with phenol/chloroform and once with chloroform.

13. Recover the DNA by ethanol precipitation. Rinse the pellet once with 70% ethanol, dry, and redissolve the DNA in a small volume of TE (pH 7.6).

[1]Girvitz et al. (1980).

Notes

i. This procedure is written for elution of a single band of DNA approximately 5–10 mm wide. For larger bands, the volumes should be scaled up approximately.

ii. Dretzen et al. (1981) have recently described a similar procedure in which the DNA is collected on strips of DEAE-cellulose paper. This procedure can be used to recover DNA from both agarose or acrylamide gels and is reported to yield DNA fragments that do not need to be purified on DEAE-Sephacel columns before being used in subsequent enzyme reactions.

Recovery of DNA from Low-melting-temperature Agarose

A number of grades of agarose are available in which hydroxyethyl groups have been introduced into the agarose molecule. This substitution causes the agarose to gel at 30°C and to melt at 65°C — well below the melting temperature for most DNAs. These properties have been exploited to develop a simple technique for the recovery of DNA from gels (Weislander 1979).

1. Dissolve the low-melting-temperature agarose by heating in electrophoresis buffer to 70°C. Cool to 37°C. Add ethidium bromide to a final concentration of 0.5 μg/ml. Pour the gel at 4°C to ensure that it sets properly.

2. Load the samples of DNA and carry out electrophoresis at 4°C to ensure that the gel does not melt during the run. The electrophoretic characteristics of low-melting-temperature agarose gels are similar to those of conventional agarose gels.

3. Cut out the desired segment(s) of gel and add about 5 volumes of 20 mM Tris · Cl (pH 8.0) and 1 mM EDTA.

4. Heat for 5 minutes at 65°C to melt the gel.

5. At *room temperature*, extract the melted gel slice with an equal volume of phenol. Recover the aqueous phase by centrifugation at 20°C and reextract with phenol/chloroform and then with chloroform.

6. Recover the DNA by ethanol precipitation. Usually the DNA is now pure enough to serve as a substrate for restriction enzymes, ligases, and so forth. If necessary, the DNA can be further purified by chromatography on DEAE-Sephacel, as described earlier.

ALKALINE AGAROSE GELS

Alkaline agarose gels (McDonnell et al. 1977) are used: (1) to analyze the size of the DNA strand in nuclease-S1-resistant DNA · RNA hybrids (see page 207 and Favaloro et al. 1980); (2) to check the size of first and second DNA strands synthesized by reverse transcriptase (see page 233); (3) to check for nicking activity in enzyme preparations used for molecular cloning; (4) to calibrate the reagents used in nick translation of DNA (see page 111).

Because addition of sodium hydroxide to hot agarose solution causes hydrolysis of the polymer, the gels are prepared in a neutral, unbuffered solution (50 mM NaCl and 1 mM EDTA) and equilibrated in alkaline electrophoresis buffer before running.

1. Add the correct amount of powdered agarose to a measured quantity of 50 mM NaCl and 1 mM EDTA.

2. Heat the slurry in a boiling-water bath or in a microwave oven until the agarose dissolves.

3. Cool the solution to 50°C, pour the gel and mount it in the electrophoresis tank as described on pages 158–159.

4. Add sufficient alkaline electrophoresis buffer to cover the gel to a depth of 3–5 mm.

   Alkaline electrophoresis buffer

   30 mM NaOH
   1 mM EDTA

   Allow the buffer to soak into the gel for at least 30 minutes before loading the samples of DNA.

5. Samples of DNA to be analyzed on alkaline agarose gels are precipitated with ethanol and dissolved in 10–20 μl of alkaline loading buffer.

   Alkaline loading buffer

   50 mM NaOH
   1 mM EDTA
   2.5% Ficoll (type 400; Pharmacia)
   0.025% bromocresol green

6. Before loading, remove the excess alkaline electrophoresis buffer from above the surface of gel. Sufficient buffer should remain to cover the gel to a depth of 1 mm.

7. Load the DNA samples. Electrophoresis is carried out at voltages up to 7.5 V/cm until the dye has migrated approximately 8 cm (about 15 V-hr/cm).

   Because bromocresol green diffuses rapidly out of alkaline gels into the electrophoresis buffer, a second glass plate should be placed directly on top of the gel after the dye has migrated out of the loading slot.

8. In many cases, DNA analyzed by alkaline gel electrophoresis is labeled with ^{32}P, which can be detected by autoradiography. At the end of the run, the gel is removed from the tank and soaked for 30 minutes at room temperature in 7% trichloracetic acid. The gel is then mounted on a glass plate and dried for several hours under many layers of Whatman 3MM paper weighted with another glass plate. The dried gel is then covered with Saran Wrap and autoradiographed at room temperature or at $-70°C$ with an intensifying screen (see Fig. A.3, page 470).

   If the DNA is unlabeled, the gel should be soaked for 1 hour in neutralizing solution (1 M Tris · Cl [pH 7.6] and 1.5 M NaCl). The DNA can then be transferred to a nitrocellulose filter as described on pages 382ff and detected by hybridization to an appropriate ^{32}P-labeled probe.

Polyacrylamide Gel Electrophoresis

Polyacrylamide gels are used to analyze and prepare fragments of DNA less than 1 kb in length (Maniatis et al. 1975). They may be cast in a variety of polyacrylamide concentrations, ranging from 3.5% to 20%, depending on the sizes of the fragments of interest.

Acrylamide (%)	Effective Range of Separation (nucleotides)
3.5	100–1000
5.0	80–500
8.0	60–400
12.0	40–200
20.0	10–100

Polyacrylamide gels are almost always poured between two glass plates that are held apart by spacers. In this arrangement, most of the acrylamide solution is shielded from exposure to the air, so that inhibition of polymerization by oxygen is confined to a narrow layer at the top of the gel. Polyacrylamide gels can range in length from 10 cm to 100 cm, depending on the separation required; they are invariably run in the vertical position.

PREPARATION OF POLYACRYLAMIDE GELS

1. Prepare stock solutions.

30% Acrylamide

acrylamide	29 g
N,N'-methylene bisacrylamide	1 g
H_2O	to 100 ml

Caution

Acrylamide is highly toxic. Use gloves and a mask.

Tris-borate electrophoresis buffer (TBE) (See Table 5.1.).

3% Ammonium persulfate

ammonium persulfate (solid should be stored at 4°C)	0.3 g
H_2O	to 10 ml

The solution may be stored at 4°C for periods of up to a week.

2. Calculate the volume of acrylamide solutions required. The recipes in Table 5.3 yield 100 ml of solution.

3. Place the required quantity of solution in a side-arm flask. Deaerate by applying vacuum, gently at first. Swirl the flask during deaeration until no more air bubbles are released.

4. Prepare the glass plates for pouring the gel. Until recently, polyacrylamide gels were poured and run in pairs of plates shaped as illustrated in Figure 5.8 (top).

   This design has two disadvantages: The "rabbit-ears" are very fragile, and the notched plates, being difficult to mill, are expensive. Currently, therefore, the gels are usually poured in plates like the one illustrated in Figure 5.8 (bottom) (plates without rabbit ears originated with W.

TABLE 5.3. POLYACRYLAMIDE GELS

Reagents	Polyacrylamide gel (%)				
	3.5	5.0	8.0	12.0	20.0
30% acrylamide	11.6	16.6	26.6	40.0	66.6
H_2O	76.3	71.3	61.3	47.9	21.3
3% ammonium persulfate	2.1	2.1	2.1	2.1	2.1
10× TBE	10.0	10.0	10.0	10.0	10.0
TOTAL	100.0	100.0	100.0	100.0	100.0

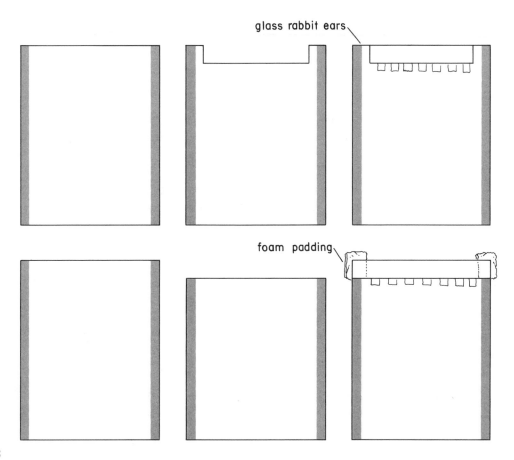

Figure 5.8

Polyacrylamide gel plates, with and without rabbit ears. (*Left*) outer plates; (*center*) inner plates; (*right*) inner and outer plates joined together, showing position of the gel slots.

Barnes and were modified, as illustrated above, by P. A. Bullock and R. E. Gelinas).

When the cast gel is clamped to the electrophoresis tank, the gap between the inner plate and the lucite support is filled by strips of a pad that is normally used as a cushion under sleeping bags. These pads may be purchased from a sports outfitter (e.g., Eastern Mountain Sports) and then cut to the desired size. When clamped tightly and compressed, this material forms a water-tight seal between the backing plate and the lucite support.

Wash the glass plates in warm detergent solution and rinse well in tap water and then deionized water. Hold the plates by the edges so that grease from your hands does not become deposited on the working surfaces of the plates. Rinse the plates with ethanol and set aside to dry.

Lay the larger (or unnotched) outer plate flat on the bench and arrange the two spacer bars in place along the sides. A couple of minute dabs of vaseline helps keep the spacer bars in position during the next steps.

Lay the inner plate in position, resting on the spacer bars. Bind the entire length of the two sides and the bottom of the plates with Whatman 3MM yellow electrical tape to make a watertight seal.

5. Remember—acrylamide is toxic and is absorbed rapidly through the skin. Wear gloves and perform the following manipulations over a tray so that any spilled acrylamide does not spread over your bench. Add 30 μl of TEMED (N,N,N',N'-tetramethylethylene diamine) to each 100 ml of deaerated acrylamide, mix, and pour immediately into the space between the two plates. Fill almost to the top. Keep the remaining acrylamide solution at 4°C to reduce the rate of polymerization. If the plates are clean, there should be no trapped air bubbles, and if they were sealed well, no leaks.

6. Insert the appropriate comb immediately, being careful not to allow air bubbles to become trapped under the teeth. The tops of the teeth should be slightly higher than the top of the glass plate.

7. Allow the acrylamide to polymerize at room temperature for 60 minutes, adding additional acrylamide if the gel retracts significantly. When polymerization is complete, a Schlieren pattern appears just beneath the teeth.

8. Remove the comb and immediately rinse out the wells with water. Remove the electrical tape from the bottom of the gel.

9. Attach the gel to the electrophoresis tank, using large bulldog clips for the sides and 3-prong clamps for the shoulders. Place foam padding as shown in Figure 5.9.

10. Fill the reservoirs with 1× electrophoresis buffer. Use a bent pasteur pipette or syringe needle to remove any air bubbles trapped beneath the bottom of the gel.

11. Use a pasteur pipette to flush out the wells with electrophoresis buffer. If this is not done, diffuse wavy bands of DNA will be observed.

12. Load the DNA samples using a glass micropipette. The capacity of polyacrylamide gels is quite high, and up to 1 μg of DNA per band can be loaded into a slot of 0.5 cm long and 0.2 cm wide.

13. Connect the electrodes to a power pack (positive outlet connected to the bottom reservoir). Polyacrylamide gels are usually run at voltage gradients between 1 V/cm and 8 V/cm. At higher voltages, heating of the gel may cause bowing of the DNA bands or even melting of the strands of small DNA fragments. The marker dyes migrate in polyacrylamide gels in 1× TBE buffer at the same rate as DNA fragments of the following sizes (in nucleotide pairs; see Table 5.4; Maniatis et al. 1975).

14. At the end of the run, detach the plates from the tank and remove the tape from the sides. Lay the gel and the two glass plates flat on the bench. Lift up a corner of the upper glass plate and pull it smoothly away, leaving the gel attached to the lower plate. Remove the spacers.

**TABLE 5.4. MIGRATION OF MARKER DYES
IN POLYACRYLAMIDE GELS**

Percentage of gel	Bromophenol blue[1]	Xylene cyanol[1]
3.5	100	460
5.0	65	260
8.0	45	160
12.0	20	70
20.0	12	45

[1]The numbers are the approximate sizes of fragments of DNA (in nucleotide pairs) with which the dyes would comigrate.

15. Submerge the gel and its attached glass plate in staining solution (0.5 μg/ml of ethidium bromide in 1× TBE buffer). Be careful! The gel is fragile and slippery. After staining for 45 minutes, remove the gel, using the glass plate as a support. Blot off excess liquid with Kimwipes. Do not use absorbent paper. Wrap the gel and glass plate in Saran Wrap.

16. The gel may now easily be photographed by inverting it on a UV-transilluminator and removing the glass plate. Polyacrylamide quenches the fluorescence of ethidium bromide so that it is not possible to detect less than about 10 ng of DNA per band.

ISOLATION OF DNA FRAGMENTS FROM POLYACRYLAMIDE GELS[2]

1. Run and stain the polyacrylamide gel as described on page 177.

2. Use a long-wave UV lamp to locate the DNA band of interest.

3. Cut out the band, using a sharp razor blade.

4. Place the gel slice on a glass plate and chop it into fine pieces with a razor blade. Transfer the pieces to a small test tube and add 1 volume of elution buffer (0.5 M ammonium acetate and 1 mM EDTA [pH8.0]).

5. Cap the tube and incubate at 37° overnight, if possible on a rotating wheel.

6. Centrifuge the sample at 10,000g for 10 minutes at 20°C. Recover the supernatant. Be careful to avoid transferring fragments of acrylamide (a drawn-out pasteur pipette works well).

7. Add an additional 0.5 volume of elution buffer to the pellet, vortex briefly, and recentrifuge. Combine the two supernatants.

8. Remove any remaining fragments of acrylamide by passing the supernatant through a small column of glass wool contained in the tip of a pasteur pipette.

9. Precipitate the DNA with ethanol.

10. Redissolve the DNA in 200 μl of TE. Add 25 μl of 3 M sodium acetate (pH 5.2) and reprecipitate the DNA.

12. Rinse the pellet once with 70% ethanol; dry briefly under vacuum and resuspend in a small volume of TE (pH 7.9).

[2]Maxam and Gilbert 1977.

Strand-separating Gels

For some purposes (e.g., sequencing by the Maxam-Gilbert procedure; hybridization to low-abundance RNAs), it is necessary to obtain separated strands of fragments of DNA. Often this can be achieved by electrophoresis of denatured DNA through neutral agarose (Hayward 1972) or polyacrylamide gels (Maxam and Gilbert 1977). Why the complementary strands of many fragments of DNA should migrate through gels at different rates is not known for certain. However, it seems likely that the two complementary strands fold into different secondary structures as a consequence of intrastrand base pairing. In many cases, this secondary structure confers different electrophoretic mobilities upon the complementary strands. Sometimes, however, the two strands migrate at the same rate even during prolonged electrophoresis. Unfortunately, there are no rules that predict whether complementary DNA strands will migrate through gels at the same or different rates, and the behavior of each fragment has to be established by experiment.

The strands of DNA fragments less than 1 kb in length are separated on polyacrylamide gels (Szalay et al. 1977; Maxam and Gilbert 1980); the strands of longer DNA fragments are separated on agarose gels (Hayward 1972; Tibbetts et al. 1974; Dunn and Sambrook 1980).

STRAND SEPARATION ON POLYACRYLAMIDE GELS

DNA Fragments Greater than 200 Nucleotides in Length

1. Make a gel containing

   5% (w/v) acrylamide
   0.1% N,N'-methylene bisacrylamide

   in a buffer consisting of

   50 mM Tris-borate (pH 8.3)
   1 mM EDTA

   After the gel has completely polymerized, which takes 1–2 hours because of the low amount of crosslinker present, it should be prerun in the same buffer at 8 V/cm for 1–2 hours before the DNA is loaded.

2. Precipitate the DNA fragment with ethanol and wash the pellet twice with 70% ethanol to remove salt.

3. Redissolve the DNA in 40 μl of

   30% (w/v) DMSO
   1 mM EDTA
   0.05% xylene cyanol
   0.05% bromophenol blue

4. Heat to 90°C for 2 minutes. Chill quickly in ice water.

5. Immediately load the sample and carry out electrophoresis at 8 V/cm until the faster-migrating, double-stranded form of the DNA fragment has traveled the length of the gel (predetermined empirically).

6. Locate the position of the DNA by staining with ethidium bromide (see page 177) or by autoradiography (see pages 470ff).

7. Recover the separated strands from the gel as described on page 164.

DNA Fragments Less than 200 Nucleotides in Length

The strands of small DNA fragments are separated on 8% polyacrylamide gels (Maxam and Gilbert 1977).

1. Make a gel containing

   8% (w/v) acrylamide
   0.24% (w/v) *N,N'*-methylene bisacrylamide

   in a buffer consisting of

   50 mM Tris-borate (pH 8.3)
    1 mM EDTA

   After the gel has completely polymerized, it should be prerun at 8 V/cm for 1–2 hours before the DNA is loaded.

2. Precipitate the DNA with ethanol and wash the pellet twice with 70% ethanol to remove salt.

3. Dissolve the DNA in 50 μl of

   0.3 M NaOH
   10% glycerol
   0.05% bromophenol blue

4. Load the sample onto the gel and immediately carry out electrophoresis at 8 V/cm until the faster-migrating, double-stranded form of the DNA has traveled the length of the gel (predetermined empirically).

5. Almost invariably, strand separation of small molecules of DNA is carried out using DNA fragments labeled with ^{32}P. At the end of the run, remove one of the glass plates, cover the gel with Saran Wrap, and autoradiograph at room temperature or at −70°C with an intensifying screen (see pages 470ff).

6. Recover the separated strands from the gel as described on page 164.

STRAND SEPARATION ON AGAROSE GELS

This method is used to separate the strands of DNA fragments >1 kb in length (Hayward 1972).

1. Cut strips of agarose containing specific DNA fragments from a preparative agarose gel.

2. Immerse the agarose strips in 0.3 M NaOH for 30 minutes at room temperature.

3. Rinse the strips briefly in several changes of ice-cold, distilled water.

4. Load the strips into slots cut to the correct size in a preformed horizontal agarose gel cast in a buffer containing

    36 mM Tris·Cl (pH 7.7)
    30 mM NaH$_2$PO$_4$
     1 mM EDTA

    Any gaps between the agarose strips and the slots are filled with melted agarose solution.

5. Add 50 μl of 1× gel-loading-dye I to any empty slot of the gel (see Table 5.2, page 160).

6. Carry out electrophoresis at 1.5 V/cm until the bromophenol blue has migrated about half the length of the gel.

7. If the DNA is to be recovered from the gel and if it is unlabeled, the gel should be stained for 45 minutes in electrophoresis buffer containing 1.0 μg/ml of ethidium bromide. Because the affinity of ethidium bromide for single-stranded DNA is relatively low and the fluorescent yield is poor, it is difficult to detect less than 0.1 μg of single-stranded DNA by staining.

   If the DNA is labeled with ^{32}P, the gel should be covered with Saran Wrap and exposed for autoradiography either at room temperature or at −70°C with intensifying screens. Once the DNA has been located, it can be recovered from the gel by electroelution, as described on page 164.
   Alternatively, the gel can be soaked in 0.5 M Tris (pH 7.6) and 1.5 M NaCl for 1 hour, and the DNA can then be transferred onto a sheet of nitrocellulose in preparation for hybridization (Southern 1975; see pages 382ff).

Notes

i. Each DNA fragment usually gives rise to three bands in strand-separating gels. One of these is native, double-stranded DNA, which forms by the reannealing of the two single strands during loading of the gel. The other two bands (usually much fainter and fuzzier than the first) are the two separated strands.

ii. Native, double-stranded DNA runs considerably slower than single-stranded DNA in agarose gels; in acrylamide gels, however, native DNA usually runs faster than single-stranded DNA.

iii. The concentration of single-stranded DNA applied to the gel should be kept as low as possible in order to minimize renaturation. Use thick gels (0.3–0.5 cm) with 2–3-cm slots.

iv. Markers for strand-separating gels may be prepared by end-labeling of DNA fragments with recessed 3′-hydroxyl termini, as described on pages 113ff. Just before loading onto the gel, the markers should be denatured by heat or alkali, exactly like the samples under investigation.

v. The preferred method of loading an agarose, strand-separating gel is by electrophoresis of denatured DNA from a chunk of agarose gel. This procedure minimizes the opportunity for renaturation of complementary strands of DNA. However, loading by conventional procedures also works, albeit less efficiently.

    a. Precipitate the DNA with ethanol and wash the pellet twice with 70% ethanol.

    b. Redissolve the DNA in

        0.1 M NaOH
         1 mM EDTA
        8% sucrose
        0.05% bromophenol blue

    c. Load the sample onto the gel and immediately carry out electrophoresis as described above.

STRAND SEPARATION OF SHORT DUPLEX DNA BY ELECTROPHORESIS THROUGH DENATURING POLYACRYLAMIDE GELS

As pointed out above, separation of the strands of denatured duplex DNA cannot always be achieved by electrophoresis through nondenaturing gels. An alternative strategy is to use restriction enzymes to generate a duplex DNA fragment with strands that are unequal in length. These strands are then separated on the basis of size by electrophoresis through denaturing polyacrylamide gels.

For example, enzymes can be used that generate one blunt end and one end containing a 3' or 5' extension

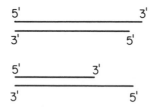

or, alternatively, two enzymes that leave a 3' extenstion at one end and a 5' extension at the other end of the DNA.

Depending on the enzymes used, DNA strands that differ in size by two to eight nucleotides can be generated. Because denaturing polyacrylamide gels used for DNA sequencing can resolve fragments that differ in size by only one nucleotide, it is possible to separate the strands of duplex DNA in which one of the strands is larger than the other.

1. Digest the DNA with the appropriate enzymes and label the 3' or 5' ends with ^{32}P using the procedures described on pages 113ff. If the DNA is well characterized and all of the fragments can be resolved by gel electrophoresis, extract the solution with phenol/chloroform and precipitate the DNA with ethanol. Resuspend the DNA in 20 μl of

   90% formamide (purified as described on page 448)
   1× TBE (see page 454)
   0.02% bromophenol blue
   0.02% xylene cyanol

   If a complex pattern of restriction fragments is expected, it may be necessary to purify the specific fragment of interest from a nondenaturing polyacrylamide gel before separating the strands by electrophoresis through a denaturing polyacrylamide gel.

2. Heat the solution to 90°C to denature the DNA, then cool quickly in ice water. Immediately layer the DNA solution into the slot of a polyacrylamide gel containing 7–8 M urea (see below). Up to 2 μg of DNA may be applied to a slot 3 cm wide formed in a gel that is 0.7 mm thick.

3. The gel is cast as described on page 174, using a solution that contains

    20 parts of acrylamide to every 1 part of bisacrylamide
    8 M urea
    1× TBE (see page 454)

    Use the following rule of thumb to estimate the concentration of polyacrylamide required to separate strands of DNA fragments of different sizes.

% Polyacrylamide	DNA Fragment Size (in nucleotides)
4	>200
5	80–200
8	40–100
12	10–50

4. Electrophoresis is carried out at 20 V/cm until the marker dyes have migrated a suitable distance (see Table 5.5). In general, the DNA fragments should be allowed to migrate almost the full length of the gel in order to obtain maximum separation of strands.

    Note. When a significant amount of salt is present in the sample, a salt front is generated that moves with the bromophenol blue dye.

5. Locate the DNA by autoradiography. Recover the DNA from the gel as described on page 164.

Note

The use of this procedure has been described in Akusjarvi and Pettersson (1978), Treisman (1980), and Plucienniczak and Streeck (1981).

TABLE 5.5. SIZE OF DNA FRAGMENTS (IN NUCLEOTIDES) THAT COMIGRATE WITH MARKER DYE IN DENATURING POLYACRYLAMIDE GELS

% Polyacrylamide	Bromophenol Blue	Xylene Cyanol
5	35	130
6	26	106
8	19	70–80
10	12	55
20	8	28

6

Extraction, Purification, and Analysis of mRNA from Eukaryotic Cells

A typical mammalian cell contains about 10^{-5} μg of RNA, 80–85% of which is ribosomal (chiefly, 28S, 18S, and 5S), whereas 10–15% is made up of a variety of low-molecular-weight species (transfer RNAs, small nuclear RNAs, etc.). All of these RNAs are of defined size and sequence and can be isolated in virtually pure form by gel electrophoresis, density gradient centrifugation, or ion-exchange chromatography. By contrast, mRNA, which makes up between 1% and 5% of the total cellular RNA, is heterogeneous both in size (from a couple of hundred bases to several kilobases in length) and sequence. However, virtually all mammalian mRNAs carry at their 3' ends a poly(A) tract that is generally long enough to allow mRNAs to be purified by affinity chromatography on oligo(dT) cellulose. The resulting, physically heterogeneous population of molecules collectively encodes all of the polypeptides synthesized by the cell.

The keys to obtaining good preparations of eukaryotic mRNA are to minimize ribonuclease activity during the initial stages of extraction and to avoid the accidental introduction of trace amounts of ribonuclease from the glassware and solutions. The following four elements are therefore important in the successful isolation of mRNA from mammalian cells.

1. Use of Exogenous Inhibitors of RNases

Two types of inhibitors of RNase are currently in widespread use: vanadyl-ribonucleoside complexes and RNasin.

Vanadyl-ribonucleoside complexes. The complexes formed between the oxovanadium ion and any of the four ribonucleosides are transition-state analogs that bind to many RNases and inhibit their activity almost completely (Berger and Birkenmeier 1979). The four vanadyl-ribonucleoside complexes are added to intact cells and used at a concentration of 10 mM during all stages of RNA extraction and purification. The resulting mRNA is isolated in high yield and in a form that can be translated efficiently in cell-free, protein-synthesizing systems. If necessary, the vanadyl-ribonucleoside complexes can be removed from the RNA preparation by multiple extractions with phenol containing 0.1% 8-hydroxyquinoline.

Vanadyl-ribonucleoside complexes are prepared as follows (Berger and Birkenmeier 1979):

1. Dissolve 0.5 mmole of each of the four ribonucleosides in 8 ml of water in a boiling-water bath.

2. Purge the solution with nitrogen gas, while adding 1 ml of 2 M vanadyl sulfate.

3. Adjust the pH to ~6 by dropwise addition of 10 M NaOH.

4. Adjust the pH carefully to 7.0 by dropwise addition of 1 M NaOH under nitrogen gas in the water bath. As the complexes form, the solution changes color from blue to green-black.

5. Adjust the volume to 10 ml; divide the solution into small aliquots and store at −20°C under nitrogen gas. The concentration of the complex is 200 mM. Dilute 1:20 for use.

Vanadyl-ribonucleoside complexes are commercially available from Bethesda Research Laboratories, P.O. Box 577, Gaithersburg, MD 20760.

RNasin. RNasin is a protein ($M_r \simeq 40,000$), isolated from rat liver and human placenta, that is a potent inhibitor of RNase. RNasin is effective during cell-free translation (Scheel and Blackburn 1979) and reverse transcription of mRNA (de Martynoff et al. 1980). Like vanadyl-ribonucleoside complexes, RNasin can be included in enzymatic reactions. Its advantage over vanadyl-ribonucleoside complexes is that it can be easily extracted with phenol.

RNasin is commercially available from Biotec, Inc., 2800 Fish Hatchery Rd., Madison, WI 53711.

2. Methods That Disrupt Cells and Inactivate Nucleases Simultaneously

Proteins readily dissolve in solutions of potent chaotropic agents such as guanidinium chloride and guanidinium isothiocyanate (Cox 1968). Cellular structures disintegrate and nucleoproteins dissociate rapidly from nucleic acids as ordered secondary structure is lost. Even RNase, an enzyme that is resistant to many forms of physical abuse (such as boiling), is inactive in the presence of 4 M guanidinium isothiocyanate and reducing agents like β-mercaptoethanol (Sela et al. 1957). This combination of reagents can therefore be used to isolate intact RNA from tissues, such as the pancreas, that are rich in RNase (Chirgwin et al. 1979).

Guanidinium isothiocyanate is made up according to the following steps.

1. To a 100-g bottle of guanidinium isothiocyanate (Eastman Laboratory and Specialty Chemicals or Fluka), add 100 ml of deionized H_2O, 10.6 ml of 1 M Tris · Cl (pH 7.6), and 10.6 ml of 0.2 M EDTA. Stir overnight at room temperature.

2. Warm the solution while stirring to 60–70°C for 10 minutes to assist dissolution. Often there is a residue of insoluble material that is removed by centrifugation at 3000g for 10 minutes at 20°C.

3. Add 21.2 ml of 20% Sarkosyl (sodium lauryl sarkosinate) and 2.1 ml of β-mercaptoethanol to the supernatant and bring the volume to 212 ml with sterile H_2O.

4. Filter through a disposable Nalgene filter and store at 4°C in a tightly sealed, brown glass bottle.

3. Use of RNase-free Glassware and Plasticware

Sterile, disposable plasticware is essentially free of RNase and can be used for the preparation and storage of RNA without pretreatment. General laboratory glassware, however, is often a source of RNase contamination and should be treated by baking at 250°C for 4 or more hours. In addition, some workers treat glassware for 12 hours at 37°C with a solution of 0.1% diethylpyrocarbonate, which is a strong but not absolute inhibitor of RNase (Fedorcsak and Ehrenberg 1966). Before using glassware treated in this way, it is important to remove traces of diethylpyrocarbonate by heating to 100°C for 15 minutes (Kumar and Lindberg 1972) or by autoclaving. Otherwise, there is a danger that the remaining traces of diethylpyrocarbonate will inactivate the RNA by carboxymethylation.

It is a good idea to set aside items of glassware and batches of plasticware that are to be used only for experiments with RNA, to mark them distinctively, and to store them in a designated place.

A potentially major source of contamination with RNase is the hands of investigators. There is little use in going to great lengths to rid glassware of contamination if no care is taken to keep one's fingers out of harm's way. Gloves should therefore be worn at all stages during the preparation of materials and solutions used for the isolation and analysis of RNA and during all manipulations involving RNA.

4. Careful Preparation of Solutions

All solutions should be prepared using baked glassware, glass-distilled autoclaved water, and dry chemicals that are reserved for work with RNA and that are handled with baked spatulas. Wherever possible, the solutions should be treated with 0.1% diethylpyrocarbonate for at least 12 hours and autoclaved (15 minutes, liquid cycle). Note that diethylpyrocarbonate cannot be used to treat solutions containing Tris. It is highly unstable in the presence of Tris buffers and decomposes rapidly into ethanol and carbon dioxide.

ISOLATION OF mRNA FROM MAMMALIAN CELLS

The method given below is a modification of that of Favaloro et al. (1980) for the isolation of mRNA from monolayers of mammalian cells grown in tissue culture. However, it can also be used to isolate mRNA from mammalian cells grown in suspension or from any mammalian tissue that can be dispersed into single cells.

1. Remove the medium from the cells and wash the monolayers three or four times with ice-cold, phosphate-buffered saline (PBS). Add 2 ml of ice-cold PBS to each of the plates and stand them on a bed of ice.

2. Scrape off the cell sheet from each plate in turn by using a rubber policeman. With a wide-mouthed pipette, transfer the cell suspension into a Corex centrifuge tube and store on ice until the cells have been removed from all of the plates.

3. Centrifuge at 2000g for 5 minutes at 4°C.

4. Remove the supernatant by aspiration and resuspend the cell pellet in ice-cold lysis buffer (0.25 ml per 85-mm dish).

   Lysis buffer
      0.14 M NaCl
      1.5 mM $MgCl_2$
      10 mM Tris · Cl (pH 8.6)
      0.5% NP-40
      1000 units/ml RNasin
         or 10 mM vanadyl-ribonucleoside complexes

5. Vortex for 10 seconds and then underlay the cell suspension with an equal volume of lysis buffer containing sucrose (24% w/v) and NP-40 (1%). Store on ice for 5 minutes.

6. Centrifuge at 10,000g for 20 minutes at 4°C in a swing-out rotor.

7. Recover the turbid, upper (cytoplasmic) layer and add an equal volume of 2 × PK buffer.

   2 × PK buffer
      0.2 M Tris · Cl (pH 7.5)
      25 mM EDTA
      0.3 M NaCl
      2% w/v SDS

   Add proteinase K to a final concentration of 200 μg/ml. Mix and incubate at 37°C for 30 minutes.

Note. If nuclear RNA is also to be prepared, discard the clear sucrose phase and resuspend the nuclear pellet at the bottom of the centrifuge tube in lysis buffer (0.25 ml per 85-mm dish). Add an equal volume of 2 × PK buffer and proteinase K to a final concentration of 200 μg/ml. Disrupt the nuclei and shear the liberated DNA by repeatedly squirting the viscous solution through a sterile 19-gauge hypodermic needle. Incubate at 37°C for 30 minutes.

8. Remove the proteins by extracting once with phenol/chloroform.

9. Recover the aqueous phase, add 2.5 volumes of ethanol, and store at −20°C for at least 2 hours.

10. Centrifuge for 10 minutes at 5000g at 0°C. Discard the supernatant and wash the pellet with 75% ethanol containing 0.1 M sodium acetate (pH 5.2).

11. Redissolve nucleic acids in a small volume (50 μl per 85-mm dish) of:

    50 mM Tris · Cl (pH 7.5)
    1 mM EDTA

    Then add MgCl$_2$ to 10 mM and RNasin or vanadyl-ribonucleoside complexes to final concentrations of 2000 units/ml or 2 mM, respectively. Add pancreatic DNase I (Worthington RNase-free DPFF) from which RNase has been removed either by chromatography on agarose-5′-(4-aminophenyl-phosphoryl) uridine 2′(3′) phosphate (Miles) as described by Maxwell et al. (1977) (see page 451) or by adsorption to macaloid (Schaffner 1982) (see page 452). Incubate for 30 minutes at 37°C.

12. Add EDTA and SDS to final concentrations of 10 mM and 0.2%, respectively.

13. Extract the solution once with phenol/chloroform.

14. Add sodium acetate (pH 5.2) to 0.3 M and precipitate the nucleic acids with 2 volumes of ethanol. Store the RNA in 70% ethanol at −70°C.
    Cytoplasmic RNA may be further purified and freed from any contaminating oligodeoxyribonucleotides by chromatography on oligo(dT)-cellulose (see page 197).

Note

i. Oligodeoxyribonucleotides contaminating preparations of nuclear RNA may be removed as follows:

   a. Resuspend the pellet in 20% sodium acetate by repeated pipetting up
      and down.

   b. Centrifuge for 10 minutes in an Eppendorf centrifuge.

   c. Discard the supernatant and redissolve the pellet in TE. Precipitate
      with ethanol.

ii. Cytoplasmic RNA can usually be purified by oligo(dT) chromatography
    without DNase treatment.

ISOLATION OF TOTAL CELLULAR RNA

The two methods given below are used to isolate RNA from cells that cannot be fractionated easily into cytoplasm and nuclei (e.g., frozen fragments of tissue) or from cells that are particularly rich in RNase (e.g., pancreatic cells).

Guanidinium/Hot Phenol Method[1]

1. Add 4 M guanidinium isothiocyanate mixture (prepared as described on page 189) to a tissue fragment or washed cell pellet in a plastic, disposable centrifuge tube. Use 1 ml of guanidinium isothiocyanate mixture for 10^7 cells or 5 ml for every gram of tissue.

   Note. Tissue fragments may require disruption by homogenization in an omnimixer or polytron mixer.

2. Bring the mixture to 60°C and, while maintaining this temperature, draw the slimy suspension into a syringe fitted with an 18-gauge needle. Forcefully eject the suspension back into the tube. Repeat until the viscosity of the suspension is reduced by shearing of the liberated chromosomal DNA.

3. Add an equal volume of phenol preheated to 60°C and continue to pass the emulsion through the syringe.

4. Add 0.5 volume of:

   0.1 M sodium acetate (pH 5.2)
   10 mM Tris·Cl (pH 7.4)
   1 mM EDTA

5. Add an equal volume of a 24:1 solution of chloroform and isoamyl alcohol and shake vigorously for 10–15 minutes while maintaining the temperature at 60°C.

6. Cool on ice and centrifuge at 2000*g* for 10 minutes at 4°C.

7. Recover the aqueous phase and reextract with phenol/chloroform.

8. Centrifuge and recover the aqueous phase and reextract twice with chloroform.

9. Add 2 volumes of ethanol. Store at −20°C for 1–2 hours.

10. Recover the RNA by centrifugation at 12,000*g* for 20 minutes at 4°C.

[1]Feramisco et al. (1982)

11. Dissolve the pellet in the original starting volume (step 1) of:

    0.1 M Tris·Cl (pH 7.4)
    50 mM NaCl
    10 mM EDTA
    0.2% SDS

    Add proteinase K to a final concentration of 200 μg/ml. Incubate for 1–2 hours at 37°C.

12. Heat to 60°C. Add 0.5 volume of phenol preheated to 60°C and mix. Add 0.5 volume of chloroform and mix vigorously at 60°C for 10 minutes.

13. Cool in ice and centrifuge at 2000g for 10 minutes at 4°C.

14. Extract once more with phenol/chloroform at 60°C and then extract twice with chloroform at room temperature.

15. Precipitate the nucleic acids with ethanol by centrifugation and rinse the pellet in 70% ethanol. Store the RNA in 70% ethanol at −70°C.

16. Dissolve the RNA in sterile water. 5×10^8 cells should yield approximately 5–10 mg of RNA.

Guanidinium/Cesium Chloride Method[2]

1. To a fragment of tissue or a cell pellet, add 5 volumes of:

   6 M guanidinium isothiocyanate
   5 mM sodium citrate (pH 7.0)
   0.1 M β-mercaptoethanol
   0.5% Sarkosyl

   Disperse the tissue by homogenization or the cell pellet by vortexing.

2. Add 1 g of cesium chloride to each 2.5 ml of homogenate.

3. Layer the homogenate onto a 1.2-ml cushion of 5.7 M CsCl in 0.1 M EDTA (pH 7.5) in a Beckman SW50.1 polyallomer tube (or its equivalent). (Other types of centrifuge tubes can be used; see Chirgwin et al. [1979]).

4. Centrifuge at 35,000 rpm for 12 hours at 20°C. This procedure takes advantage of the fact that the buoyant density of RNA in cesium chloride is much greater than that of other cellular macromolecules. During centrifugation, the RNA forms a pellet on the bottom of the tube while most of the DNA and protein floats upward in the cesium chloride solution.

5. Discard the supernatant, dry the walls of the centrifuge tubes thoroughly, and dissolve the pellet of RNA in:

   10 mM Tris · Cl (pH 7.4)
   5 mM EDTA
   1% SDS

6. Extract once with a 4:1 mixture of chloroform and 1-butanol and transfer the aqueous phase to a fresh tube. Reextract the organic phase with an equal volume of:

   10 mM Tris · Cl (pH 7.4)
   5 mM EDTA
   1% SDS

   Combine the two aqueous phases.

7. Add 0.1 volume of 3 M sodium acetate (pH 5.2) and 2.2 volumes of ethanol. Store at −20°C for at least 2 hours. Recover the RNA by centrifugation.

8. Dissolve the pellet in 1 ml of H_2O and reprecipitate with ethanol. Store the RNA in 70% ethanol at −70°C.

[2]Glisin et al. (1974); Ullrich et al. (1977).

SELECTION OF POLY(A)$^+$ RNA

Several techniques have been developed to separate polyadenylated RNA from nonpolyadenylated RNAs. The method of choice is chromatography on oligo(dT)-cellulose (Edmonds et al. 1971; Aviv and Leder 1972), which can be prepared as described by Gilham (1964) or obtained commercially. Up to 10 mg of RNA can be processed per ml of oligo(dT)-cellulose.

1. Equilibrate the oligo(dT)-cellulose in sterile loading buffer. To sterilize the loading buffer, mix appropriate amounts of RNase-free stock solutions of the Tris, sodium chloride, and EDTA and autoclave. Add SDS from a 20% stock solution that has been treated at 65°C for 1 hour. Alternatively, 0.05 M sodium citrate can be substituted for Tris, and the loading buffer and SDS can then be treated with diethylpyrocarbonate.

   Loading buffer
   20 mM Tris·Cl (pH 7.6)
   0.5 M NaCl
   1 mM EDTA
   0.1% SDS

2. Pour a 1.0-ml column in a Dispocolumn or pasteur pipette. Wash the column with 3 column-volumes each of:

   a. sterile H$_2$O
   b. 0.1 M NaOH and 5 mM EDTA
   c. sterile H$_2$O

3. Check that the pH of the column effluent is less than 8.

4. Wash the column with 5 volumes of sterile loading buffer.

5. Dissolve the RNA in sterile water and heat to 65°C for 5 minutes. Add an equal amount of 2× loading buffer, cool the sample to room temperature, and apply to the column. Collect the flow-through, again heat to 65°C, cool, and reapply to the column.

6. Wash the column with 5–10 column-volumes of loading buffer, followed by 4 column-volumes of loading buffer containing 0.1 M NaCl.

   Note. Read the OD$_{260}$ of each column-volume fraction collected. Initially, the OD$_{260}$ will be very high as the poly(A)$^-$ RNA comes through the column. The later fractions should have no or very little OD$_{260}$ absorbing material.

7. Elute the poly(A)$^+$ RNA with 2–3 column-volumes of sterile

   10 mM Tris-Cl (pH 7.5)
   1 mM EDTA
   0.05% SDS

   If desired, the eluted poly(A)$^+$ mRNA can be selected again by oligo(dT)-cellulose chromatography by adjusting the sodium chloride concentration of the eluted mRNA to 0.5 M and repeating steps 5, 6, and 7.

8. Add sodium acetate (3 M, pH 5.2) to a final concentration of 0.3 M. Precipitate the RNA with 2.2 volumes of ethanol at $-20°C$. Rinse the pellet in 70% ethanol.

9. Dissolve the pellet in sterile water. The yield from 10^7 cells should be 1–5 μg of poly(A)$^+$ RNA. Regenerate the column by sequential washing in sodium hydroxide, water, and loading buffer as described in steps 2, 3, and 4 above.

Notes

i. When many RNA samples are to be processed, it may be more efficient to carry out a batch absorption and elution with oligo(dT)-cellulose. After dissolving the RNA in loading buffer (step 5), add 0.3 g (dry weight) of oligo(dT)-cellulose for each 0.5 mg of RNA. Centrifuge at $1500g$ for 4 minutes at 15°C and wash the oligo(dT)-cellulose four to five times with 5 ml of loading buffer at room temperature. Elute the poly(A)$^+$ RNA with four 1-ml washes of sterile

   10 mM Tris · Cl (pH 7.5)
   1 mM EDTA
   0.05% SDS

   Continue with step 8 above.

ii. The sodium salt of SDS is relatively insoluble and therefore may tend to impede the flow of the column. This can be avoided by using lithium chloride instead of sodium chloride in the loading buffer.

iii. RNA should be stored in 70% ethanol at $-70°C$.

GEL ELECTROPHORESIS OF RNA

The two systems most frequently used to measure the molecular weight of RNA and to separate RNAs of different sizes for "Northern" transfer or in vitro translation are:

1. Electrophoresis through agarose gels after denaturation of the RNA with glyoxal and dimethylsulfoxide.

2. Electrophoresis through agarose gels containing methylmercuric hydroxide or formaldehyde.

In each case, the RNA is fully denatured, and its rate of migration through the gel is in linear proportion to the $\log_{10}$ of its molecular weight.

Electrophoresis of RNA after Denaturation with Glyoxal and Dimethylsulfoxide [3]

1. Mix in a sterile Eppendorf tube:

6 M glyoxal	2.7 μl
dimethylsulfoxide (DMSO)	8.0 μl
0.1 M NaH$_2$PO$_4$ (pH 7.0)	1.6 μl
RNA (up to 20 μg)	3.7 μl

   Glyoxal is usually obtained as a 40% solution (6 M). Because it readily oxidizes in air, the glyoxal solution must be deionized before use by passage through a mixed-bed resin (Bio-Rad AG 501-X8) until its pH is neutral. It is then stored at −20°C in small aliquots in tightly capped tubes.

   The sodium phosphate solution is made up as follows:

   a. Dissolve 0.1 moles of NaH$_2$PO$_4$ in a minimum volume of water.

   b. Adjust the pH to 7.0 with concentrated phosphoric acid.

   c. Adjust volume to 1 liter and autoclave.

2. Incubate the RNA solution at 50°C for 60 minutes in a tightly closed tube.

3. While the RNA is incubating, pour a horizontal agarose gel. For RNAs up to 1 kb in length, use 1.4% agarose; for larger RNAs, use 1.0% agarose. The gels are poured and run in 0.01 M NaH$_2$PO$_4$ (pH 7.0).

   Note. Because glyoxal reacts with ethidium bromide, the gels are poured and run in the absence of the dye.

4. Cool the RNA sample to 20°C and add 4 μl of *sterile* loading buffer and load the sample immediately. As molecular-weight markers, use glyoxylated RNAs (e.g., 18S and 28S ribosomal RNAs; globin 9S mRNA).

   Loading buffer
   50% glycerol
   0.01 M NaH$_2$PO$_4$ (pH 7.0)
   0.4% bromophenol blue

5. The gel is run submerged in buffer at 3–4 V/cm. Constant recirculation of the buffer (see Fig. 6.1) is required in order to maintain the pH within acceptable limits (glyoxal dissociates from RNA at pH >8.0). Alternatively, the buffer may be changed every 30 minutes during the run.

6. At the end of the run, the gel may be stained with ethidium bromide (0.5 μg/ml in H$_2$O) and photographed as described on page 162.

[3]McMaster and Carmichael (1977).

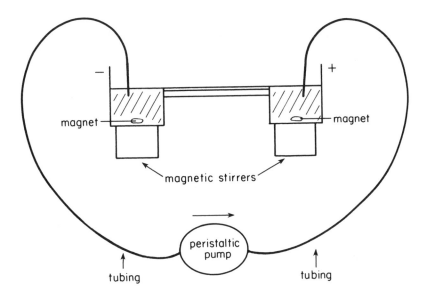

Figure 6.1

Transfer of Glyoxylated RNA to Nitrocellulose Filters

Glyoxylated RNA may be transferred immediately after electrophoresis from agarose gels to nitrocellulose filters; no further treatment of the gel is necessary (Thomas 1980). The gel is placed in contact with the nitrocellulose filter and blotted essentially as described on pages 382ff.

Notes

i. The blotting buffer is 20× SSC (3 M NaCl and 0.3 M trisodium citrate).

ii. The nitrocellulose filter should be wetted in water and then soaked for 5 minutes in 20× SSC just before use.

iii. Transfer of RNA is less efficient if the gel is presoaked in 20× SSC or if the RNA has been stained with ethidium bromide.

iv. Transfer is complete in 15–24 hours.

v. Do not wash the blot with more dilute salt solutions before baking, otherwise most of the RNA will be lost.

vi. The blots are dried at room temperature and baked in a vacuum oven at 80°C for 2 hours.

Electrophoresis of RNA through Gels Containing Formaldehyde[4]

1. Prepare gel-running buffer and formaldehyde.

   Gel-running buffer

   > 0.2 M morpholinopropanesulfonic acid (MOPS) (pH 7.0)
   > 50 mM sodium acetate
   > 1 mM EDTA (pH 8.0)

   This buffer yellows with age if exposed to light or autoclaved. Discoloration does not affect its performance appreciably.

   Formaldehyde (F.W. = 30.03) is usually obtained as a 37% solution in water (12.3 M). Check that the pH of the concentrated solution is greater than 4.0. The concentrated solution should be handled and stored in a chemical hood.

2. Prepare the gel by melting agarose in water, cooling to 60°C, and adding 10× gel buffer and formaldehyde to give 1× and 2.2 M final concentrations, respectively. (One part of stock formaldehyde solution should be diluted with 4.6 parts of agarose solution.)

   Note. The fractionation properties of formaldehyde gels with different agarose concentrations have been determined by Lehrach et al. (1977).

3. Prepare the sample by mixing the following in a sterile Eppendorf tube:

RNA (up to 20 μg)	4.5 μl
10× gel-running buffer	2.0 μl
formaldehyde	3.5 μl
formamide	10.0 μl

   Incubate at 55°C for 15 minutes.

   Note. Formamide oxidizes readily in air and should be deionized by passage through a mixed-bed resin (Bio-Rad AG 501-X8) until its pH is neutral. It is then recrystallized at 0°C and stored at −20°C in small aliquots in tightly capped tubes.

4. Add 2 μl of sterile loading buffer.

   Loading buffer

   > 50% glycerol
   > 1 mM EDTA
   > 0.4% bromophenol blue
   > 0.4% xylene cyanol

[4]Lehrach et al. (1977); Goldberg (1980); B. Seed and D. A. Goldberg (unpubl.).

5. Load the RNA samples onto the gel. Restriction fragments of DNA are convenient molecular-weight markers. They should be treated and run exactly as the RNA samples. Labeled DNA markers or markers that will be detected by a labeled hybridization probe are preferred because the ethidium fluorescence of formaldehyde-denatured nucleic acids is weak. If such markers are impossible to obtain, the following protocol may be used:

a. Apply to the gel sufficient DNA to give at least 50–100 ng per band.

b. Cut the lanes containing the markers from the gel before the alkaline hydrolysis step (see below), and wash with four or five changes of water for 2 hours.

c. Wash with two changes of 0.1 M ammonium acetate for 1 hour.

d. Stain for 1 hour with 0.5 μg/ml of ethidium bromide in 0.1 M ammonium acetate, and 0.1 M β-mercaptoethanol.

e. Destain for 45 minutes with a solution of 0.1 M ammonium acetate and 0.01 M β-mercaptoethanol.

Transfer of Formaldehyde-denatured RNA to Nitrocellulose

1. After electrophoresis is complete, soak the gel for 5 minutes in several changes of water.

   Note. Gels containing formaldehyde are less rigid than nondenaturing agarose gels. Care must be exercised in handling them.

2. Soak the gel in an excess of 50 mM NaOH and 10 mM NaCl for 45 minutes at room temperature.

   Note. The partial alkaline hydrolysis improves the transfer of high-molecular-weight RNA.

3. Neutralize the gel by soaking for 45 minutes at room temperature in 0.1 M Tris · Cl (pH 7.5).

4. Soak the gel for 1 hour in 20× SSC.

5. Transfer the RNA to nitrocellulose by the method described on page 201. The transfer is complete in 3–4 hours.

6. After transfer is complete, wash the filter in 3× SSC, dry in air for 1–2 hours, and bake for 3–4 hours at 80°C under vacuum.

Electrophoresis of RNA through Gels Containing Methylmercuric Hydroxide[5]

Methylmercuric hydroxide reacts primarily with the amino bonds of uridine and guanosine in RNA. Because these bonds are normally involved in Watson-Crick pairing, methylmercuric hydroxide is an effective denaturing agent that disrupts all secondary structure in the RNA.

Methylmercuric hydroxide also reacts reversibly with various small molecules; for example:

$$CH_3HgOH + RSH \rightarrow CH_3HgSR + H_2O$$

Thus, sulfhydryl compounds can be used to reverse the binding of CH_3Hg^+ to nucleic acids.

Caution

Methylmercuric hydroxide is extremely toxic. It is also volatile. Therefore all manipulations of pouring, running, cutting, and processing of gel slices should be carried out in a chemical hood. All solid and liquid wastes should be treated as toxic materials and disposed of accordingly.

1. Prepare the gel (1–1.5% agarose, depending on the size of RNA to be analyzed) in gel-running buffer. No methylmercuric hydroxide is added to the running buffer. The compound is uncharged and does not migrate rapidly out of the gel.

   Gel-running buffer

   50 mM boric acid
   5 mM sodium borate
   10 mM sodium sulfate

   Note. Buffers that contain nitrogen bases, EDTA, or chloride ions should not be used because these compounds form complexes with methyl hydroxide.

2. Mix equal volumes of the RNA solution (up to 10 μg may be loaded per standard 0.6-cm slot) and 2× loading buffer.

   2× Loading buffer

methylmercuric hydroxide	25 μl
4× running buffer	500 μl
100% glycerol	200 μl
H₂O	275 μl
bromophenol blue	0.2% w/v

---

[5]Bailey and Davidson (1976).

3. Load the samples and run the gel at 1.5 V/cm for 12–16 hours. Recirculate the buffer to avoid generating a pH gradient in the gel.

4. After electrophoresis, RNA may be stained by incubating the gel for 30 minutes in 0.5 M ammonium acetate and 0.5 μg/ml of ethidium bromide. The ammonium salt complexes methylmercury and enhances binding of the dye to RNA.

5. The RNA may be transferred from methylmercury gels to nitrocellulose filters. The transfer of RNA is inefficient if the gel is presoaked in 20× SSC or if the RNA has been stained with ethidium bromide.

Recovery of RNA from Agarose Gels Containing Methylmercuric Hydroxide[6]

1. The gel is prepared and run as described above, except that low-melting-temperature agarose is used (Wieslander 1979).

2. After electrophoresis, soak the gel in 0.1 M dithiothreitol for 30–40 minutes.

3. Cut the gel into slices approximately 3 mm in width. Stained tracks containing 18S and 28S ribosomal RNAs and 9S globin RNAs may be used as rough molecular-weight guides.

4. To each gel slice, add approximately 4 volumes of 0.5 M ammonium acetate preheated to 65°C. Be sure to use a large volume of extraction buffer so that the gel is completely melted. Otherwise, agarose will be carried over into the aqueous phase during subsequent extraction with phenol and chloroform.

5. Heat at 65°C until the gel is melted. Vortex well.

6. Extract with phenol at room temperature. Centrifuge at 2000g for 10 minutes at 4°C. During extraction with phenol and chloroform, agarose becomes a powder and forms a layer at the interface upon centrifugation.

7. Reextract the aqueous phase twice more with chloroform. Repeated chloroform extractions are required to remove the agarose.

8. Precipitate the RNA with ethanol and wash the precipitate with 70% ethanol and 0.05 M ammonium acetate.

Note

The extracted RNA translates well in in vitro protein-synthesizing systems and is an efficient template for cDNA synthesis using reverse transcriptase.

[6]See, for example, Lemischka et al. (1981).

Staining RNA after Transfer to Nitrocellulose Filters

This procedure may be used to check the size of the RNA transferred to nitrocellulose and to estimate the efficiency of its transfer (A. Efstratiadis, unpubl.). A lane may be cut off the filter after baking, or the entire filter may be stained after hybridization and exposure to X-ray film.

1. Soak the dried filter in 5% acetic acid for 15 minutes at room temperature.

2. Transfer the filter to a solution of 0.5 M sodium acetate (pH 5.2) and 0.04% methylene blue for 5–10 minutes at room temperature.

3. Rinse the filter in water for 5–10 minutes. Marker RNAs (28S and 18S ribosomal RNAs and 9S globin mRNA) appear as bands. Total poly(A)$^{+}$ mammalian mRNA appears as a smear.

NUCLEASE-S1 MAPPING OF RNA

Nuclease-S1 mapping is used to map the locations of the ends of RNA molecules and of any splice points within them in relation to specific sites (e.g., positions of restriction endonuclease cleavage) within the template DNA. The procedure is based on the observation of Casey and Davidson (1977) that hybridization conditions can be established that minimize the formation of DNA · DNA hybrids while promoting the formation of DNA · RNA hybrids. These hybrids are then digested with the single-strand-specific nuclease S1 and analyzed by gel electrophoresis (Berk and Sharp 1977).

Although more-refined techniques are available (e.g., cDNA copies of RNAs can be prepared by primer extension and sequenced) (Ghosh et al. 1978; Reddy et al. 1978), the coordinates determined by nuclease-S1 analysis are accurate enough for most purposes. Furthermore, the method is rapid, easy to perform, and extremely sensitive—the equivalent of as little as one molecule of RNA per cell can be detected.

The method given below describes the nuclease-S1 technique in its simplest and most-basic form. A more-detailed description, together with a thorough discussion of the pitfalls of nuclease-S1 mapping, is presented in an excellent review by Favaloro et al. (1980).

1. Mix in a sterile Eppendorf tube:

   DNA   0.1–1.0 μg
   RNA   0.5–500 μg

   If necessary, add tRNA carrier to bring the total amount of RNA to 100–200 μg.

   Note. The amount of DNA used in the hybridization reaction depends on its molecular weight. For fragments approximately 5 kb in length, 0.1 μg is required; for smaller fragments, proportionately less DNA should be used.

   The amount of RNA required depends on the concentration of the sequences of interest. To detect sequences present in low amounts, up to 250 μg of RNA may be used per 50 μl of hybridization reaction. These quantities of RNA and DNA are suitable for hybridization in a volume of 50 μl. If reagents are in short supply, the hybridization reactions can be scaled down to 10 μl. For ease of manipulation in subsequent steps, it is advisable to keep the hybridization volume to 50 μl or less.

2. Precipitate the mixed DNA and RNA with ethanol (−70°C for 15 minutes).

3. Resuspend the pellet in 30 μl of hybridization buffer (see page 208). Pipette up and down many times to make sure that the pellet is dissolved.

Hybridization buffer

40 mM PIPES (pH 6.4)
1 mM EDTA (pH 8.0)
0.4 M NaCl
80% formamide

Notes

i. Use the disodium salt of PIPES.

ii. Use formamide deionized by passage through a mixed-bed resin (Bio-Rad AG 501-X8), recrystallized at 0°C, and stored in small aliquots in tightly capped tubes.

4. Immerse the tubes in a water bath at 72°C for 10–15 minutes to denature the DNA. (Higher temperatures [75–85°C] have been used successfully.)

5. Transfer the tubes *rapidly* to a water bath set at the desired hybridization temperature. Do not allow the tubes to cool below the hybridization temperature during transfer.

    The hybridization temperature, which depends on the G + C content of the DNA, is chosen so as to minimize the formation of DNA · DNA hybrids while allowing DNA · RNA hybrids to form. The following table gives the approximate hybridization temperatures for DNAs of different G + C content. It is advisable to carry out the hybridization reactions at different temperatures to find the optimal temperature for your RNA.

G + C	Hybridization temperature
41%	49°C
49%	52°C
58%	60°C

6. Hybridize for 3 hours. Open the lid of the hybridization tube, but keep the body of the tube submerged. Rapidly add 0.3 ml of ice-cold nuclease-S1 buffer (usually containing 100–1000 units/ml of nuclease S1). Mix well and incubate for the appropriate time and temperature (see note below).

Nuclease-S1 buffer

0.28 M NaCl
0.05 M sodium acetate (pH 4.6)
4.5 mM $ZnSO_4$
20 μg/ml carrier ssDNA

Note. A variety of temperatures and nuclease-S1 concentrations have been used with particular RNA · DNA hybrids. Incubation at 20°C will minimize the digestion of RNA across from DNA loops (this is critical for neutral gel analysis). Higher temperatures (37–45°C) may be required for complete digestion for denaturing gel analysis. In general, the nuclease-S1 digestion and the incubation temperature should be optimized for each RNA · DNA hybrid.

7. Chill to 0°C. Add 50 μl of 4.0 M ammonium acetate and 0.1 M EDTA to stop the reaction.

8. Extract once with phenol/chloroform. Add 20 μg of carrier tRNA. Precipitate with an equal volume of isopropanol (−70°C for 15 minutes).

9. Dissolve the precipitate in 40 μl of TE (pH 7.4).

10. Add 10 μl of loading buffer (50% glycerol and 0.2% bromocresol green) and mix well.

11. Load half of the sample on a neutral agarose gel. Adjust the remaining sample to 0.05 M NaOH and run on an alkaline agarose gel.

12. Following electrophoresis, transfer the DNA from both gels to nitrocellulose filters as described on pages 382ff. (The alkaline gel does not require soaking in denaturation solution).

13. Hybridize the filters to an appropriate ^{32}P-labeled DNA probe.

7

Synthesis and Cloning of cDNA

The enzymatic conversion of poly(A)$^+$ mRNA to double-stranded cDNA and the insertion of this DNA into bacterial plasmids has become a fundamental tool of eukaryotic molecular biology (for reviews, see Efstratiadis and Villa-Komaroff 1979; Williams 1981). Although a number of different approaches to synthesizing double-stranded DNA copies of mRNA have been reported, the most commonly used procedure involves synthesis of the first cDNA strand with reverse transcriptase (RNA-dependent DNA polymerase), removal of the RNA template by alkaline degradation, synthesis of the second DNA strand with *E. coli* DNA polymerase I or reverse transcriptase (using a hairpin loop at the 3′ end of the first DNA strand as primer), and finally, digestion of the loop connecting the first and second cDNA strands with the single-strand-specific nuclease S1. In this chapter, we summarize important technical points for each step of the synthesis of double-stranded cDNA, and we then discuss procedures necessary for joining this DNA to plasmid cloning vectors.

Synthesis of cDNA

SYNTHESIS OF THE FIRST cDNA STRAND

A number of papers describing optimization of conditions for producing "full-length" cDNA transcripts have been published (Efstratiadis et al. 1976; Buell et al. 1978; Retzel et al. 1980). Different mRNAs are copied into DNA with different efficiencies; thus conditions that are optimal for copying one species of mRNA may not work as well for another. In general, when dealing with heterogeneous populations of mRNA, conditions are used that lead to the greatest overall yield of cDNA. The following parameters are important.

Reverse Transcriptase

The most important factor in the synthesis of long cDNAs is the quality of the reverse transcriptase used in the reaction. Until recently, the major producer of reverse transcriptase was Dr. J. W. Beard (Life Sciences, Inc., 1509½ 49th Street South, St. Petersburg, FL 33707), who provided the enzyme on contract to the National Institutes of Health. After the NIH program was terminated, Life Sciences, Inc., began selling the enzyme directly. Reverse transcriptase is also available commercially from Bethesda Research Laboratories (Gaithersburg, MD) and Boehringer Mannheim Biochemicals (Indianapolis, IN).

Although the quality of these enzymes is generally good, the amount of contaminating RNase varies from batch to batch. (Some suppliers assay for and provide information about contaminating RNase.) This problem can be circumvented by additional purification of the enzyme (Marcus et al. 1974; Faras and Dibble 1975; Kacian 1977; Myers et al. 1980) or by including potent inhibitors of RNase, such as vanadyl-ribonucleoside complexes or RNasin, in the reverse transcription reaction. Many factors previously thought to be important for efficient synthesis of full-length cDNA transcripts actually work by protecting the RNA template from RNases (Buell et al. 1978; Retzel et al. 1980). For example, the addition of sodium pyrophosphate or ribonucleoside triphosphates was originally thought to increase the efficiency with which reverse transcriptase copied RNA (Kacian et al. 1972). However, with highly purified reverse transcriptase, the addition of these compounds has no effect (Retzel et al. 1980).

The ratio of reverse transcriptase to mRNA template is also important in optimizing the yield of full-length cDNA (Friedman and Rosbash 1977). With a given amount of template, the yield and the size of the cDNA transcript increases with increasing amounts of reverse transcriptase. In one study, maximum yield of full-length transcripts was reached at 80 units of enzyme per microgram of template, a 30-fold to 60-fold molar excess of enzyme to template (Friedman and Rosbash 1977). Such a high ratio of enzyme to template requires the use of highly purified enzyme and the inclusion of inhibitors of RNase in the reaction.

pH

A pH of 8.3 is optimal for efficient incorporation and production of full-length transcripts. A deviation of ± 0.5 pH units will result in a 5-fold decrease in the production of full-length transcripts. A number of buffer systems have been tested but none are better than Tris.

Monovalent Cation

Ionic conditions substantially affect the transcriptional efficiency of various templates. Longer transcripts are obtained with potassium than with sodium ions. The optimum potassium-ion concentration for both total synthesis and length of cDNA is 140–150 mM.

Divalent Cation

Divalent cations are an absolute requirement for reverse transcriptase activity. No activity is observed below 4 mM Mg^{++}; the optimum concentration for the production of full-length transcripts is 6–10 mM.

Deoxynucleoside Triphosphates

The use of high concentrations of each of the four deoxynucleoside triphosphates (dNTPs) is particularly important for efficient cDNA synthesis (Efstratiadis et al. 1976; Retzel et al. 1980). If the concentration of only one of them drops below 10–50 μM, the yield of full-length transcripts decreases significantly. Using avian myeloblastosis virus (AMV) RNA as a template, maximum production of full-length cDNAs was achieved at a concentration of 75 μM of all four dNTPs (Retzel et al. 1980). However, since little or no inhibition of transcription is observed in the 100 μM to 1 mM range, dNTP concentrations of 200–250 μM are generally used.

SYNTHESIS OF THE SECOND cDNA STRAND

For reasons that are not yet understood, the 3' ends of single-stranded cDNAs are capable of forming hairpin structures and therefore can be used to prime the synthesis of the second cDNA strand by *E. coli* DNA polymerase I or reverse transcriptase (see Fig. 7.1). Although there has been a great deal of speculation regarding the structure of the hairpin loops at the end of cDNAs and the mechanism by which they are generated, the phenomenon has not been systematically studied.

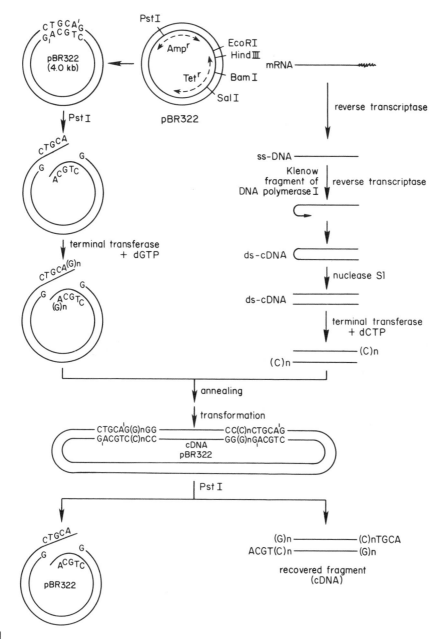

Figure 7.1

The conditions first used to achieve full-length, second-strand cDNA synthesis by DNA polymerase I (Efstratiadis et al. 1976) are still widely used (Wickens et al. 1978). In brief, the reaction is carried out at pH 6.9 to minimize the $5' \rightarrow 3'$ exonuclease activity of DNA polymerase I and at 15°C to minimize the possibility of synthesizing "snapback" DNA. The Klenow fragment of DNA polymerase I, which lacks the $5' \rightarrow 3'$ exonuclease activity, has also been successfully employed to synthesize the second cDNA strand.

Many investigators have utilized reverse transcriptase to synthesize the second cDNA strand using conditions similar to those already described for the first cDNA strand synthesis. Although there is one report that AMV reverse transcriptase could not be used to synthesize the second strand of an immunoglobulin cDNA (Rougeon and Mach 1976), the success of a large number of experiments that used reverse transcriptase for second-strand synthesis indicates that this is not a general problem. We recommend using both enzymes in succession. The rationale of this procedure, which was suggested by A. Efstratiadis, is that DNA polymerase I and reverse transcriptase may pause or stop at different sequences. Thus, partially synthesized second strands produced by one enzyme may be extended to completion by the other.

CLEAVAGE OF THE HAIRPIN LOOP WITH NUCLEASE S1

After synthesis of cDNA is complete, the first and second strands are covalently joined by the hairpin loop that was used to prime the second-strand synthesis (Efstratiadis et al. 1976). This loop is susceptible to cleavage by the single-strand-specific nuclease S1. The resulting termini are not always perfectly blunt-ended, and the efficiency of cloning is improved if they are repaired with the Klenow fragment of *E. coli* DNA polymerase I (Seeburg et al. 1977). The duplex DNA is then either fractionated according to size and the largest molecules inserted into bacterial plasmids, or an entire spectrum of sizes of double-stranded DNA is cloned to generate a cDNA library.

Molecular Cloning of Double-stranded cDNA

A variety of methods has been used to link double-stranded cDNA to plasmid vectors (Efstratiadis and Villa-Komaroff 1979; Maniatis 1980; Williams 1981). The most commonly used procedures are:

1. The addition of complementary homopolymer tracts to double-stranded cDNA and to the plasmid DNA. The vector and double-stranded cDNA are then joined by hydrogen bonding between the complementary homo-polymeric tails to form open circular, hybrid molecules capable of transforming *E. coli*. The formation of closed circular DNA by in vitro enzymatic ligation is not necessary to establish the recombinant plasmids in *E. coli*.

2. The addition of synthetic linkers to the termini of double-stranded cDNA. After cleavage with the appropriate restriction enzyme, the cDNA molecules are inserted into plasmid DNA that has been cleaved with a compatible enzyme.

HOMOPOLYMERIC TAILING

dA · dT Tailing

Calf-thymus terminal deoxynucleotidyl transferase, which catalyzes the addition of deoxynucleotides to the 3'-hydroxyl ends of single- or double-stranded DNA, was first used by Wensink et al. (1974) to introduce recombinant DNA into *E. coli* by a dA · dT joining procedure (Jackson et al. 1972; Lobban and Kaiser 1973). In the original procedure, a small number of nucleotides were removed from the 5' ends of the duplex DNAs to leave protruding, single-stranded, 3'-hydroxyl termini, which served as efficient templates for terminal transferase. The need for this step was obviated when it was shown that terminal transferase could utilize recessed 3' termini in the presence of cobalt ions (Roychoudhury et al. 1976). Usually, 50 to 150 dA residues are added to the linearized vector DNA and a corresponding number of dT residues to the double-stranded cDNA.

Double-stranded cDNA inserted into plasmids via the dA · dT joining procedure can be excised and recovered in one of three ways. The first method involves digestion of the recombinant plasmid DNA with nuclease S1 under moderately denaturing conditions, which cause preferential melting of the dA · dT linkers (Hofstetter et al. 1976). This is achieved by including formamide in the digestion buffer (25–50%) and carrying out the digestion at an elevated temperature (37–55°C). The efficiency of this reaction depends on the length of the dA · dT linkers: Inserts with short linkers are difficult to excise. The optimal conditions (including DNA and enzyme concentrations) should be determined empirically for each cDNA clone. In some cases, higher yields of the insert and fewer extraneous cleavage products are obtained when the single-strand-specific, mung-bean nuclease is used rather than nuclease S1 (M. R. Green, unpubl.). This may be related to the relatively high specificity of mung-bean nuclease for AT-rich duplex DNA (Johnson and Laskowski 1970).

An alternative but seldom-used procedure involves the conversion of plasmid DNA to linear duplex molecules using a restriction enzyme that does not cleave within the inserted DNA (Goff and Berg 1978). The linear DNA is then denatured and briefly renatured to allow snapback structures to form between the dA and dT residues that flank the cDNA insert on each strand. The vector DNA is then removed by treatment with *E. coli* exonuclease VII, which digests single-stranded DNA in both the 5′ → 3′ and 3′ → 5′ directions. The dA · dT duplex is then melted, and the two strands of the insert are reannealed to form duplex DNA.

Another strategy is to insert the double-stranded cDNA into a site that is closely flanked by two hexanucleotide restriction sites: In pBR322, for example, the sites for *Eco*RI, *Cla*I, and *Hin*dIII occur within a 30-bp region. Thus, by inserting the double-stranded cDNA into the *Cla*I site via dA · dT tailing, the insert can be recovered by digesting with *Eco*RI and *Hin*dIII.

In principle, *Hin*dIII sites can be regenerated by digesting plasmid DNA with *Hin*dIII, tailing with oligo(dT), and annealing with dA-tailed, double-stranded cDNA. The cDNA insert can then be recovered by *Hin*dIII digestion. For reasons that are not clear, this method has been used only rarely.

dC · dG Tailing

Currently, the most widely used procedure for cloning cDNAs by homopolymeric tailing involves addition of dG tails to the plasmid and complementary dC tails to the cDNA (Villa-Komaroff et al. 1978; Rowekamp and Firtel 1980). This method yields clones from which inserts can be easily removed. A plasmid such as pBR322 or pAT153 is digested with the enzyme *Pst*I, which cleaves the sequence

$$5' \quad C\text{-}T\text{-}G\text{-}C\text{-}A\text{-}G \quad 3'$$
$$3' \quad G\text{-}A\text{-}C\text{-}G\text{-}T\text{-}C \quad 5'$$

leaving protruding 3′ tails. The addition of a short stretch of dG residues to the linear plasmid DNA results in regeneration of a *Pst*I site at each end of the insert, which can therefore be recovered from the plasmid by digestion with *Pst*I (see Fig. 7.1). In practice, the efficiency of regenerating the *Pst*I site depends on the quality of *Pst*I used to linearize pBR322 DNA and the quality of the terminal transferase. If the penultimate residue is removed from the protruding 3′ tail by trace amounts of exonuclease, subsequent addition of dG residues will not recreate a PstI recognition sequence. Given reasonable care, however, as high as 80–90% of the recombinant plasmids constructed by this method contain inserts flanked by PstI sites.

Recently, the number of dA · dT and dG · dC residues required for optimal efficiencies of DNA transformation was determined (Peacock et al. 1981). In general, the number of residues on the plasmid and the cDNA should be approximately equal, with approximately 100 residues being added to each DNA for dA · dT joining and approximately 20 for dG · dC joining (Peacock et al. 1981). Interestingly, the bacterial strain can make a significant difference to the transformation efficiency. RR1, a recA$^+$ strain of E. coli, yielded 10 times as many recombinant cDNA clones made by the dA · dT tailing procedure as did the recA$^-$ host HB101. In the same experiment, untreated pBR322 DNA transformed the two strains with equal efficiency (Peacock et al. 1981). It would therefore appear that the bacterial recA system is involved in repairing open circular, hybrid DNA molecules that contain homopolymer tails.

SYNTHETIC DNA LINKERS

Synthetic linkers containing one or more restriction sites provide an alternative method to join double-stranded cDNA to plasmid vectors. Double-stranded cDNA, generated as described earlier, is treated with bacteriophage T4 DNA polymerase or E. coli DNA polymerase I, enzymes that remove protruding, 3′, single-stranded termini with their 3′ → 5′ exonucleolytic activities and fill in recessed 3′ ends with their polymerizing activities. The combination of these activities therefore generates blunt-ended cDNA molecules, which are then incubated with a large molar excess of linker molecules in the presence of bacteriophage T4 DNA ligase, an enzyme that is able to catalyze the ligation of blunt-ended DNA molecules. Thus, the products of the reaction are cDNA molecules carrying polymeric linker sequences at their ends. These molecules are then cleaved with the appropriate restriction enzyme and ligated to a plasmid vector that has been cleaved with a compatible enzyme.

The double-stranded cDNA molecules containing the synthetic cohesive ends will, of course, ligate to each other as well as to the vector DNA. In addition, the vector can recircularize by self-ligation and increase the background of nonrecombinant plasmids. These problems can be circumvented to a large extent by treating the linearized plasmid with phosphatase (see page 133) and/or by ligating different linkers to each end of the cDNA. In the original description of this method (Kurtz and Nicodemus 1981), two different linkers were simultaneously ligated to cDNA. During this process, it would be expected that 50% of the cDNA molecules would receive the same linker at each end; such molecules could not be inserted into plasmid DNA by directional cloning. In practice, this figure is even greater because one linker almost always has a higher rate of ligation to cDNA than the other.

This problem can be solved by adding one linker to the cDNA before cleaving the hairpin loop with nuclease S1 and the second linker after the S1 treatment. The double-linkered cDNA can then be treated with the appropriate restriction enzymes and inserted into a plasmid vector by directional cloning (see Fig. 7.2).

This opens the possibility of inserting cDNA in the correct orientation into vectors that allow expression of the inserted sequences in bacteria (see Chapter 12) and of identifying clones of interest by screening bacterial colonies for the presence of material that reacts with specific antisera to a particular gene product. This technique could be of great value when cloning rare mRNAs, for which no nucleic acid probes are available.

One problem with this approach is that the double-stranded, cDNA linker–DNA hybrids must be digested with the appropriate restriction enzymes to generate cohesive ends (Scheller et al. 1977). If the double-stranded cDNA contains one or more recognition sites for either one of the enzymes, it will be cleaved and subsequently cloned as two or more DNA fragments, making the structural analysis of the full-length cDNA difficult. This problem can be alleviated by using synthetic linkers carrying recognition sequences for restriction enzymes that cleave mammalian DNA very rarely (e.g., *Sal*I), by using *Eco*RI methylase to protect the DNA from cleavage with *Eco*RI, or by using synthetic adapters rather than linkers. Adapters are short, synthetic, double-stranded cDNAs that are blunt at one end and cohesive at the other (e.g., a *Hin*dIII cohesive end). By placing a 5′ phosphate on the blunt end of the adapter and a 3′ hydroxyl on the sticky end, the adapter will ligate to blunt-ended, double-stranded cDNA but not to itself. Unlike linkers, adapters do not have to be digested with restriction enzymes prior to ligation to double-stranded cDNA.

OTHER METHODS OF CLONING cDNA

Most of the cDNA clones thus far characterized have been constructed by using one of the techniques described above. Below we briefly describe three alternative procedures for cDNA cloning. The first procedure, mRNA · cDNA hybrid cloning, has limited applicability because of its low efficiency. The second procedure involves second-strand cDNA synthesis primed by oligonucleotides, while the third method involves plasmid-primed, first- and second-strand cDNA synthesis. Although the latter two procedures have not yet been widely applied and we ourselves have no direct experience with them, the published reports indicate that both provide an efficient means of obtaining full-length cDNA clones.

mRNA · cDNA Cloning

Another method for cDNA cloning involves transformation of *E. coli* with mRNA · cDNA hybrids that have been joined to plasmid vectors (Wood and Lee 1976; Zain et al. 1979). The bacterial host removes the mRNA and replaces it with DNA. After the first strand of cDNA has been synthesized in

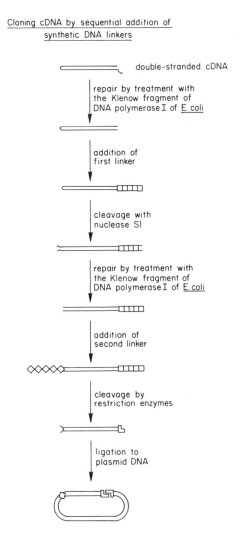

Cloning cDNA by sequential addition of synthetic DNA linkers

double-stranded cDNA

repair by treatment with the Klenow fragment of DNA polymerase I of E. coli

addition of first linker

cleavage with nuclease S1

repair by treatment with the Klenow fragment of DNA polymerase I of E. coli

addition of second linker

cleavage by restriction enzymes

ligation to plasmid DNA

Figure 7.2

Cloning cDNA by sequential addition of synthetic DNA linkers.

the usual way, dA residues are added to the mRNA · cDNA hybrid, and the tailed hybrid is then annealed to a plasmid tailed with dT. Because the efficiency of tailing the 3′-hydroxyl group of RNA is at least 10 times less than the homologous reaction with DNA, most of the dA residues added to the hybrid are incorporated at the 3′ end of the DNA strand. Joining of the other end of the hybrid to the vector is probably accomplished by hydrogen bonding between the tract of natural poly(A) at the 3′ end of the mRNA and the dT-tailed plasmid. The two practical advantages of this procedure are (1) that no synthesis of second cDNA strand is required and (2) that cleavage of the DNA hairpin by nuclease S1 is not necessary. Furthermore, the procedure should in theory allow the sequences at the 5′ end of the mRNA (which are normally lost during nuclease-S1 cleavage) to be cloned. Its major disadvantage, however, is that it is at least 10 times less efficient than double-stranded cDNA cloning and is therefore unsuitable for constructing large numbers of cDNA clones.

Second-strand cDNA Synthesis Primed by Oligonucleotides

Synthesis of the second strand of cDNA is usually primed by hairpin structures at the 3′ terminus of the first strand. An alternative procedure is to tail the first strand of cDNA directly with dT (Rougeon et al. 1975) or dC (Land et al. 1981). The second strand is then synthesized using an oligo(dA) or oligo(dG) primer, respectively, producing duplex cDNA flanked by duplex homopolymeric tracts at each end. The duplex DNA is then tailed with dC and inserted into a plasmid that has been cleaved with *Pst*I and tailed with dG.

The chief advantage of this procedure is that it eliminates the difficult step in which nuclease S1 is used to cleave the hairpin loop in double-stranded cDNA and thus facilitates the efficient cloning of full-length, double-stranded cDNA. One potential pitfall in this procedure is that even highly purified preparations of terminal transferase are contaminated with single-strand-specific nucleases. Presumably, this latter problem could be circumvented by tailing the first cDNA strand as a DNA · RNA hybrid.

Plasmid-primed, First- and Second-strand cDNA Synthesis

Recently, a novel method for high-efficiency cloning of full-length, double-stranded cDNA was published by Okayama and Berg (1982). The steps in their protocol are as follows (see Fig. 7.3A,B,C):

1. A plasmid primer for cDNA synthesis is prepared by dT tailing with terminal transferase. A fragment containing one of the dT tails, the bacterial origin of replication, and the ampicillin-resistance gene is prepared by digestion with a second enzyme, followed by agarose gel electrophoresis and oligo(dA) cellulose chromatography (Fig. 7.3A).

2. An oligo(dG)-tailed linker DNA is prepared by dG tailing a *Pst*I DNA fragment with terminal transferase, followed by digestion with a second enzyme to separate the two ends. The desired end fragment is purified by agarose gel electrophoresis (Fig. 7.3B).

3. The dT-tailed vector-primer is annealed with poly(A) mRNA at a molar ratio of 1.5–3 (mRNA:vector-primer), and a first cDNA strand is synthesized with reverse transcriptase (Fig. 7.3C).

4. dC tails are added to the 3′ end of the cDNA copy while it is still hydrogen bonded to the mRNA template. The dC tail added at the other end of the vector is then removed by restriction endonuclease digestion.

5. The oligo(dG)-tailed cDNA · mRNA plasmid is annealed and ligated to the oligo(dG)-tailed linker DNA.

6. The mRNA strand is replaced by DNA using the combined activities of RNase H, which degrades the RNA strand in an RNA · DNA hybrid, *E. coli* DNA polymerase I, which carries out a nick-translation repair of the second cDNA strand, and DNA ligase, which covalently closes the circular DNA molecule.

Okayama and Berg find that full-length or nearly full length cDNA copies are preferentially converted to duplex cDNA, and an efficiency of approximately 100,000 transformants per microgram of starting mRNA is obtained. The preferential cloning of long cDNA transcripts is thought to be a consequence of the preferential utilization of full-length reverse transcription by terminal transferase. They speculate that shortened or truncated cDNA strands in the mRNA · DNA duplex are not efficiently recognized by the terminal transferase and are therefore selected against. Although the rabbit α- and β-globin mRNA was used to establish this cDNA cloning procedure, Okayama and Berg indicate that other cDNA clones representing both rare and long (6500-nucleotide) mRNAs have been obtained with this procedure.

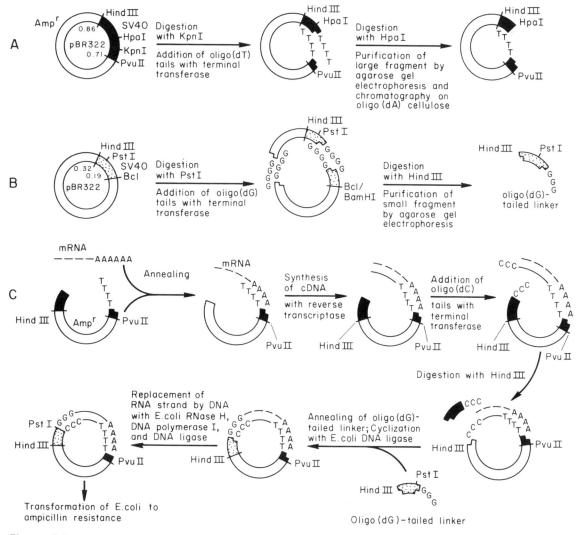

Figure 7.3

Preparation of (*A*) plasmid primer and (*B*) oligo(dG)-tailed linker DNA. (*C*) Steps in the construction of plasmid-cDNA recombinants. pBR322 DNA is represented by the open sections of each ring; SV40 DNA is indicated by the darkened or stippled segments. The numbers next to the restriction site designations are the corresponding SV40 DNA map coordinates.

Strategies for cDNA Cloning

ABUNDANT mRNAs

Initially, cDNA cloning was used to obtain copies of abundant mRNAs such as globin and ovalbumin. In these cases, the RNA of interest comprises as much as 50–90% of the total poly(A)$^+$ cytoplasmic RNA isolated from certain specific cell types. Consequently, no further purification of the particular mRNA is required before double-stranded cDNA is synthesized and cloned.

To identify cDNA clones of abundant mRNAs, transformed bacteria are assayed by nucleic acid hybridization for the presence of the desired DNA sequences. The probes consist either of ^{32}P-labeled, single-stranded cDNA synthesized in vitro by reverse transcriptase, using as template mRNA preparations that are rich in the sequences of interest, or of a partially fragmented, end-labeled preparation of the mRNA itself. As a good approximation, the mRNA sequences of interest will be represented both in the cloned, double-stranded cDNAs and in the probe in proportion to their abundances in the starting population. In cases like ovalbumin and globin, the chances are high that any colony hybridizing strongly to the probe will contain the desired DNA sequences.

Proof of the identity of the clone can be obtained in one of three ways:

1. By showing that the cloned cDNA is able to select the mRNA of interest from the starting population of mRNA. Usually the cloned cDNA is immobilized on a nitrocellulose filter and hybridized to mRNA in solution. After extensive washing, the mRNA is released from the hybrid and translated in a cell-free, protein-synthesizing system (hybridization/selection) (Goldberg et al. 1979).

2. By showing that the cloned cDNA is able to hybridize to the mRNA of interest and thereby inhibit its translation in vitro (hybrid-arrested translation) (Paterson et al. 1977).

3. By direct DNA sequencing. When the amino acid sequence of the protein product is known, it is a simple matter to establish that the cloned cDNA and the protein are colinear. Rapid methods have recently been developed to apply the Maxam-Gilbert (1977) or the Maat-Smith (1978) techniques to obtain the sequence of DNA fragments cloned in plasmids (Frischauf et al. 1980).

LOW-ABUNDANCE mRNAs

With refinements of methods for the efficient introduction of recombinant cDNA plasmids into *E. coli* (Hanahan and Meselson 1980) and for screening large numbers of transformed bacterial colonies for foreign DNA sequences (D. Hanahan, unpubl.), the cloning of mRNAs of relatively low abundance is possible.

The strategy currently employed involves the construction of large numbers of cDNA clones from total poly(A)$^+$ mRNA and the identification of the cDNA clones of interest. The entire collection of cDNA clones from a particular preparation of poly(A)$^+$ RNA is called a cDNA library.

A typical mammalian cell contains between 10,000 and 30,000 different mRNA sequences (Davidson 1976). Williams (1981) has determined the number of clones necessary to obtain a complete cDNA library from a human fibroblast cell that contains approximately 12,000 different mRNA sequences. The low-abundance class of mRNAs ($<$ 14 copies/cell) comprises approximately 30% of the mRNA, and there are about 11,000 different mRNAs in this class. The minimum number of clones required to obtain a complete representation of low-abundance mRNA sequences is therefore 11,000/.30 = $\sim$ 37,000. Of course, because of sampling variation and of preferential cloning of certain sequences, a much larger number of recombinants must be obtained to increase the chance that any given clone will be represented in the library. The number of clones required to achieve a given probability that any given low-abundance sequence will be present in a library is:

$$N = \frac{\ln\,(1 - P)}{\ln\,(1 - {}^1/_n)}$$

where

N = number of clones required;
P = probability desired (usually 0.99);
n = fractional proportion of the total mRNA population that a single type of low-abundance mRNA represents.

Therefore, to achieve a 99% probability of obtaining a particular low-abundance mRNA from the human fibroblast described above:

P = 0.99
n = 1/37,000
N = 170,000

This number is within reach of existing techniques, since between 1×10^5 and 6×10^5 colonies per μg of double-stranded cDNA can be obtained either by homopolymeric tailing or by double-linker procedures.

A major problem, however, is the detection of extremely low abundance mRNA sequences by in situ colony hybridization (Gergen et al. 1979; Williams and Lloyd 1979; Dworkin and Dawid 1980). Several authors have calculated that clones representing as little as 0.05%–0.1% of the total mRNA molecules can be detected when in vitro labeled mRNA (or cDNA) is used as a probe. In practice, however, with the concentrations of probe that are usually available and with hybridization and autoradiographic exposure times that are reasonable, it is extremely difficult to detect clones containing cDNA complementary to mRNA species that are present in the initial population at less than 1 part in 200.

So far, no general method has been developed to clone such molecules. However, there are several techniques that may be used singly or in combination to deal with the problems encountered in identifying cDNA clones of RNAs that are only minor components of the total population and for which no hybridization probes are available.

Size Fractionation

The simplest technique is to fractionate the mRNA by size, for example, by density gradient centrifugation or gel electrophoresis under denaturing contions (see pages 199–206). Each fraction of the mRNA is then translated in vitro and the protein product of interest is identified by a combination of immunoprecipitation and SDS-polyacrylamide gel electrophoresis. The degree of enrichment obviously varies from mRNA to mRNA, depending on its size relative to the bulk of the mRNA population. At best, an enrichment of perhaps 10-fold may be attained; however, this may be sufficient to bring the mRNA within cloning range.

An alternative strategy is to construct a cDNA library from a partially enriched mRNA population obtained, perhaps, by sucrose gradient centrifugation, and then to screen the library by hybridization to probes synthesized by reverse transcription of a still more highly enriched mRNA population obtained by fractionation of mRNA through density gradients *and* denaturing gels. The aim is to reduce the library to a manageable number of cDNA clones that can be screened individually or in small batches by hybrid selection.

Synthetic Oligodeoxynucleotides

Purification of an mRNA present in low concentrations can be arduous and difficult. If a partial or complete amino acid sequence of the protein of interest is available, the method of choice involves the chemical synthesis of oligonucleotides complementary to the mRNA. The sequence of such oligonucleotides can be deduced from favorable short sequences of amino acids (Wu 1972). In essence, one scans the known protein sequence for areas rich in amino acids specified either by a single codon (e.g., methionine, AUG; tryptophan, UGG) or by two codons (e.g., phenylalanine, UUU, UUC; tyrosine, UAU, UAC; histidine, CAU, CAC). Knowing the frequency with which different degenerate codons are used (e.g., glutamine is usually specified by CAG) and by taking advantage of G · T base-pairing, it is often possible to narrow down the candidate oligonucleotides to a manageable number. These oligonucleotides are then synthesized in vitro by either the phosphodiester method (Agarwal et al. 1972) or the more-efficient phosphotriester method (Hsiung et al. 1979). When these oligonucleotides are incubated under carefully defined annealing conditions with total poly(A)$^+$ mRNA, they form hybrids only with those species of mRNA to which they are exactly complementary. They can therefore be used as primers in reverse transcription reactions with unfractionated poly(A)$^+$ RNA to synthesize single-stranded probes for screening cDNA libraries (Chan et al. 1979; Noyes et al. 1979; Goeddel et al. 1980a; Houghton et al. 1980). If the synthetic oligodeoxynucleotides are sufficiently long (14–20 nucleotides), they can be used directly as probes to screen cDNA libraries for the clones containing sequences of interest (Montgomery et al. 1978; Goeddel et al. 1980a; Suggs et al. 1981).

A useful approach is to synthesize chemically a mixture of oligonucleotides that represent all possible coding combinations for a small portion of the amino acid sequence of the protein of interest (Wallace et al. 1979, 1981). One of these oligonucleotides will form a perfectly base-paired duplex with the double-stranded DNA, whereas the other oligonucleotides will form mismatched duplexes. If hybridization conditions of the appropriate stringency are chosen, only the perfectly matched duplex will be stable. This approach was recently employed to isolate cloned cDNA sequences for human β_2-microglobulin (Suggs et al. 1981). Note that the conditions used to screen colonies by hybridization are considerably more stringent than the conditions used to anneal primers to mRNA. Thus, when oligonucleotides are used as probes, they are much more specific than when used to prime cDNA synthesis on an mRNA template.

Oligonucleotides complementary to the coding region of an mRNA can never prime the synthesis of full-length cDNA molecules in reverse transcription reactions. Such oligonucleotides are therefore hardly ever used as primers to synthesize cDNA for cloning purposes. Their outstanding virtue is that they are (or can be used to generate) highly specific probes. The sensitivity of screening cDNA libraries is thereby increased to the level where clones synthesized from extremely rare mRNAs can easily be detected.

Differential Hybridization

This method has been used when two mRNA preparations are available that contain many sequences in common but that are different from each other in the presence and absence of a few species of interest. Examples of such sibling pairs might be mRNAs extracted from cells before and after exposure to heat shock, drugs, or hormones. In the simplest application of this technique, ^{32}P-labeled cDNA is synthesized in vitro from both preparations of poly(A)$^+$ RNA. Most of the cDNA sequences will be shared by the two preparations. However, the cDNA synthesized from the induced-cell RNA should contain additional sequences complementary to any new species of poly(A)$^+$ RNA. The two probes are then used to screen replicas of a cDNA library constructed from mRNA extracted from the induced-cell population. Those colonies hybridizing specifically to the induced-cell cDNA probe are likely to contain cloned copies of the induced mRNAs. Examples of inducible genes cloned in this way include the galactose-inducible genes of yeast (St. John and Davis 1979) and human fibroblast interferon (Taniguchi et al. 1980a). This procedure has also been used to identify cDNA clones of developmentally regulated mRNAs from *Xenopus laevis* (Dworkin and Dawid 1980), *Dictyostelium discoidum* (Williams and Lloyd 1979; Rowekamp and Firtel 1980), and sea urchins (Lasky et al. 1980).

cDNA clones corresponding to developmentally regulated mRNAs can also be identified using another type of differential hybridization. A population of cDNA molecules enriched in sequences characteristic for a particular developmental stage is used to probe a cDNA or genomic library (Timberlake 1980; Zimmerman 1980). This enrichment is accomplished by "cascade hybridization" in which cDNA prepared from mRNA obtained at one developmental stage (stage 1) is hybridized to a 20-fold excess of mRNA obtained from another stage (stage 2). The mRNA $\cdot$ cDNA hybrid is then removed by binding to hydroxyapatite. This procedure is repeated twice more using a 50-fold to 100-fold excess of stage-2 mRNA. The final, unbound cDNA fraction is then hybridized to a 100-fold excess of stage-1 mRNA, and the hybrid is recovered from hydroxyapatite. After removing the mRNA by alkaline hydrolysis, the cDNA that is highly enriched in stage-1-specific sequences is used to probe a stage-1 cDNA library.

Immunopurification of Polysomes

One approach to enriching specific mRNAs is to purify particular polysomes by virtue of the reaction between antibodies and nascent polypeptide chains (Cowie et al. 1961). The technique, which originally involved immunoprecipitation of polysomes, was limited to mRNAs that encode abundant proteins such as ovalbumin (Palacios et al. 1972) and immunoglobulin (Schechter 1973). Attempts to apply the method to mRNAs of lesser abundance were disappointing (Flick et al. 1978). However, recently the use of immunoaffinity columns (Schutz et al. 1977) and protein A–Sepharose columns (Shapiro and Young 1981) has resulted in significant improvements of the technique. For example, a relatively abundant trypanosome surface-antigen mRNA was purified by reacting polysomes with a heterogeneous antiserum to the surface antigen and trapping the complex on a protein A–Sepharose column (Shapiro and Young 1981). Lower-abundance mRNAs now can be isolated by combining the use of protein A–Sepharose with the use of monoclonal antibodies. Korman et al. (1982) used a monoclonal antibody to the heavy chain of the human HLA-DR antigen to purify the corresponding mRNA, which represents only 0.01–0.05% of the total mRNA. These investigators report a 2000-fold to 3000-fold purification of the HLA-DR mRNA. The purified mRNA can then be used to prepare a cDNA probe for screening a total cDNA library, or it can be used directly to prepare a double-stranded cDNA clone.

Procedures for cDNA Cloning

On the following pages, we describe in detail two methods for cDNA cloning using either dG · dC homopolymer tailing or the double-linker technique. We have used both methods successfully to produce cDNA libraries that appear to reflect the complexity of mRNA populations extracted from several types of mammalian cells.

The method for homopolymer tailing is a synthesis of protocols published by a number of different groups, in particular Efstratiadis and Villa-Komaroff (1979), Rowekamp and Firtel (1980), and B. Roberts (pers. comm.). The method utilizing sequential addition of linkers is an unpublished modification by J. Fiddes of a protocol devised by Kurtz and Nicodemus (1981).

SYNTHESIS OF DOUBLE-STRANDED cDNA

The conditions given below are optimal for synthesis of cDNA from heterogeneous populations of mRNA. However, individual species of mRNA may be copied by reverse transcriptase at different efficiencies.

First-strand Synthesis

1. Purify poly(A)$^+$ mRNA from the cells of interest using the methods described in Chapter 6. For optimal results, you will need about 10 μg of poly(A)$^+$ RNA to synthesize enough double-stranded cDNA for a library. However, the reactions will work (albeit less efficiently) if less template is available. Before proceeding, the integrity of the poly(A)$^+$ RNA should be checked by gel electrophoresis (see Chapter 6), using as markers 18S and 28S ribosomal RNAs and purified 9S globin mRNA. As visualized by ethidium-bromide staining of gels or methylene-blue staining of nitrocellulose filters, the poly(A)$^+$ RNA should form a continuous smear ($\sim$ 10S – $\sim$ 30S) with most of the molecules migrating at about 16S–18S. There is usually ribosomal RNA present in the poly(A)$^+$ RNA even after two cycles of selection on oligo(dT) columns. The sharpness of the ribosomal RNA bands provides a rough indication of whether the mRNA is degraded.

2. Prepare sterile stock buffers and solutions for first-strand synthesis:

   1 M Tris · Cl (pH 8.3) at 42° (the pH should be measured at 42°C since the pH of Tris changes with temperature)
   1 M KCl
   250 mM MgCl$_2$
   700 mM β-mercaptoethanol (add 50 μl of a concentrated [14 M] solution to 950 μl of H$_2$O
   dNTP solution (containing all four dNTPs [20 mM] in 0.01 M Tris · Cl [pH 8.0])
   oligo(dT)$_{12-18}$ primer (1 mg/ml in H$_2$O)
   100 mM methylmercuric hydroxide (see Chapter 6 for preparation and precautions to be taken in handling)

3. Estimate the volume of the AMV reverse transcriptase required. For 10 μg of poly(A)$^+$ mRNA, you will need approximately 40 units of reverse transcriptase. Most enzyme preparations contain 5–20 units/μl.

   The contribution of the storage buffer to the final composition of the reaction mixture must be considered. For example, the standard reverse transcriptase storage buffer contains 200 mM potassium phosphate (pH 7.2). Therefore, to obtain the optimum monovalent cation concentration (140–150 mM K$^+$), the amount of stock potassium chloride solution included in the reaction mixture must be reduced appropriately if large volumes of reverse transcriptase are added. Moreover, to prevent the added phosphate buffer (pH 7.2) from lowering the final pH of the reaction (optimally, pH 8.3), a relatively high concentration of Tris (100 mM) is used.

For prolonged storage, reverse transcriptase should be kept in small aliquots at $-70°C$. The enzyme subunits dissociate with time at $-20°C$. Therefore, only the working solution should be kept at $-20°C$.

4. Because many batches of reverse transcriptase are contaminated with RNase, potent inhibitors of RNase (RNasin or vanadyl-ribonucleoside complexes) are routinely included in the reaction. Although both types of inhibitor are effective, RNasin has a slight advantage in that it is readily removed by a single extraction with phenol/chloroform.

   RNasin should be used at a final concentration of 0.5 units/μl of reaction mixture.

   Vanadyl-ribonucleoside complexes are prepared as follows. Thaw the stock solution (200 mM; see page 188) immediately before use, centrifuge for 2 minutes (in an Eppendorf centrifuge), and dilute to 10 mM with water. The final concentration in the reaction mix is 1 mM; higher concentrations inhibit reverse transcriptase.

5. In an autoclaved Eppendorf tube, dry down approximately 50 pmoles (~ 40 μl) of each of the four [α-^{32}P]dNTPs (sp. act. = 800 Ci/mM; supplied in ethanol/water [50% v/v]).

   In this case, [α-^{32}P]dNTPs supplied in ethanol have some advantage over those supplied as stabilized aqueous solutions. The latter, which contain Tricene buffer at pH 6.0, would occupy about a third of the reaction volume and would change the pH of the reaction.

   If only aqueous [α-^{32}P]dNTPs are available, the following changes should be made to the reaction mixture:

   a. Make up a solution that contains three unlabeled dNTPs at a concentration of 20 mM and one unlabeled dNTP at a concentration of 10 mM. Use 2.5 μl of this composite solution per 50 μl of reaction mixture.

   b. Add to the reaction 10 μl (100 μCi) of the [α-^{32}P]dNTP present in the composite solution at low concentration.

6. Set up the reaction mixture. A reasonable reaction volume is 50 μl. Smaller volumes are more difficult to handle, and the presence of impurities in the radioactive triphosphates (especially after storage for more than one half-life) may lead to the inhibition of the reaction.

   A larger reaction volume requires more [α-^{32}P]dNTP to achieve the same amount of incorporation into DNA and is unnecessarily expensive.

   a. To the dried down radioactive triphosphates add:

1 mg/ml mRNA	10 μl (10 μg)
100 mM methylmercuric hydroxide	1 μl

Let stand at room temperature for 10 minutes. This treatment denatures the RNA and increases the yield of full-length cDNA from some mRNA templates.

b. Add 2 μl of 700 mM β-mercaptoethanol and 5 μl of 10 mM vanadyl-ribonucleoside complexes (or 2 μl of RNasin, 25 units). Let stand at room temperature for 5 minutes. The β-mercaptoethanol, which is necessary for the stability of reverse transcriptase, is added at this point to sequester the mercury ions since these ions otherwise would inhibit reverse transcription.

c. Add:

1 mg/ml oligo(dT)$_{12-18}$	10 μl (10 μg)
1 M Tris·Cl (pH 8.3)	5 μl
1 M KCl	7 μl
250 mM MgCl$_2$	2 μl
20 mM dNTPs	2.5 μl
H$_2$O to a final volume of	50 μl

d. Add 2 μl (40 units) of reverse transcriptase, mix well by vortexing, and centrifuge briefly in an Eppendorf centrifuge to eliminate the bubbles that are generated by the presence of Triton X-100 in the enzyme storage buffer. Incubate at 42°C for 1–3 hours.

The final reaction conditions for first-strand synthesis are:

100 mM Tris·Cl (pH 8.3)
 10 mM MgCl$_2$
140 mM KCl
100 μg/ml oligo(dT)$_{12-18}$
  2 mM methylmercuric hydroxide
 20 mM β-mercaptoethanol
  1 mM vanadyl-ribonucleoside complexes, *or*
0.5 units/μl RNasin
  1 mM each dNTPs
100 μg/ml poly(A)$^+$ RNA
400–800 units/ml reverse transcriptase

7. Stop the reaction by adding 2 μl of 0.5 M EDTA (pH 8.0), followed by 25 μl of 150 mM NaOH.

Note. It is important that the concentration of EDTA is sufficiently high to chelate all the divalent magnesium ions; otherwise an insoluble magnesium hydroxide–DNA complex will form when the sodium hydroxide is added.

8. Incubate for 1 hour at 65° or for 8 hours at 37°C to hydrolyze the mRNA template.

9. Neutralize the solution by adding

1.0 M Tris · Cl (pH 8.0)	25 μl
1.0 N HCl	25 μl

10. Measure the total amount of radioactivity in the reaction and the amount of material incorporated into TCA-precipitable material, as described on page 473.

11. Calculate the yield of cDNA from the percent of dNTPs incorporated. In theory, it is possible to synthesize an amount of cDNA equal in weight to the RNA template. In practice, the yield of the first-strand reverse transcriptase reaction is usually no more than 10–30% of the weight of poly(A)$^+$ RNA added.

12. Extract the remainder of the reaction with an equal volume of phenol/chloroform. After centrifugation, transfer the aqueous phase to a fresh Eppendorf tube. Reextract the organic phase with an equal volume of 10 mM Tris · Cl (pH 8.0), 100 mM NaCl, and 1 mM EDTA. Combine the two aqueous phases.

13. Separate the cDNA from unincorporated dNTPs and the products of alkaline hydrolysis of the template by chromatography on Sephadex G-100 as follows. Layer the combined aqueous phases on a column (~ 2 ml bed volume) of Sephadex G-100. Collect 0.2-ml fractions and measure the amount of radioactivity by Cerenkov counting. Pool the fractions in the excluded volume that contain radioactivity (see Fig. 7.4). Remove an aliquot (20,000 cpm) of the cDNA for analysis by gel electrophoresis. Precipitate the remainder of the cDNA with ethanol. Alternatively, the unincorporated dNTPs can be removed by spun-column chromatography (see page 466).

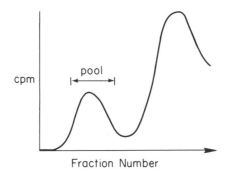

Figure 7.4

14. Measure the size of the first-strand cDNA by electrophoresis through a 1.4% alkaline agarose gel (see page 171).

Apply 20,000 cpm of the cDNA to the gel. For molecular-weight markers, use a mixture of end-labeled restriction endonuclease fragments of pBR322 DNA (see page 115). Apply 3000 cpm of the markers to the gel. Continue electrophoresis until the bromocresol green has migrated half the length of the gel.

Fix the DNA by immersing the gel for 30 minutes in each of two changes of 7% trichloroacetic acid.

Wash the gel briefly in water and blot off any excess fluid. Cover with Saran Wrap and expose for autoradiography (Kodak XR film or equivalent) at room temperature for several hours without intensifying screens.

If synthesis of the first strand was successful, a smear of radioactivity will be seen (from 100 nucleotides to the size of the largest species in the RNA preparation). Unless the poly(A)$^+$ RNA has been prepared from a differentiated cell type that contains one or more species of highly abundant mRNA, specific bands will not be detected.

Second-strand Synthesis

1. Recover the first-strand cDNA by centrifugation (10 minutes at 4°C in an Eppendorf centrifuge).

2. Resuspend the cDNA in 50 μl of H_2O. Add 50 μl of 2× second-strand buffer. Set 4.0 μl aside for later analysis by nuclease S1 and gel electrophoresis.

   2× Second-strand buffer
   0.2 M HEPES (pH 6.9)
   20 mM $MgCl_2$
   5 mM dithiothreitol
   0.14 M KCl
   1 mM of each of the four dNTPs

3. Add 20–50 units of the Klenow fragment of *E. coli* DNA polymerase I for every microgram of first-strand cDNA in the reaction. The volume of enzyme added should not exceed 15% of the total volume of the reaction, otherwise synthesis of the second cDNA strand may be inhibited by glycerol and phosphate in the enzyme storage buffer. The concentration of enzyme in many commercial preparations is quite low (1–2 units/μl), and it is often necessary to increase the volume of the second-strand reaction in order to accommodate the amount of enzyme required.

   Incubate at 15°C for 20 hours. The long incubation period allows the enzyme to find first-strand cDNA molecules with hairpin loops at their 3′ ends. Presumably, these structures are quite unstable and transient, and the enzyme must wait for and catch molecules in this unlikely configuration in order to begin synthesis of the second DNA strand. Some workers denature the first strand of cDNA by treatment with methylmercuric hydroxide before beginning second-strand synthesis. We have not found this procedure to make any difference to either the efficiency of the second-strand synthesis or to the size of the double-stranded cDNA product.

4. Stop the reaction by adding 2.0 μl of 0.5 M EDTA.

5. Remove two 2.0-μl aliquots from the reaction mixture for later analysis by gel electrophoresis. Extract the remainder of the sample with an equal volume of phenol/chloroform. Separate the double-stranded cDNA from unincorporated dNTPs by chromatography on Sephadex G-50, as described on pages 464–467. Precipitate the DNA with ethanol.

6. Even if the length-distribution of the population of double-stranded cDNA appears to be correct, it is highly unlikely that second-strand synthesis has been completed in all of the molecules. Truncated double-stranded cDNA is thought to arise because of the presence in the template strand of sequences that cause the Klenow polymerase to pause or stop (strong-stop sequences). Because such stopping points are different for Klenow polymerase and reverse transcriptase, it is possible to obtain a greater yield of full-length, double-stranded cDNA by carrying out a reaction with reverse transcriptase after the reaction with the Klenow enzyme has been completed.

Dissolve the cDNA in 20 μl of H_2O. Add:

1 M Tris·Cl (pH 8.3)	5	μl
1 M KCl	7	μl
250 mM $MgCl_2$	2	μl
a solution containing all four		
dNTPs at a concentration of 20 mM	2.5	μl
700 mM β-mercaptoethanol	2	μl
H_2O	to 48	μl
reverse transcriptase	2	μl (40 units)

Incubate at 42°C for 1 hour.

8. Stop the reaction by adding 2.0 μl of 0.5 M EDTA. Remove two 1-μl aliquots for analysis by gel electrophoresis. Extract the remainder of the sample with an equal volume of phenol/chloroform. Separate the double-stranded cDNA from unincorporated dNTPs by chromatography on Sephadex G-50 as described on pages 464–467. Precipitate the double-stranded cDNA with ethanol.

9. Apply samples from steps 2 (2 μl of the 4-μl aliquot), 5, and 8 to an alkaline, 1.4% agarose gel. Run the gel and locate the position of the cDNA by autoradiography, as described on page 171. For molecular-weight markers, use a set of end-labeled fragments of pBR322 DNA (see pages 115ff).

   If synthesis of the second strand is successful, the length of the double-stranded cDNA calculated from its rate of migration through the alkaline gel should be approximately twice that of the first strand. This is because the first and second strands are covalently joined by the hairpin loop.

DIGESTION WITH NUCLEASE S1

The amount of nuclease S1 required is determined by carrying out a set of pilot-scale reactions, each containing approximately 2000 cpm of ^{32}P-labeled, double-stranded cDNA.

1. Dissolve double-stranded cDNA in 50 μl of a solution of 1 mM Tris · Cl (pH 7.6) and 0.1 mM EDTA. Measure the amount of radioactivity by Cerenkov counting.

2. From the solution, remove an aliquot containing 10,000 cpm of ^{32}P. Add 10 μl of 10× nuclease-S1 buffer and sufficient water to bring the volume to 100 μl. Freeze the remainder of the double-stranded cDNA.

   Dispense 20-μl aliquots in five Eppendorf tubes. To each aliquot, add 0, 1, 2, 4, or 6 units of nuclease S1. Incubate at 37°C for 30 minutes.

   10× Nuclease-S1 buffer

   2   M NaCl
   0.5 M sodium acetate (pH 4.5)
   10   mM ZnSO$_4$
   5% glycerol

3. Add 1 μl of 0.5 M EDTA to stop the reactions. Analyze each sample on a 1.4% alkaline gel, using a set of end-labeled fragments of pBR322 DNA as molecular-weight markers (see page 115). Be sure to include on the gel a sample of the first-strand cDNA that was set aside for this purpose.

   Locate the position of the DNA by autoradiography. To obtain the result as quickly as possible:

   a. Fix the gel in 7% trichloroacetic acid (30 minutes, 2 changes).
   b. Wash the gel briefly with water.
   c. Dry down the gel onto Whatman 3MM paper.

   Alternatively:

   a. Soak the gel for 45 minutes in 0.5 M Tris · Cl (pH 7.5) and 1.0 M NaCl.
   b. Transfer the DNA to a nitrocellulose filter by Southern blotting (see pages 382ff).

   Expose the dried-down gel or nitrocellulose filter for autoradiography using Kodak XR film, or its equivalent, with intensifying screens. An overnight exposure at −70°C should be sufficient. Digestion with increasing amounts of nuclease S1 should yield populations of molecules of decreasing modal size. Choose the concentration of enzyme that yields molecules whose modal distribution is the same as that of the first-strand cDNA.

Note. S. Zeitlin and A. Efstratiadis (pers. comm.) have suggested an alternative procedure for calibrating the nuclease-S1 reaction. The procedure is based on the observation that the plasmid PML-21 (Hershfield et al. 1974) contains an inverted-repeat sequence of 1050 bp that is part of the kanamycin-resistance element Tn*903* (Sim et al. 1979). The assay is carried out by linearizing the plasmid DNA by digestion with a restriction enzyme, denaturing by boiling and quick cooling, digesting with nuclease S1, and analyzing the product on native and denaturing agarose gels. A successful digest is indicated by the presence of a discrete band of 1050 bp in both the native and denaturing gels.

a. Digest 10 μg of PML-21 DNA with *Eco*RI in 100 μl of *Eco*RI buffer. Note that *Eco*RI does not cleave within the inverted-repeat sequence (1 μg plasmid DNA ≡ 100 ng of hairpin DNA).

b. Add 900 μl of H₂O, mix, and divide the sample into 100-μl aliquots.

c. Denature the DNA by boiling and then cool the samples rapidly by plunging them into a dry-ice/ethanol bath.

d. Allow the DNA solutions to thaw in ice and then to each tube add 100 μl of ice-cold 2× nuclease-S1 buffer containing 0, 10, 20, 40, 60, 80, or 100 units of nuclease S1. Incubate at 37°C for 30 minutes. Usually, approximately 50 units of nuclease S1 are required to digest 1 μg of denatured PML-21 DNA.

e. Analyze 25 μl of each sample by electrophoresis through neutral and denaturing 2.0% agarose gels. Visualize the DNA by staining with ethidium bromide.

4. Digest the remainder of the double-stranded cDNA with the appropriate amount of nuclease S1.

5. Stop the reaction by addition of 2 μl of 0.5 M EDTA. Add 2 M Tris base to a final concentration of 0.05 M. Extract the solution once with phenol/chloroform. Precipitate the DNA with ethanol.

6. Redissolve the cDNA in 18 μl of TE (pH 8.0). Add 2 μl of 3 M NaCl. Fractionate the cDNA into size classes by passage through a 1-ml column of Sepharose CL·4B (see pages 464–465) equilibrated in 10 mM Tris·Cl (pH 8.0), 0.3 M NaCl, and 1 mM EDTA. Collect 50-μl fractions. Assay an aliquot of each fraction by electrophoresis through an alkaline agarose gel (1.4%). For molecular-weight markers, use a set of end-labeled fragments of pBR322 DNA. Locate the position of the cDNA by autoradiography as described in step 2 above. Pool the fractions that contain cDNA molecules greater than 500 bp in length. Precipitate the cDNA with ethanol.

CLONING DOUBLE-STRANDED cDNA

Homopolymeric Tailing of Vector DNA with Poly(dG)

1. Digest 55 μg of vector DNA (pBR322, pAT153, or pXf3) with *Pst*I. Check that the digestion is complete by analyzing a small sample by electrophoresis through a 1% agarose minigel.

2. Purify the linear DNA by electrophoresis through a preparative 1% agarose gel. (A slot 4 cm long and 4 mm deep will be required to avoid overloading).

3. Extract the DNA from the gel by electroelution, as described on pages 164ff. This purification step is important for two reasons. First, it removes any RNA or low-molecular-weight DNA contaminating the plasmid DNA or the restriction enzyme. Second, it separates the linear, plasmid DNA from any circular molecules that have not been digested with *Pst*I. Such molecules contribute significantly to the background of nonrecombinant transformants when the vector DNA preparation is used to transform *E. coli.*

4. Dissolve the linear plasmid DNA in 55 μl of H_2O. Add 55 μl of 2× tailing buffer.

   2× Tailing buffer

   0.4 M potassium cacodylate
   50 mM Tris · Cl (pH 6.9)
   4 mM dithiothreitol
   1 mM $CoCl_2$
   2 mM [^{3}H]dGTP (sp. act. = 12 Ci/mmole)
   500 μg/ml bovine serum albumin

   (The potassium cacodylate should be diluted from a 1 M solution that has been passed through a Chelex column equilibrated with potassium ions.)

5. Transfer 10 μl to a fresh tube and incubate for 10 minutes at 37°C. Store the remainder of the sample at −20°C.

6. Add 2 units of terminal transferase to the 10-μl aliquot. Mix and continue incubation at 37°C.

7. Remove 1-μl aliquots after 0, 1, 2, 5, 10, 20, 30, and 40 minutes of incubation. Spot the aliquots onto DE-81 filter discs. Wash the discs and count the radioactivity as described on page 473. Calculate how many dG residues have been added per end.

8. Incubate the remaining 100 μl of linear plasmid DNA for 10 minutes at 37°C. Add 20 units of terminal transferase and incubate for the time that results in the addition of 15–20 dG residues per end.

9. Stop the reaction by chilling to 0°C. Add 10 μl of 0.5 M EDTA (pH 8.0). Extract once with phenol/chloroform.

10. Separate the homopolymerically tailed DNA from low-molecular-weight contaminants by chromatography on a column of Sephadex G-100 equilibrated in 1× annealing buffer. Store the tailed DNA in aliquots at −20°C.

Note. Sephadex G-100 cannot be used in spun columns because the centrifugation crushes the beads.

10× Annealing buffer

  1 M NaCl
  0.1 M Tris · Cl (pH 7.8)
  1 mM EDTA

11. To check that the tailed vector is functional and is not contaminated by uncut or unit-length, untailed plasmid DNA, a trial annealing and transformation of *E. coli* should be carried out:

  a. Add 20–30 dC residues to a small (200–500 bp) fragment of DNA using the procedure described above for dG tailing.

  b. Set up the following annealing reactions:

Tube A:

uncut plasmid DNA	0.1 μg
H$_2$O	to 18 μl
10× annealing buffer	2 μl

Tube B:

dG-tailed vector	0.1 μg
H$_2$O	to 18 μl
10× annealing buffer	2 μl

Tube C:

dG-tailed vector	0.1 μg
dC-tailed insert	0.01 μg
H$_2$O	to 18 μl
10× annealing buffer	2 μl

  c. Heat to 65°C for 5 minutes and allow the DNAs to reanneal by incubating at 57°C for 1–2 hours. Transform *E. coli* strain RR1 (see Chapter 8). The efficiency of transformation by dG-tailed vector alone should be reduced at least 100-fold compared with circular plasmid.
  The efficiency of transformation by the recombination plasmid (tube C) should be at least 10-fold greater than that of dG-tailed vector alone.

Homopolymeric Tailing of Double-stranded cDNA with Poly(dC)

1. Calculate the quantity of double-stranded cDNA synthesized from the amount of $[\alpha\text{-}^{32}P]dCTP$ incorporated during first-strand synthesis. Estimate the total number of molecules synthesized from the size distribution of the double-stranded cDNA. Because an accurate measurement of size is not usually possible and the amount of double-stranded cDNA is limited, the rate of addition of homopolymeric dC tails is tested using 5 μg of plasmid DNA linearized with *Pst*I. This is not as irrational as it sounds. Terminal transferase reactions are carried out with the enzyme in vast excess, so that the number of residues added is essentially independent of DNA concentration.

   Set up a series of pilot reactions with terminal transferase by using 0.5 μg of linearized plasmid DNA and 2 units of terminal transferase, exactly as described on page 238 except that $[^3H]dCTP$ is used instead of dGTP. Take samples after 30 seconds, 1 minute, 2 minutes, and 4 minutes of incubation. Spot the aliquots onto DE-81 filter discs. Wash the discs and count the radioactivity as described on page 473. Calculate how many dC residues have been added per end.

2. Recover the double-stranded cDNA by centrifugation for 10 minutes at 4°C in an Eppendorf centrifuge. Dry the DNA pellet briefly under vacuum.

3. Dissolve the double-stranded cDNA in 25 μl of H$_2$O. Add 25 μl of 2 × tailing buffer prepared with $[^3H]dCTP$ (see page 239). Incubate for 10 minutes at 37°C.

4. Add 5 units of terminal transferase for every microgram of double-stranded cDNA in the reaction. Incubate for the time calculated from the pilot reactions to allow addition of 15–20 dC residues.

5. Stop the reaction by chilling to 0°C. Add 10 μl of 0.5 M EDTA (pH 8.0). Extract once with phenol/chloroform.

6. Separate the tailed DNA from low-molecular-weight contaminants by chromatography through a column of Sephadex G-100 equilibrated in annealing buffer or by spun-column chromatography using Sephadex G-50 (see page 466). Store the tailed DNA at −20°C.

ANNEALING VECTOR AND DOUBLE-STRANDED cDNA

1. Mix equimolar amounts of dC-tailed cDNA and dG-tailed vector in annealing buffer at a final concentration of 1 ng/μl.

   Annealing buffer
   0.1 M NaCl
   10 mM Tris · Cl (pH 7.8)
   1.0 mM EDTA

2. Heat to 65°C for 5 minutes and allow the DNAs to anneal by incubating at 57°C for 1–2 hours. Store the reannealed DNAs at −20°C.

3. Carry out the transformation of *E. coli* strain RR1 by using the protocol described on page 254.

CLONING DOUBLE-STRANDED cDNA BY SEQUENTIAL ADDITION OF LINKERS

1. Synthesize the first and second strands of cDNA, as described previously. Do not digest the double-stranded hairpin DNA with nuclease S1.

2. Prepare two sets of kinased linkers, as described on page 396.

3. To maximize the number of molecules with a perfectly blunt end, the hairpin double-stranded cDNA is treated with the Klenow fragment of *E. coli* DNA polymerase I in the presence of all four dNTPs.

   Dissolve approximately 2 μg of double-stranded cDNA in 11 μl of TE (pH 7.4). Add:

10× repair buffer	2 μl
1.0 mM dATP	1.25 μl
1.0 mM dCTP	1.25 μl
1.0 mM dGTP	1.25 μl
1.0 mM dTTP	1.25 μl
Klenow fragment of DNA polymerase I	1 unit ($\sim$ 1 μl)

   Incubate for 30 minutes at room temperature.

   10× Repair buffer

   0.5 M Tris · Cl (pH 7.4)
   70 mM $MgCl_2$
   10 mM dithiothreitol

4. The first linker is added to the blunt end of the hairpin double-stranded cDNA, i.e., at the end corresponding to the 3′ terminus of the original mRNA. Enough kinased linkers should be added to achieve a 1:1 mass ratio with double-stranded cDNA.

   At the end of the repair reaction (step 3), add 30 μl of 2× blunt-end ligation buffer. Then add:

2 μg kinased linkers in a volume of	4 μl
10 Weiss units T4 polynucleotide ligase	$\sim$ 1 μl
20 units RNA ligase	2 μl

   Incubate for 12–16 hours at 4°C.

2× Blunt-end ligation buffer

50 mM Tris·Cl (pH 7.4)
10 mM MgCl$_2$
10 mM dithiothreitol
0.5 mM spermidine
2 mM ATP
2.5 mM hexamine cobalt chloride
20 μg/ml BSA

Blunt-end ligation buffer should be stored in small aliquots at −20°C.

5. Stop the reaction by addition of 2 μl of 0.5 M EDTA. Extract once with phenol/chloroform. Precipitate the DNA with ethanol.

6. Dissolve the double-stranded cDNA in 45 μl of a solution of 1 mM Tris·Cl (pH 7.6) and 0.1 mM EDTA. Cleave the hairpin loop with nuclease S1, as described on pages 237ff.

7. Stop the reaction by addition of 2 μl of 0.5 M EDTA and 2.5 μl of 1 M Tris base. Extract once with phenol/chloroform.

8. Separate the double-stranded cDNA from low-molecular-weight contaminants by chromatography on Sephadex G-100. Precipitate the double-stranded cDNA with ethanol.

9. Repair the double-stranded cDNA with the Klenow fragment of *E. coli* DNA polymerase I, as described in step 3 (page 243).

10. Add the second kinased linker as described in step 4 (page 243).

11. Dilute the ligation reaction so that the composition of the buffer is suitable for digestion of the linkered DNA by the appropriate restriction enzymes. Add 50 units of each enzyme for every microgram of linker used in the ligation reactions. Incubate for 6–8 hours at the appropriate temperature.

12. Terminate the reaction by addition of EDTA to a final concentration of 10 mM. Extract once with phenol/chloroform. Precipitate the DNA with ethanol.

13. Redissolve the double-stranded cDNA in 10 μl of H$_2$O. Add 10 μl of

0.6 M NaCl
20 mM Tris·Cl (pH 8.0)
2 mM EDTA

Fractionate the cDNA into size classes by passage through a 1-ml column of Sepharose CL-4B equilibrated in the same buffer. Collect 50-μl fractions. Assay an aliquot of each fraction by electrophoresis through an alkaline agarose gel (1.4%), using as molecular-weight markers a set of end-labeled fragments of pBR322 DNA. Locate the position of the cDNA by autoradiography (see page 470). Pool the fractions that contain cDNA molecules greater than 500 bp in length. Precipitate the cDNA with ethanol.

14. Prepare the vector DNA as follows. Digest 50 μg of plasmid with the appropriate restriction enzymes. Purify the desired fragment of DNA either by gel electrophoresis (see Chapter 5) or by sucrose gradient centrifugation, essentially as described by Kurtz and Nicodemus (1981). The gradient (10–40% [w/v] sucrose in 10 mM Tris·Cl [pH 7.9], 1 mM EDTA, and 1 M NaCl) can be poured in the conventional way in a Beckman SW41 centrifuge tube, or it can be made by three cycles of freezing at $-70°C$ and thawing at 4°C of a 20% (w/v) solution of sucrose in the same buffer.

Up to 100 μg of DNA may be loaded onto a single gradient, which is centrifuged for 34 hours at 40,000 rpm in an SW41 rotor at 4°C. Fractions (0.4 ml) are collected from the bottom of the tube and 15-μl aliquots are analyzed by electrophoresis on an agarose gel. Fractions containing the vector DNA are pooled, diluted threefold with water to reduce the sucrose concentration, and precipitated with ethanol.

To check that the vector DNA is functional and is not contaminated by uncut or unit-length, linear plasmid DNA, trial ligation and transformation of *E. coli* are carried out.

a. Prepare a small DNA fragment (200–500 bp) with ends that are compatible with those of the vector.

b. Set up the following ligation reactions:

ligation tube A:

uncut plasmid DNA	0.1 μg
H$_2$O	to 18 μl
10× ligation buffer	2 μl
ligase	5 Weiss units

Ligation tube B:

vector DNA	0.1 μg
H₂O	to 18 μl
10× ligation buffer	2 μl
ligase	5 Weiss units

Ligation tube C:

vector DNA	0.1 μg
small fragment of DNA	0.01 μg
H₂O	to 18 μl
10× ligation buffer	2 μl
ligase	5 Weiss units

Incubate each sample at 4°C for 12–16 hours.

10× Ligation buffer

0.5 M Tris (pH 7.4)
0.1 M MgCl₂
0.1 M dithiothreitol
10 mM spermidine
10 mM ATP
1 mg/ml BSA

  c. Transform *E. coli* strain DH1 or HB101 (see Chapter 8) with 10 ng of the ligated DNA.

     The efficiency of transformation of *E. coli* by the vector DNA (ligation tube B) should be reduced 10^4-fold compared with undigested plasmid. The efficiency of transformation of *E. coli* by the reconstructed plasmid (ligation tube C) should be at 10-fold to 100-fold greater than that of the vector alone.

15. Mix the appropriate amount of vector with double-stranded cDNA to achieve a molar ratio of vector to cDNA of 5:1. Heat to 68°C for 10 minutes. Chill in ice. Add water and 10× ligase buffer so that the final concentration of vector DNA is 1.5 μg/ml in 1× ligation buffer.

16. Add 10 Weiss units of T4 polynucleotide ligase for every microgram of vector DNA in the reaction. Incubate for 12–16 hours at 12°C.

17. Add EDTA to a final concentration of 10 mM. Extract once with phenol/chloroform and precipitate the DNA with ethanol.

18. Carry out transformation of *E. coli* DH-1 by using one of the protocols given in Chapter 8.

8

Introduction of Plasmid and Bacteriophage λ DNA into *Escherichia coli*

The central act of molecular cloning is the introduction of recombinant DNA into bacterial cells. At one time, this was an erratic and inefficient process; now, however, reproducible and effective methods are available to transform bacteria with plasmid DNA molecules and to package bacteriophage λ DNA in vitro into infectious virus particles.

Transformation of *Escherichia coli* by Plasmid DNA

Most methods for bacterial transformation are based on an observation of Mandel and Higa (1970), who demonstrated that uptake of bacteriophage λ DNA is enhanced by treatment of bacterial cells with calcium chloride. In 1973, their method was shown also to work for plasmid DNA by Cohen, Chang, and Hsu. Many variations in this basic technique have since been described, all directed toward optimizing the efficiency of transformation for different bacterial strains. Most protocols yield 10^5-10^7 transformants per microgram of intact pBR322 DNA. However, conditions have been established (D. Hanahan, unpubl.; see page 254) in which the *E. coli* strain χ1776 reproducibly yields 10^7-10^8 transformants per microgram of intact pBR322 DNA. Impressive though this efficiency is, it is important to realize, first, that only a very small proportion of the cells are competent to incorporate plasmid DNA in a stable fashion, and second, that only 1 DNA molecule in approximately 10,000 is successful at transformation.

Once inside the bacterium, the plasmid DNA replicates and expresses the drug-resistance markers that allow the transformed cell to survive in the presence of an antibiotic.

The ability of bacteria to take up DNA is short-lived. After exposure to agents that enhance uptake, most strains of bacteria remain in a competent state for only 1–2 days. However, competent cells of the strain χ1776 can be prepared in large quantities, tested, and stored frozen in aliquots. These advantages are partially offset by the fastidious requirements and expensive media needed to grow χ1776. For these reasons, it is common practice to use χ1776 for initial transformation and screening and then to use the plasmid DNA obtained from minipreparations (see pages 366ff) to transform more manageable hosts.

TRANSFORMATION BY THE CALCIUM CHLORIDE PROCEDURE

This is a frequently used procedure to transform bacteria with plasmid DNA (Mandel and Higa 1970). It yields 10^5–10^7 transformants per microgram of intact pBR322 DNA when used with *E. coli* strain HB101.

1. Inoculate 100 ml of L broth in a 500-ml flask with 1 ml of an overnight bacterial culture. Grow the cells with vigorous shaking at 37°C to a density of $\sim 5 \times 10^7$ cells/ml. This usually takes 2–4 hours. For each transformation assay, 3 ml of cells will be needed.

   Note. The relationship between optical density and the number of viable bacteria per milliliter of culture varies from strain to strain. For example, for rec^+ strains (χ1776, MM294) 1 $OD_{550} = 0.2$ ($\sim 5 \times 10^7$ cells/ml); whereas for rec^- strains (DH1, HB101) 1 $OD_{550} = 0.5$ ($\sim 5 \times 10^7$ cells/ml). A curve calibrating the OD_{550} and the number of bacteria per milliliter of culture should be constructed for each new strain of *E. coli* used.

2. Chill the culture on ice for 10 minutes. Centrifuge the cell suspension at 4000g for 5 minutes at 4°C.

3. Discard the supernatant. Resuspend the cells in half of the original culture volume of an ice-cold, sterile solution of 50 mM $CaCl_2$ and 10 mM Tris·Cl (pH 8.0).

4. Place the cell suspension in an ice bath for 15 minutes and then centrifuge the suspension at 4000g for 5 minutes at 4°C.

5. Discard the supernatant. Resuspend the cells in $1/15$ of the original volume of an ice-cold, sterile solution of 50 mM $CaCl_2$ and 10 mM Tris·Cl (pH 8.0). Dispense 0.2-ml aliquots into prechilled tubes. Store the cells at 4°C for 12–24 hours.

   Note. For maximum transformation efficiency, it is very important (1) that the bacterial culture is in the logarithmic phase of growth and that the cell density is low at the time of treatment with calcium chloride; and (2) that the cells are maintained at 4°C for 12–24 hours. During this period, the efficiency of transformation increases fourfold to sixfold (Dagert and Ehrlich 1979). After an additional 24 hours, the efficiency decreases to the original level.

6. Add DNA in ligation buffer or TE. Mix and store on ice for 30 minutes. Up to 40 ng of DNA (dissolved in up to 100 μl of ligation buffer or TE) can be used for each transformation reaction. Addition of more DNA or a greater volume of buffer leads to a reduction in transformation efficiency.

7. Transfer to a water bath, preheated to 42°C, for 2 minutes.

8. Add 1.0 ml of L broth to each tube and incubate at 37°C for 30 minutes (tetracycline selection) or 1 hour (ampicillin or kanamycin selection) without shaking. This period allows the bacteria to recover and to begin to express antibiotic resistance.

9. Spread an appropriate quantity of cells onto selective media by using either the spreading or the top agar procedure (page 60). Usually the top agar procedure yields slightly more transformants.

    The entire transformation mixture may be spread on a single plate or plated in top agar if the selection is for tetracycline resistance. If ampicillin resistance is required, only a portion of the culture (empirically determined) should be spread on a single plate since the number of transformants obtained does not increase in linear proportion to the volume applied to the plate, perhaps because of toxic substances released by cells killed by the antibiotic. Furthermore, when selecting for ampicillin resistance, the density of cells plated should be low and the plates should be removed from the incubator after 16–24 hours and placed at 4°C. β-Lactamase, secreted into the medium from ampicillin-resistant transformants, rapidly depletes the antibiotic in regions surrounding the colonies. Thus, plating cells at high density or incubating for long periods results in the appearance of ampicillin-sensitive satellite colonies.

10. Leave the plates at room temperature until the top agar has hardened or until the liquid has been absorbed.

11. Invert the plates and incubate at 37°C. Colonies should appear in 12–16 hours.

TRANSFORMATION BY THE CALCIUM CHLORIDE/RUBIDIUM CHLORIDE PROCEDURE

This protocol (Kushner 1978) works best with *E. coli* strain SK1590 (see Appendix C). However, the efficiency of transformation of a number of other *E. coli* strains is higher than can be achieved with the standard calcium chloride procedure.

1. Inoculate 100 ml of L broth in a 500-ml flask with 0.5 ml of an overnight bacterial culture. Grow the cells with vigorous shaking at 37°C until a cell density of $\sim 5 \times 10^7$ cells/ml is reached (see note to step 1, page 250). For each transformation assay, 2 ml of cells will be needed.

2. Centrifuge 2-ml aliquots of the culture (1×10^8 cells) in sterile 15-ml Corex tubes for 10 minutes at 4000g at 4°C.

3. Discard the supernatant. Gently resuspend the cell pellet in 1 ml of a sterile solution of

   10 mM MOPS (morpholinopropane sulfonic acid) (pH 7.0)
   10 mM RbCl

4. Recover the cells by centrifugation at 4000g for 10 minutes at 4°C.

5. Discard the supernatant. Gently resuspend the cell pellet in 1 ml of a sterile solution of

   0.1 M MOPS (pH 6.5)
   50 mM CaCl$_2$
   10 mM RbCl

6. Place the bacterial suspension on ice for 15 minutes.

7. Recover the cells by centrifugation at 4000g for 10 minutes at 4°C.

8. Drain off as much of the supernatant as possible. Gently resuspend the pellet in 0.2 ml of sterile

   0.1 M MOPS (pH 6.5)
   50 mM CaCl$_2$
   10 mM RbCl

9. Add 3 μl of dimethylsulfoxide (Spectroquality DMSO; Matheson, Coleman and Bell) and 1–200 ng of DNA in a volume of 10 μl or less of TE (pH 8.0).

Note. The oxidation products of DMSO are very inhibitory to transformation. Therefore, 0.5-ml aliquots should be taken from a fresh bottle of Spectroquality DMSO and frozen at −70°C. An aliquot should be used only once and then discarded.

10. Place on ice for 30 minutes.

11. Transfer to a water bath at 43–44°C for 30 seconds.

12. Add L broth to a volume of 5 ml. Incubate for 60 minutes at 37°C without shaking.

13. Plate on selective medium as described on page 60. If necessary, the cells can be concentrated by centrifugation before plating.

TRANSFORMATION OF *E. coli* χ**1776**

This is the most efficient procedure that is currently available (D. Hanahan, unpubl.). The full version of the protocol, given below, can be used only with *E. coli* χ1776. However, with some minor changes (see note on page 255), the method can be used with several other *E. coli* strains (MM294, C600, DH1, but *not* HB101) with only a slight reduction in efficiency.

1. Dilute 1 ml of an overnight culture of *E. coli* strain χ1776 into 100 ml of χ1776 broth in a 500-ml flask. Incubate at 37°C with vigorous aeration until a cell density of 5×10^7 cells/ml is reached (1 $OD_{550} = 0.2$; see note to step 1, page 250). This usually takes 3–4 hours. For each transformation assay, 5 ml of the culture are needed.

   Note. χ1776 is very sensitive to detergents. Therefore, only glassware that has been rinsed very well or plasticware should be used. Also, note that χ1776 requires diaminopimelic acid and thymidine for growth (see Appendix C, page 442).

2. Centrifuge the cells at 4000*g* for 5 minutes at 4°C.

3. Discard the supernatant. Gently resuspend the cells in 0.2× the original culture volume of ice-cold χ1776 transformation buffer. Place on ice for 5 minutes.

   χ1776 Transformation buffer (pH 5.8)
   100 mM RbCl
   45 mM MnCl₂
   10 mM CaCl₂
   5 mM MgCl₂
   0.5 mM LiCl
   35 mM potassium acetate (pH 6.2)
   15% sucrose (Schwartz-Mann; Ultrapure)

   Buffer preparation: Make a solution containing all the components except the rubidium chloride and manganese chloride. Add solid rubidium chloride and manganese chloride and carefully adjust the pH of the solution to 5.8 with dilute acetic acid. Sterilize the solution by filtration and freeze in aliquots at −20°C.

4. Centrifuge the cells at 4000*g* for 5 minutes at 4°C.

5. Discard the supernatant. Gently resuspend the cells in 0.04× the original culture volume of ice-cold transformation buffer. Dispense 0.2-ml aliquots into prechilled tubes. Place on ice for 5 minutes.

6. To each tube add 7 μl of dimethylsulfoxide (Spectroquality DMSO; Matheson, Coleman and Bell) (see note to step 9, page 253). Place on ice for 15 minutes.

Note. Competent cells can be prepared and stored at $-70°C$ using this procedure. Prepare the cells in bulk as described in steps 1-6, and then dispense 200-μl aliquots into sterile Eppendorf tubes. Freeze each aliquot as soon as it is made in dry ice/ethanol and store at $-70°C$. Thaw each aliquot as needed by placing it on ice for 30 minutes and continue the transformation protocol at step 9. When stored at $-70°C$, the cells remain competent for transformation for at least 6 months (this is not true for DH1, C600, or MM294).

7. Freeze-shock the cells by placing the tube at $-55°C$ in a dry-ice/ethanol bath for 20 seconds.

8. Rapidly thaw the cells by placing the tube in water at room temperature. When the cells are thawed, transfer the tube to an ice bath for 5 minutes.

9. Add 7 μl of DMSO, mix, and place the tube on ice for 5-10 minutes.

10. Add 1-25 ng of DNA in 1-10 μl of ligation buffer or TE (pH 7.4). Place the cells in an ice bath for 10-30 minutes.

11. Freeze-shock the cells by immersing the tube for 30-60 seconds in a dry-ice/ethanol bath.

12. Rapidly thaw the cells by placing the tube in water at room temperature. Immediately after thawing, place the tube in an ice bath.

13. Heat-shock the cells by immersing the tube in a water bath at $42°C$ for 1 minute *or* at $37°C$ for 2 minutes.

14. Add 1 ml of χ1776 broth and incubate for 30 minutes (tetracycline selection) or 1 hour (ampicillin or kanamycin selection) at $37°C$ without agitation.

15. Plate the bacteria onto selective medium as described on page 60.

Notes

i. The method given above can be used with other strains of *E. coli* (e.g., DH1, C600, MM294) if magnesium chloride and lithium chloride are left out of the transformation buffer and if the freeze-shock treatments (steps 7 and 11) are omitted. Under these circumstances, transformation efficiencies of 10^7-10^8 transformants per microgram of superhelical pBR322 DNA are obtained.

ii. For unknown reasons, this protocol does not work well with *E. coli* HB101.

In Vitro Packaging of Bacteriophage λ DNA

Packaging of bacteriophage λ DNA in vitro was initially developed by Becker and Gold (1975) using mixtures of extracts prepared from bacteria infected with mutants of bacteriophage λ that map in genes required for assembly of bacteriophage particles. The procedure has been improved and modified in a number of laboratories to the point where efficiencies of 10^7–10^8 pfu/μg of intact λ DNA can now be reproducibly attained. Approximately 0.05–0.5% of the λ DNA molecules present in the reaction can be packaged into infectious virions.

A diagrammatic representation of the various stages of bacteriophage λ DNA packaging and the stages at which various mutations affect the process is shown in Figure 1.8.

The E protein is the major component of the bacteriophage head and is required for assembly of the earliest identifiable precursor. Mutants in protein E accumulate all the components of the viral capsid.

The D protein is localized on the outside of the bacteriophage head and is involved in the coupled process of insertion of λ DNA into the head precursor and subsequent maturation of the head. Mutants in the D gene accumulate the immature "prehead" but do not allow insertion of λ DNA into the head.

The A protein is involved in the insertion of λ DNA into the bacteriophage head and cleavage of the concatenated, precursor DNA at the cohesive end (*cos*) sites. Mutants in the A gene also accumulate empty preheads. Complementing extracts have been prepared from cells infected with A^- and E^- or D^- and E^- strains; alternatively, extracts prepared from cells infected with A^- mutants can be complemented by addition of purified wild-type A protein.

Extracts are usually prepared from cells containing bacteriophage λ lysogens of the appropriate genotype (amber mutations in the A, D, or E genes). The lysogens also carry the following mutations:

cIts857—specifies a temperature-sensitive bacteriophage λ repressor molecule. This mutation causes λ DNA to be maintained in the lysogenic state when the host bacteria are grown at 32°C; bacteriophage growth is induced by transiently raising the temperature to 42–45°C to inactivate the repressor specified by the cI gene.

Sam7—an amber mutation in the bacteriophage S gene that is required for cell lysis. This mutation causes capsid components to accumulate within SuIII$^-$ bacterial cells for 2-3 hours following induction of the cIts857 lysogen.

b-region deletion (*b*2 or *b*1007)—a deletion in the bacteriophage genome that effectively removes the λ DNA attachment site *(att)*. This mutation reduces, but does not entirely eliminate, the packaging of endogenous λ DNA molecules in extracts made from the induced cells.

*red*3 (in λ) and *recA* (in *E. coli*)—mutations that inactivate the generalized recombination systems of bacteriophage λ and the host, thereby minimizing recombination between the endogenous λ DNA in the extract and the exogenously added recombinant genomes.

Packaging extracts are usually prepared by growing the lysogenic bacteria at 30-32°C to midlog phase, inducing lytic functions by inactivating the *c*I repressor protein at 45°C, and growing the cultures for an additional 2-3 hours at 38-39°C to allow packaging components to accumulate. Cell extracts are then prepared.

Of the many protocols available for the preparation of packaging extracts, the two presented below are perhaps the simplest to perform. Both yield highly efficient packaging extracts. The particular advantage of protocol I is its lack of bias in packaging recombinant molecules of different sizes. Its main disadvantage is that a relatively high level of background plaques is sometimes observed, resulting from packaging of endogenous λ DNA. The advantages of protocol II are simplicity, high efficiency, low background, and size selectivity.

Two alternative but slightly more complicated procedures for preparing packaging extracts are also available. The protocol described by Sternberg et al. (1977) is efficient but involves two different extracts and a complicated packaging procedure. The protocol described by Faber et al. (1978) depends on the purification of the bacteriophage λ *A* protein and the preparation of two separate extracts.

MAINTENANCE AND TESTING OF BACTERIOPHAGE λ LYSOGENS

Because successful preparation of packaging extracts is dependent on the presence of specific mutations in the bacteriophage λ genome, considerable care should be exercised in growing and maintaining the bacteriophage λ lysogens. We recommend taking the following precautions.

1. Master stocks of *E. coli* BHB2690 and BHB2688 should be stored in glycerol at −70°C as described in Chapter 1. Extracts should be made from cultures prepared directly from these master stocks as described immediately below.

2. Check for the presence of the mutation that renders the *c*I-gene product temperature-sensitive by streaking from the master stocks of *E. coli* strains BHB2690 and BHB2688 onto LB plates (two plates for each bacterial strain). Incubate one plate of each strain at 30–32°C and the other at 42°C. The bacteria should grow only on the plates incubated at 32°C. Pick a single colony of each strain from the plates incubated at 32°C and grow small overnight cultures at 32°C. Test these cultures once again by plating aliquots at 32°C and 42°C, as described below. If the bacteria should grow only on the plates incubated at 32°C, the residue of the small overnight cultures may be used as inocula for larger cultures from which extracts are prepared.

3. Pick small, rather than large, colonies. The *recA*⁻ mutation causes the bacteria to grow slowly. Revertants thus give rise to larger colonies. If necessary, the *recA* function can be tested as follows:

   a. Streak the strain being tested, a known *recA*⁻ strain, and a wild-type *E. coli* strain onto a plate as shown in Figure 8.1 (top).

   b. Use a piece of cardboard to cover about three quarters of each streak. Expose the remaining portion of each streak for 10 seconds to the UV light of a Biogard hood. Move the cardboard so that half of each streak is now visible and repeat the exposure to UV light. Move the cardboard once again so three quarters of the streak is exposed and repeat the exposure to UV light. The remaining one quarter of each streak should receive no exposure. Following overnight growth, the plate should look like the one illustrated in Figure 8.1 (bottom).
   Mutations in *recA* prevent repair of UV-induced damage so that irradiated, *recA*⁻ cells do not grow. Different *recA*⁻ mutants are killed by different amounts of UV irradiation (compare *E. coli* strains HB101 and 1046).

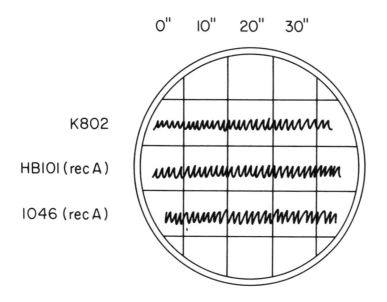

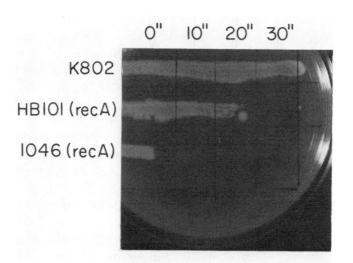

Figure 8.1

PREPARATION OF PACKAGING EXTRACTS—PROTOCOL I

In this protocol *E. coli* lysogens BHB2690 (*D*am) and BHB2688 (*E*am) are grown separately, induced, mixed, concentrated, and frozen in aliquots. The DNA to be packaged is added directly to a single tube of packaging extract (Hohn 1979).

1. Make subcultures of the master stocks of *E. coli* BHB2690 and BHB2688 and verify their genotypes as described on page 258.

2. Read the OD_{600} of a 50-ml overnight culture of each strain. Set up a fresh culture of each strain by inoculating approximately 500 ml of medium M9 prewarmed to 32°C in a 2-liter flask with sufficient cells to give an initial $OD_{600} \simeq 0.1$.

3. Grow the cells with vigorous shaking at 32°C until the OD_{600} of each culture equals 0.3 (2–3 hours if the initial $OD_{600} \simeq 0.1$). It is important that the cultures be in the midlog phase of growth at the time of induction.

4. Induce the lysogens by placing the flasks in a water bath preheated to 45°C. Swirl the cultures continuously for the next 15 minutes.

   An alternative protocol is to induce the culture by immersing the flask in a shaking water bath set at 65°C. As soon as the temperature inside the flask reaches 45°C, the culture should be transferred for 15 minutes to a water bath set at 45°C.

5. Incubate the induced cells at 38–39°C for 2–3 hours with vigorous aeration. Check for successful induction by adding a drop of chloroform to a small sample of culture; it should clear in a few minutes (see page 78).

6. Mix the two cultures, chill in an ice-water bath, and then recover the bacteria by centrifugation at 4000*g* for 10 minutes at 4°C.

7. Discard the supernatant. Wash the cells in 300 ml of ice-cold medium M9 lacking casamino acids. Recover the bacteria by centrifugation as described above.

8. Discard the supernatant. All of the remaining steps should be carried out in the cold room. Invert the centrifuge bottle and let the fluid drain away from the bacterial pellet. Remove any remaining medium with a pasteur pipette and Q-tips. Dry the walls of the centrifuge bottle with Kimwipes.

9. Use a siliconized pasteur pipette to disrupt and resuspend the cell pellet in $0.004\times$ the original volume (i.e., 4 ml for each liter of culture of the mixed lysogens) of freshly prepared CH buffer. This buffer differs from that used for packaging cosmids (see page 298) in that it contains putrescine and therefore allows DNA molecules of a greater size-range to be packaged (Hohn 1979).

CH buffer
40 mM Tris·Cl (pH 8.0)
1 mM spermidine
1 mM putrescine
0.1% β-mercaptoethanol
7% DMSO (Spectropure; Matheson, Coleman and Bell)

10. Transfer the bacterial suspension to a small glass tube. When the suspension is homogeneous, use an automatic pipette to dispense 50-μl aliquots into precooled 1.5-ml Eppendorf tubes. Because of the viscosity of the suspension, the disposable pipette tip should be cut to widen its diameter. Work quickly, and periodically swirl the bacterial suspension.

11. After each aliquot is transferred, cap the tube and plunge it into liquid nitrogen.

12. Use forceps to recover the tubes from the liquid nitrogen and immediately transfer the aliquots to a freezer at −70°C.
    Extracts made and stored in this way are stable for at least 6 months.

PACKAGING IN VITRO—PROTOCOL I

1. Mix:

   DNA dissolved in 5 μl of
      66 mM Tris·Cl (pH 7.9) and 10 mM $MgCl_2$ 0.1–1.0 μg
   CH buffer (see note i below)   appropriate volume
   0.1 M ATP (pH 7.5)   1 μl

   Store on ice.

   Notes

   i. Each batch of extracts should be titrated to determine the optimal volume for packaging. The titration is carried out by adding various quantities of CH buffer (0–50 μl) containing 1.5 mM ATP to a standard packaging reaction containing 0.5 μg of intact bacteriophage λ DNA and 50 μl of extract. By choosing the optimal dilution, the efficiency of packaging can sometimes be increased by an order of magnitude or more. Use the same preparation of λ DNA to test different batches of extracts. This DNA should be stored in aliquots at −20°C.

   ii. 0.1 M ATP is prepared as follows: Dissolve 60 mg of ATP in 0.8 ml of H_2O. Adjust the pH to 7.0 with 0.1 M NaOH. Adjust the volume to 1.0 ml with H_2O. Dispense the solution into small aliquots and store at −70°C.

2. Remove up to three tubes containing packaging extracts from storage at −70°C and place on ice. Do not try to carry out more than three packaging reactions at a time. While the extracts are still frozen, add the mixture prepared in step 1. Using a sealed capillary pipette, blend together the contents of the tube as the extract thaws. Mix very well.

   Note. The DNA *must* be mixed into the extracts just as they thaw because the phage components will start to assemble immediately following the cell lysis, which accompanies thawing. This is a critical step. If the extract thaws before mixing is completed, the viscosity of the liberated bacterial DNA prevents efficient mixing of the added bacteriophage λ DNA.

3. Incubate for 60 minutes at 37°C.

4. Remove another set of packaging extracts from storage at $-70°C$. Place on ice. Add to each tube 5 μl of pancreatic DNase (100 μg/ml) and 2.5 μl of 0.5 M $MgCl_2$. Mix the contents of the tube as the extract thaws and then store on ice for 5–10 minutes. Add 20 μl of the second packaging extract (containing pancreatic DNase and magnesium chloride) to each packaging reaction. Mix well and incubate for a further 30 minutes at 37°C. Flick the tube every few minutes during the incubation.

Notes

i. The addition of a second portion of extract improves the efficiency of packaging twofold to fivefold, presumably by supplying additional phage components required for steps in morphogenesis after assembly of the head.

ii. The addition of DNase should result in a rapid reduction in viscosity of the packaging reaction. If the viscosity does not decrease markedly, add another aliquot of pancreatic DNase and incubate for a further 5 minutes.

5. Add 1 ml of SM and 5 μl of chloroform to each packaging reaction. Mix. Remove debris by centrifuging for 30 seconds in an Eppendorf centrifuge. Titrate the number of viable bacteriophage particles in the supernatant as described on page 64. The phage particles are stable in the packaging mixture containing SM and chloroform for several weeks.

Note

Because components in the packaging extracts can inhibit infection by bacteriophage particles, the amount of a packaging reaction that can be plated on a single plate is limited. The severity of inhibition can be determined empirically for each batch of extracts by setting up a mock packaging reaction (no added DNA) and mixing a known number of infectious bacteriophage particles with increasing amounts of the packaging extracts and 1 ml of SM. After incubating for 10 minutes at room temperature, measure the titer of the bacteriophages. Generally, inhibition only becomes a problem when more than one tenth of a packaging reaction is plated on a single 85-mm plate.

PREPARATION OF PACKAGING EXTRACTS—PROTOCOL II[1]

Sonicated Extract from Induced BHB2690 (Prehead Donor)

1. Make a subculture of the master stock of *E. coli* BHB2690. Verify its genotype, as described on page 258.

2. Read the OD_{600} of a 100-ml overnight culture and inoculate 500 ml of NZM broth prewarmed to 32°C in a 2-liter flask with sufficient cells to give an initial $OD_{600} \leqslant 0.1$. Incubate with aeration at 32°C until an $OD_{600} = 0.3$ is reached (2–3 hours).

   It is important that the cultures be in the midlog phase of growth before induction.

3. Induce the lysogen by placing the flask into a water bath preheated to 45°C. Swirl the culture continuously for 15 minutes.

   Alternatively, induce the culture by immersing the flask in a shaking water bath set at 65°C. As soon as the temperature inside the flask reaches 45°C, the culture should be transferred for 15 minutes to a water bath set at 45°C.

4. Incubate the induced cells at 38–39°C for 2–3 hours with vigorous aeration. Check for successful induction by adding a drop of chloroform to a small sample of culture; it should clear within a few minutes.

5. Recover the cells by centrifugation at 4000*g* for 10 minutes at 4°C.

6. Drain off as much liquid as possible. Remove any remaining medium with a pasteur pipette and Q-tips. Dry the walls of the centrifuge bottle with Kimwipes.

7. Add 3.6 ml of freshly prepared sonication buffer. Resuspend the pellet thoroughly and transfer the resulting homogeneous suspension to a small, clear plastic tube (Falcon no. 2054 or 2057).

   Sonication buffer

   20 mM Tris·Cl (pH 8.0)
   1 mM EDTA
   5 mM β-mercaptoethanol

8. Sonicate in short bursts (10 seconds) at maximum power using a microtip probe. The tube should be immersed in ice-water and the temperature of the sonication buffer should not be allowed to exceed 4°C. Allow the sample to cool for 20–30 seconds between each burst of sonication. Patience is critical!

   Sonicate until the solution clears and its viscosity decreases.

[1]Scalenghe et al. (1978); B. Hohn (unpubl.).

Note. The amount of sonication is critical, and the clearing and change in viscosity of the solution are not always readily apparent. When preparing these extracts for the first time, you should remove aliquots of the suspension after sonicating for various times. The aliquots are then processed and used in separate packaging reactions to determine the optimal sonication time.

9. Transfer the sonicated sample to a centrifuge tube and remove debris by centrifugation at 12,000g for 10 minutes at 4°C.

10. To the supernatant (3 ml) add an equal volume of cold sonication buffer and $\frac{1}{6}$ volume of freshly prepared packaging buffer. Dispense 15-μl aliquots into precooled (4°C), 1.5-ml Eppendorf tubes. Immediately close the caps of the tubes, immerse them briefly in liquid nitrogen, and transfer them to −70°C for long-term storage.

Packaging buffer

    6 mM Tris · Cl (pH 8.0)
    50 mM spermidine
    50 mM putrescine
    20 mM $MgCl_2$
    30 mM ATP (see note ii, page 262)
    30 mM β-mercaptoethanol

Freeze/Thaw Lysate from Induced BHB2688 (Packaging Protein Donor)

1. Make a subculture of the master stock of *E. coli* BHB2688. Verify its genotype as described on page 258.

2. Read the OD_{600} of a 100-ml overnight culture and inoculate 500 ml of NZM broth prewarmed to 32°C into each of three 2-liter flasks with sufficient cells to give an initial $OD_{600} \leqslant 0.1$. Incubate with aeration at 32°C until the $OD_{600} \simeq 0.3$ (2–3 hours).

3. Induce the lysogen by placing the flasks into a water bath preheated to 45°C. Swirl the flasks continuously for 15 minutes (see step 4, page 260).

4. Incubate the induced cells at 38–39°C for 2–3 hours with vigorous aeration. Check for successful induction by adding a drop of chloroform to a small sample of the culture; it should clear within a few minutes.

5. Recover the cells by centrifugation at 4000*g* for 10 minutes at 4°C.

6. Drain off as much liquid as possible. Remove any remaining medium with a pasteur pipette and Q-tips. Dry the walls of the centrifuge bottle with Kimwipes.

7. Resuspend the cells in a total of 3 ml of ice-cold sucrose solution. Distribute 0.5 ml of the suspension into each of six precooled (4°C) Eppendorf tubes. Add 25 μl of fresh, ice-cold lysozyme solution to each tube. Mix gently. Quickly close the cap of the tube and plunge it into liquid nitrogen.

   Sucrose solution

   10% sucrose
   in 50 mM Tris·Cl (pH 8.0)

   Lysozyme solution

   2 mg/ml lysozyme
   in 0.25 M Tris·Cl (pH 8.0)

8. Use forceps to remove the tubes from the liquid nitrogen. Thaw the extracts in ice. Add 25 μl of freshly prepared packaging buffer (see step 10, page 265) to each tube and mix.

9. Combine the thawed extracts in a centrifuge tube and centrifuge at 48,000*g* for 1 hour at 4°C.

10. Dispense 10 μl of the supernatant into precooled (4°C) Eppendorf tubes. Immediately close the caps of the tubes and immerse them in liquid nitrogen. After all the aliquots have been frozen, remove the tubes from the liquid nitrogen and immediately transfer them to long-term storage at −70°C.

Note

The sonicated extract and the freeze/thaw extract can be combined at the time of preparation, if desired. Prepare the sonicated extract first, and freeze 15-μl aliquots in open Eppendorf tubes arranged in a rack in liquid nitrogen. Add 10 μl of freeze/thaw lysate directly into each tube. Close the cap of the tubes and immerse them in liquid nitrogen. Store at −70°C.

The major problem in preparing combined extracts is handling the frozen tubes and attempting to pipette 10 μl of freeze/thaw lysate into a tube at −70°C. The efficiencies of packaging of extracts prepared by combining before or after freezing are similar.

PACKAGING IN VITRO—PROTOCOL II

1. Remove tubes from storage at −70°C and allow the packaging extracts to thaw on ice. The freeze/thaw lysate will thaw first. Transfer the freeze/thaw lysate to the still-frozen, sonicated extract.

2. Mix gently. When the combined extracts are almost totally thawed, add the DNA to be packaged (up to 1 μg dissolved in 5 μl of 10 mM Tris · Cl [pH 7.9] and 10 mM MgCl₂). Mix with a sealed capillary pipette and incubate for 1 hour at room temperature.

3. Add 0.5-1 ml of SM and a drop of chloroform and mix. Remove debris by centrifugation in an Eppendorf centrifuge for 30 seconds and measure the titer of the viable bacteriophage particles as described on page 64.

Notes

i. Each batch of extracts should be tested with a standardized preparation of intact bacteriophage λ DNA.

ii. These extracts exhibit a high degree of selectivity in the size of the DNA that is packaged (Sternberg et al. 1977). Recombinant DNAs that are 90% or 80% of wild-type bacteriophage λ in length are packaged with efficiencies 20-fold to 50-fold lower, respectively, than wild-type λ DNA.

iii. The same packaging buffer may be used for packaging of both bacteriophage λ and cosmids.

iv. See note to Packaging In Vitro—Protocol I, page 263.

9

Construction of
Genomic Libraries

Construction of Genomic Libraries in Bacteriophage λ Vectors

Two strategies have been used to clone specific sequences of eukaryotic, genomic DNA in bacteriophage λ vectors. The first, developed before efficient methods became available to package bacteriophage λ DNA in vitro, involved cloning of DNA preparations that had been highly enriched for the sequences of interest. Total genomic DNA was digested to completion with a restriction endonuclease and the resulting fragments were separated by preparative gel electrophoresis (Tilghman et al. 1977; Tonegawa et al. 1977; Edgell et al. 1979), reverse-phase column chromatography (Hardies and Wells 1976; Landy et al. 1976), or both (Tilghman et al. 1977). The fractions containing the DNA sequences of interest were identified by hybridization to appropriate probes and cloned in bacteriophage λ vectors. Because these fractionation procedures may yield an overall enrichment of as much as 100-fold to 300-fold, the number of recombinant bacteriophages that had to be generated and screened for the DNA sequences of interest could be significantly reduced.

Subsequently, the development of efficient in vitro packaging systems (Hohn and Murray 1977; Sternberg et al. 1977) and in situ hybridization techniques (Benton and Davis 1977) eliminated the need to enrich the starting DNA for the sequences of interest. Nowadays, the usual strategy is to construct complete libraries of eukaryotic DNA and then to identify by hybridization those recombinant bacteriophages that contain the desired sequences.

Libraries of eukaryotic DNA may be prepared in two ways. The first approach involves digestion of genomic DNA to completion with a restriction enzyme and insertion of the resulting fragments in an appropriate bacteriophage λ vector (Smithies et al. 1978). This method suffers from two drawbacks. First, if the sequence of interest contains recognition site(s) for the particular restriction enzyme chosen, it will be cloned in two or more pieces. There is also a chance that the sequence may not be cloned at all if, for example, it is contained in a larger DNA fragment than the bacteriophage vector can accept. Second, the average size of the fragments generated by cleavage of eukaryotic DNA with many of the restriction enzymes that recognize hexanucleotide sequences is relatively small (~ 4 kb). An entire library therefore contains a very large number of recombinant bacteriophages, and screening by hybridization becomes a very laborious and expensive proposition.

Both of these problems can be avoided by cloning large (~ 20 kb) DNA fragments that are generated by random shearing of eukaryotic DNA

(Maniatis et al. 1978). This method ensures that there is no systematic exclusion of sequences from the cloned library merely because of an unfortunate distribution of restriction sites. Other advantages gained by constructing libraries of randomly sheared DNA are:

1. It gives an opportunity to "walk" along the eukaryotic chromosome in a way that is impossible with a nonoverlapping library of fragments obtained by complete digestion of eukaryotic DNA with a restriction enzyme. The number of potentially different clones of randomly sheared DNA fragments is essentially infinite. It is therefore possible to use DNA fragments derived from one recombinant clone as hybridization probes to identify overlapping clones and thus to "walk" along the chromosome.

2. Because the randomly sheared DNA fragments chosen for cloning are relatively large, less than a million recombinant bacteriophages need be generated and screened in order to have a good chance of isolating a particular single-copy sequence of eukaryotic DNA.

   Note. The exact probability of having any DNA sequence represented in the library can be calculated from the formula:

   $$N = \frac{\ln(1-P)}{\ln(1-f)}$$

   where P is the desired probability, f is the fractional proportion of the genome in a single recombinant, and N is the necessary number of recombinants (Clarke and Carbon 1976). For example, to achieve a 99% probability ($P = 0.99$) of having a given DNA sequence represented in a library of 17-kb fragments of a mammalian genome (3×10^9 bp):

   $$N = \frac{\ln(1-0.99)}{\ln\left(1 - \left[\frac{1.7 \times 10^4}{3 \times 10^9}\right]\right)}$$

   $$= 8.1 \times 10^5$$

3. Knowledge of the distribution of restriction sites in or around the sequence of interest is not required before cloning is attempted.

The basic steps used in the construction of this type of genomic DNA library are shown in Figure 9.1 and are summarized as follows:

1. The DNA of a bacteriophage λ substitution vector that will accept inserts up to 20 kb in length is digested with a restriction endonuclease. The left and right arms, which carry all of the genetic information essential for

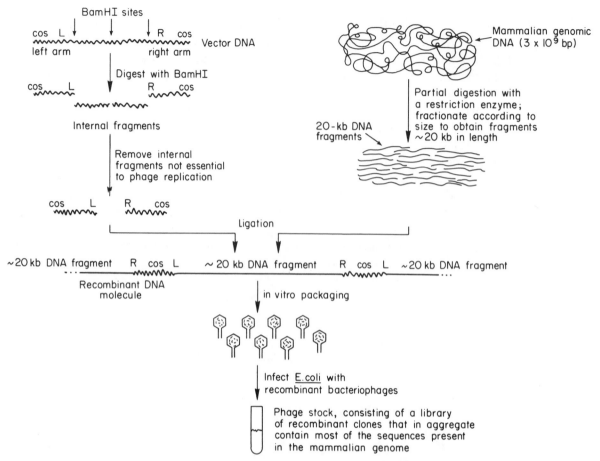

Figure 9.1

A strategy used to construct libraries of random fragments of eukaryotic DNA. (*Left*) Preparation of the vector DNA fragments; (*right*) preparation of eukaryotic DNA fragments. A concatameric, recombinant DNA molecule is produced by the action of bacteriophage T4 DNA ligase. This concatamer is the substrate for the in vitro packaging reaction during which a different recombinant DNA molecule is inserted into each bacteriophage particle. Following amplification by growth in *E. coli*, a lysate is obtained, consisting of a library of recombinant clones that, in aggregate, contain most of the sequences present in the mammalian genome.

lytic growth of the virus (see Chapter 1), are separated from the internal stuffer fragments by size fractionation.

2. The eukaryotic DNA is prepared for cloning by random fragmentation of high-molecular-weight DNA so that a population of molecules is generated with an average size of 20 kb. The only method by which DNA can be fragmented in a truly random fashion, irrespective of its base composition and sequence, is mechanical shearing. However, DNA prepared in this way requires several additional enzymatic manipulations (repair of termini, methylation, ligation to linkers, digestion of linkers) to generate

cohesive ends compatible with those of the vector arms (Maniatis et al. 1978). On the other hand, partial digestion with restriction enzymes that recognize frequently occurring tetranucleotide sequences within eukaryotic DNA frequently yields a population of fragments that is close to random and yet can be cloned directly.

However, not all sequences of the eukaryotic genome are equally represented in such populations, for three reasons. First, the termini of the fragments are not randomly distributed along the DNA but are defined by the locations of their restriction sites. The number of potential termini is almost equal to the total number of restriction sites in the DNA. Second, although restriction sites are randomly distributed in the bulk of eukaryotic DNA, this is not true for highly repeated DNA or satellite DNA (Botchan et al. 1974). Such factors as local variation in G + C content and the presence of repeated sequences of DNA can cause distortion in the distribution of sites. Third, in bacteriophage λ DNA and some eukaryotic viral DNAs (J. Sambrook, unpubl.), not all restriction sites are cleaved with equal efficiency. It is possible that the same phenomenon occurs in eukaryotic, cellular DNA, although definitive experiments have not been carried out. Thus, the termini of the fragments are distributed nonrandomly among the potential cleavage sites, and the potential sites themselves are distributed nonrandomly along the DNA. However, the number of potential sites in a eukaryotic genome is so great and the size of the cloned fragments ($\sim$ 20 kb) is so large relative to the average distance between the potential sites (256 bp) that the population of fragments, though not random in a strictly mathematical sense, can be treated as such for most practical purposes. The fragmented DNA is fractionated by sucrose density gradient centrifugation or gel electrophoresis to obtain molecules approximately 20 kb in length. For a detailed discussion of this problem, see B. Seed (1982) and B. Seed, R. Parker, and N. Davidson (submitted).

3. The arms of the vector and the fragmented eukaryotic DNA are then ligated to form long concatameric molecules that are subsequently packaged into bacteriophage λ heads using in vitro packaging systems (see Chapter 8). The recombinant bacteriophages are amplified by growth in *E. coli*. The resulting library can be stored and used over a long period of time and thus can be screened for the presence of many different gene sequences.

The aim of the method is to produce a library of recombinants that is as complete as possible. However, for a variety of reasons, particular sequences may be underrepresented or not represented at all. For example, as discussed above, the distribution of the recognition sites in and around the DNA sequence of interest will affect its representation. If the sites are very far apart or extremely close together, the chances of obtaining a fragment of clonable size are small. Furthermore, particular regions of the eukaryotic chromosome may contain DNA sequences that are detrimental to the growth of bacteriophage λ and hence may be lost from the library. Finally, in several cases, the presence of tandemly repeated sequences in the cloned eukaryotic DNA is known to lead to deletion by recombination during phage propagation (Arnheim and Kuehn 1979; Fritsch et al. 1980; Lauer et al. 1980).

The steps used in construction of a recombinant DNA library are as follows:

1. Preparation of in vitro packaging extracts.
2. Preparation of vector DNA.
3. Preparation of fragmented, eukaryotic DNA.
4. Ligation and packaging.
5. Amplification.

Preparation of in vitro packaging extracts is discussed in Chapter 8. Steps 2-5 are discussed in detail on the following pages.

PREPARATION OF VECTOR DNA

The bacteriophage λ vectors that are most useful in library construction require that the middle stuffer segment of their genomes be removed in order to accommodate 15–20 kb of foreign DNA. This process is generally referred to as "preparation of arms." Depending on the vector used, any of several approaches are possible. We will describe two methods of purifying arms by density gradient centrifugation.

Arms can also be purified by preparative electrophoresis through gels made of 0.5% low-melting-temperature agarose. The DNA is extracted from the gel and purified as described on page 170. In general, however, the yields of arms obtained by this technique are much lower than those obtained by density gradient centrifugation.

Centrifugation through Sucrose Density Gradients

This is the standard method and can be used to prepare the arms of any vector DNA (Maniatis et al. 1978).

1. Digest the bacteriophage λ vector DNA at a concentration of 150 μg/ml or greater with a twofold to threefold excess of restriction enzyme for 1 hour. Stop the reaction by cooling to 0°C. Remove an aliquot ($\sim 0.5 \mu$g) and heat at 68°C for 10 minutes to disrupt the cohesive ends of bacteriophage λ DNA and analyze by electrophoresis through a 0.5% agarose gel. Use uncut λ DNA as a marker. If digestion is incomplete, warm the reaction mixture to 37°C, add more restriction enzyme, and continue the incubation.

   Note. It is sometimes possible to digest the vector DNA with another enzyme that cleaves within the stuffer fragment but not in the arms (e.g., *Sal*I for cloning in the *Bam*HI sites of λBF101). The aim of this strategy is to minimize the amount of intact bacteriophage λ DNA in the final preparation.

2. When digestion is complete, extract the sample twice with phenol/chloroform and precipitate the DNA with ethanol. Resuspend the DNA in TE at a concentration of 150 μg/ml. Remove an aliquot (0.2 μg) and store at 4°C. If necessary, the digested DNA may be stored at -20°C until you are ready to proceed to step 3.

3. Add MgCl₂ to 0.01 M and incubate at 42°C for 1 hour to allow the cohesive ends of bacteriophage λ DNA to reanneal. Analyze an aliquot (0.2 μg) by gel electrophoresis to determine whether annealing has occurred. Use intact bacteriophage λ DNA and the aliquot set aside in step 2 (and heated to 68°C for 10 minutes) as markers.

Notes

i. Do not run the analytical gels at high voltage or in electrophoresis buffers of high electrical resistance; overheating will melt the cohesive ends of bacteriophage λ DNA during electrophoresis.

ii. An alternative procedure is to ligate the cohesive ends of the vector DNA before digesting it with a restriction endonuclease. Incubate the DNA at 42°C in 0.1 M Tris · Cl (pH 8.0) and 10 mM MgCl₂ for 1 hour, adjust the reaction to 1× ligation buffer (see page 479), add ligase, and incubate for 1–2 hours at 37°C. After gentle extraction with phenol/chloroform, the DNA is recovered by ethanol precipitation, resuspended in the appropriate buffer, and digested with restriction enzyme.

4. Prepare two 38-ml sucrose (10–40% w/v) gradients in ultracentrifuge tubes (Beckman SW27 or equivalent). The sucrose solutions are made in a buffer containing:

    1 M NaCl
    20 mM Tris · Cl (pH 8.0)
    5 mM EDTA

5. Load no more than 60–70 μg of annealed, digested bacteriophage λ DNA onto each gradient in a volume of 500 μl or less. Loading more DNA than this can cause the gradient to be overloaded and can lead to poor separation between the stuffer fragments and the arms. Centrifuge the samples at 26,000 rpm for 24 hours at 15°C in a Beckman SW27 rotor (or its equivalent).

6. Collect 0.5-ml fractions through a 21-gauge needle inserted through the bottom of the centrifuge tube.

7. Dilute 15 μl of every third fraction with 35 μl of H₂O. Add 8 μl of gel-loading dye I (see page 455), heat the samples at 68°C for 5 minutes, and analyze them by electrophoresis through a 0.5% agarose gel. Use uncut vector DNA and vector DNA cleaved with the appropriate restriction enzyme as markers. Adjust the sucrose and salt concentrations of the markers to match those of the samples, otherwise their electrophoretic mobilities will not be comparable. After photographing the gel, locate and pool the fractions that contain the annealed arms (Fig. 9.2). Be careful not to take fractions that are visibly contaminated with uncut bacteriophage λ DNA or that contain significant quantities of unannealed left or right arms.

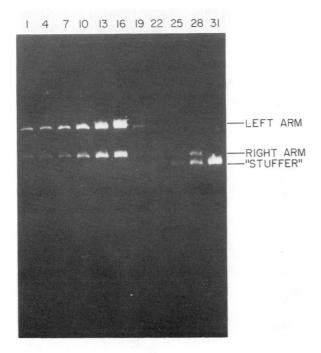

Figure 9.2

Preparation of the arms of bacteriophage λ DNA by sucrose gradient centrifugation. In this experiment, the DNA of bacteriophage λ vector Charon 28 was digested with *Bam*HI, annealed, and centrifuged through a 10-40% sucrose gradient, which was then fractionated as described in the text. Aliquots of every third fraction were heated briefly to 68°C and analyzed by electrophoresis through a 0.5% agarose gel. The positions of the left arm (23.5 kb), right arm (9 kb), and "stuffer" fragments (6.5 kb) are indicated. Fractions 1-16 containing the annealed arms were pooled.

8. Dialyze the pooled fractions against a large volume of TE at 4°C for 12-16 hours, with at least one change of buffer. Be sure to allow for a twofold to threefold increase in volume during dialysis. Extract the sample several times with 2-butanol (see page 463) to reduce the volume to about 5 ml. Precipitate the DNA with ethanol.

   Alternatively, if the pooled sample is small, the DNA can be precipitated with ethanol without prior dialysis by diluting the sample with TE so that the concentration of sucrose is reduced to about 10%.

9. Wash the precipitate with 70% ethanol. Dissolve the DNA in TE at a concentration of 300-500 μg/ml.

10. Measure the exact concentration of the DNA and analyze an aliquot by electrophoresis through a 0.5% agarose gel to assess its purity. Store the DNA at −20°C in aliquots containing 1-5 μg.

Centrifugation through Potassium Acetate Gradients

This method, which was developed comparatively recently (B. Seed, unpubl.), has the considerable advantage of avoiding the time-consuming electrophoretic analysis of gradient fractions. However, it gives the experimenter no latitude to choose which fractions to include and which to discard. Before committing a large-scale preparation to the method, it would be worthwhile to carry out a pilot experiment.

1. Digest the bacteriophage λ vector DNA with the appropriate restriction enzyme(s) and prepare the annealed arms as described in steps 1–3 on page 275.

2. Prepare 38-ml density gradients (5–20% w/v) of potassium acetate in ultracentrifuge tubes (Beckman SW27 or equivalent).

   The potassium acetate is dissolved in 10 mM Tris · Cl (pH 8.0) and 5 mM EDTA. The gradients are poured through a pipette inserted into the bottom of the centrifuge tube so that the solution of lighter density is continuously displaced by material of heavier density.

3. Load the samples of DNA onto the gradients. Each gradient will take up to 100 μg of annealed arms without overloading.

4. Centrifuge at 27,000 rpm for 16 hours at 20°C in a Beckman SW27 rotor (or its equivalent). Under these conditions, the annealed arms will form pellets on the bottom of the tube while the stuffer fragments remain in the body of the gradient.

5. Drain off as much supernatant as possible and dry the sides of the tube with Kimwipes or cotton swabs. Resuspend the pelleted arms in 0.5 ml of TE (pH 7.9). Usually, it is possible to proceed directly to the ligation step. However, if too much salt is carried through, it will be necessary to precipitate the DNA with ethanol and wash the pellet with 70% ethanol. The arms then can be redissolved in a small volume (250 μl) of TE (pH 7.9) and analyzed by electrophoresis through a 0.5% agarose gel.

6. Measure the exact concentration of the DNA and store at −20°C in aliquots containing 1–5 μg.

Note

Ethidium bromide can be included at a concentration of 2 μg/ml in both types of gradient. The position of the different species of DNA within the gradient can then be determined visually. With practice, it is possible to pool those fractions from sucrose gradients that contain the annealed arms without prior analysis by agarose gel electrophoresis. If ethidium bromide is used in this way, the purified arms should be extracted three times with isoamyl alcohol to remove the dye.

Test Ligation of Arms

The ability of the purified arms to be ligated is tested as follows.

1. Mix:

preparation of bacteriophage λ arms	1 μg
10× ligation buffer (see page 474)	1 μl
H₂O	to 10 μl

2. Remove two 1-μl aliquots (0.1 μg each) to serve as markers for gel electrophoresis. Add bacteriophage T4 DNA ligase (10–50 Weiss units) to the remainder of the reaction. Incubate for 12–16 hours at 12°C.

3. Remove two 1-μl aliquots. Add 10 μl of TE to all four aliquots. Heat one aliquot from step 2 and one from step 3 to 68°C for 10 minutes. Add 3 μl of gel-loading buffer I (see page 455) and apply all four samples to a 0.5% agarose gel. If the ligation has been successful, most of the arms should have been converted to catenates 42 kb or larger in size.

ISOLATION OF HIGH-MOLECULAR-WEIGHT, EUKARYOTIC DNA FROM CELLS GROWN IN TISSUE CULTURE[1]

1. Wash the cell monolayers twice with ice-cold, Tris-buffered saline (TBS). Using a rubber policeman, scrape the monolayers into a small amount of TBS. Centrifuge the cell suspension at 1500g for 10 minutes at 4°C.

   Tris-buffered saline

NaCl	8.0 g
KCl	0.38 g
Tris-base	3.0 g
phenol red	0.015 g
H_2O	to 800 ml

   Adjust the pH to 7.4 with 1 N HCl. Adjust the volume to 1 liter. Sterilize by autoclaving.

2. Resuspend the cell pellet at a concentration of 10^8 cells/ml in ice-cold TE.

3. Add 10 volumes of:

   0.5 M EDTA (pH 8.0)
   100 μg/ml proteinase K
   0.5% Sarcosyl

4. Place the suspension of lysed cells in a 50°C water bath for 3 hours. Swirl the viscous solution periodically.

5. Gently extract the DNA three times with an equal volume of phenol. After centrifugation, the phenol generally forms the upper phase in these extractions because of the high salt concentration in the sample. Remove the phenol and as much of the interface as possible. If necessary, reextract the interface with additional phenol and buffer. Pool the two aqueous phases.

6. After the third extraction, dialyze the DNA against 4 liters of a solution of 50 mM Tris · Cl (pH 8.0), 10 mM EDTA, and 10 mM NaCl with several changes until the OD_{270} of the dialysate is less than 0.05. Allow room in the dialysis bag for the sample to increase threefold in volume.

7. Treat the sample with 100 μg/ml of DNase-free RNase (see pages 445ff) at 37°C for 3 hours.

8. Gently extract the sample twice with phenol/chloroform.

9. Dialyze the sample extensively against TE.

[1]Blin and Stafford (1976).

10. Measure the exact concentration of the DNA and analyze an aliquot by electrophoresis through a 0.3% agarose gel. The DNA should be greater than 100 kb in size and should migrate more slowly than a marker of intact bacteriophage λ DNA. Store the DNA at 4°C.

Notes

i. DNA can also be isolated from tissues by a modification of the above procedure:

   a. Mince the tissue using a scalpel or scissors.

   b. Pour liquid nitrogen into a stainless steel Waring Blendor.

   c. Add the minced tissue to the liquid nitrogen and blend (1-5 minutes at top speed) until the tissue is ground to a fine powder.

   d. Let the liquid nitrogen evaporate and add the powder to approximately 10 volumes of lysis solution (step 3).

   e. Follow steps 4-10 as above.

ii. If further purification or concentration of the DNA is necessary, centrifuge the DNA to equilibrium in a cesium chloride density gradient ($\rho = 1.70$) in a Beckman Type-40 rotor or equivalent at 33,000 rpm for 60-70 hours at 15°C. Collect fractions from the gradient by dripping through a large needle (16-gauge) inserted into the side of the tube near the bottom (see Fig. 3.2, page 82). The viscous fractions containing DNA should be pooled and dialyzed extensively against TE (pH 7.9).

PREPARATION OF 20-KB FRAGMENTS OF EUKARYOTIC DNA

Because of its simplicity, the current method of choice for constructing libraries of eukaryotic DNA in bacteriophage λ vectors is to clone DNA fragments obtained by partial digestion of high-molecular-weight DNA with restriction enzymes that recognize a 4-bp sequence and generate a cohesive end (Maniatis et al. 1978).

Establishing Conditions for Partial Digestion of High-molecular-weight DNA

1. Prepare a reaction mixture with 10 μg of eukaryotic DNA and restriction enzyme buffer in a final volume of 150 μl. Mix well by inverting the tube several times.

2. Dispense 30 μl into an Eppendorf tube (tube 1). Dispense 15 μl into tubes 2–8. Dispense the remainder into tube 9. Chill all tubes on ice.

3. Add 4 units of restriction enzyme to tube 1 and mix well. The concentration of enzyme is thus 2 units/μg DNA. Transfer 15 μl of the reaction mixture to tube 2. The enzyme concentration has now been diluted to 1 unit/μg DNA. Mix well and continue the twofold serial dilution through to tube 8 (do not add anything to tube 9).

4. Place tubes 1–8 in a 37°C bath. Incubate for 1 hour at 37°C.

5. Stop the reactions by chilling to 0°C and adding EDTA to a final concentration of 20 mM.

6. Add 3 μl of gel-loading dye I to each sample and analyze by electrophoresis through a 0.4% agarose gel. Use accurate markers in the 10–30-kb size range on the two outside lanes (pBR322 multimers work very well; see page 474). Electrophoresis is carried out slowly (1–2 V/cm) until the bromophenol blue has just migrated off the gel.

7. Photograph the gel without overexposing the film (Fig. 9.3). Using the photograph and the markers, ascertain the amount of enzyme needed to produce the maximum intensity of fluorescence in the 15–20-kb region of the gel. This is accomplished by blocking off all regions of the gel not containing the DNA of the desired size. The various gel tracks are then compared to estimate the degree of digestion that produces the maximum amount of DNA of the desired size. The intensity of fluorescence is related to the *mass* distribution of the DNA. To obtain the maximum *number* of molecules in this size range, use half of the amount of enzyme that produces the maximum amount of fluorescence (B. Seed, R. Parker, and N. Davidson, submitted).

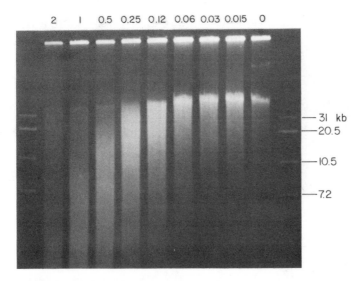

Figure 9.3

Establishing conditions for partial digestion of high-molecular-weight DNA. In this experiment, 1 μg of high-molecular-weight eukaryotic DNA was digested with varying amounts of *Mbo*I for 1 hour at 37°C. The samples were analyzed by electrophoresis using a series of bacteriophage λ DNA fragments of known molecular weight as markers. The digestion conditions that gave the greatest mass of DNA fragments in the 15–20-kb range were determined from the intensity of fluorescence. In the subsequent large-scale digestion, half the number of units of *Mbo*I/μg DNA were used to optimize sequence representation of molecules in this size range.

Note

An alternative and equally effective method for establishing optimal conditions for partial digestion is to set up the reaction with a fixed amount of enzyme for variable amounts of time. In this case, be certain to warm the reaction mixtures to the correct temperature before adding enzyme.

Large-scale Preparation of Partially Digested DNA

1. Using optimized conditions (see page 282), carry out a digestion with 250–500 μg of high-molecular-weight, eukaryotic DNA. The enzyme concentration, time, temperature, and DNA concentration should be identical to those used for the pilot reactions.

2. Analyze an aliquot of the DNA (1 μg) by electrophoresis through a 0.4% agarose gel to check that the size distribution of the digestion products is correct.

3. Meanwhile, gently extract the DNA twice with phenol/chloroform. Precipitate the DNA with ethanol. Wash the precipitate once with 70% ethanol.

4. Dissolve the DNA in 500 μl of TE.

5. If the size distribution of the DNA fragments appears satisfactory, prepare a 38-ml, 10–40% sucrose density gradient in a Beckman SW27 polyallomer tube (or its equivalent). The sucrose solutions are made in a buffer containing 1 M NaCl, 20 mM Tris·Cl (pH 8.0), and 5 mM EDTA. Heat the DNA sample for 10 minutes at 68°C, cool to 20°C, and load the gradient. Centrifuge at 26,000 rpm for 24 hours at 20°C.

6. Puncture the bottom of the centrifuge tube and collect 0.5-ml fractions.

7. Mix 10 μl of every third fraction with 10 μl of H_2O and 5 μl of gel-loading dye I (see page 455). Analyze by electrophoresis through a 0.4% agarose gel, using pBR322 multimers (see page 474) or fragments of bacteriophage λ DNA as markers. Be sure to adjust the sucrose and salt concentrations of the markers to correspond to that of the samples (Fig. 9.4).

8. Following electrophoresis, pool the gradient fractions containing DNA fragments in the 15–20-kb size range. Dialyze against 4 liters of TE (pH 7.8) at 4°C for 12–16 hours with one buffer change after at least 4–6 hours. Leave room in the dialysis bag for the samples to expand twofold to threefold.

9. Extract the dialyzed DNA several times with an equal volume of 2-butanol until the volume is reduced to about 5–8 ml (see page 463). Precipitate the DNA with ethanol.

   Alternatively, if the volume of pooled sample is sufficiently small, the DNA can be precipitated with ethanol without prior dialysis by diluting the sample with TE (pH 7.8) so that the concentration of sucrose is reduced to about 10%.

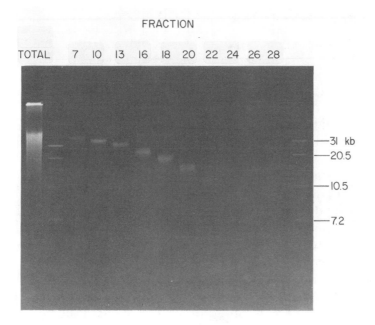

Figure 9.4

Large-scale preparation of ~20-kb fragments of eukaryotic DNA. In this experiment, eukaryotic DNA was partially digested with *Mbo*I and fractionated by centrifugation through a sucrose density gradient as described in the text. Aliquots of selected fractions were then analyzed by electrophoresis through a 0.4% agarose gel. Fractions 13–16 were pooled and used subsequently to construct a library of 15–20-kb fragments of eukaryotic DNA in the bacteriophage λ vector Charon 28.

10. Dissolve the DNA in TE at a concentration of 300–500 μg/ml. Analyze an aliquot of the DNA (0.5 μg) by electrophoresis through a 0.4% agarose gel to check that the size distribution of the digestion products is correct.

11. Carry out a test ligation of the 15–20-kb DNA as described on page 289.

Note

15–20-kb fragments of eukaryotic DNA may also be prepared by agarose gel electrophoresis, which gives better resolution but lower yields than sucrose density centrifugation. To avoid overloading, apply no more than 400–500 μg of partially digested DNA to a single slot (12 cm × 0.4 mm) of a 0.4% agarose gel. Carry out electrophoresis slowly (1–2 V/cm) and examine the stained gel under long-wave UV light. Recover the DNA by electroelution and purify it as described in Chapter 6.

LIGATION AND PACKAGING

Ligation Theory

There are two important parameters to consider in setting up a ligation reaction: the ratio of arms to potential inserts and the concentration of each of the DNA species. Optimal values for both of these parameters can be estimated theoretically. By necessity, however, such calculations assume that all the DNA molecules in the reaction are perfect. Since this is rarely the case, it is advisable also to carry out pilot reactions to check the efficiency of each new preparation of arms and potential inserts.

The best substrate for in vitro packaging is a concatamer of the form (left arm–insert–right arm)$_n$. Each arm of bacteriophage λ DNA carries two different cohesive termini: one (*cos*) that is compatible only with the complementary *cos* sequence at the end of the other arm, and one (*ct*) that is compatible with both termini of the insert and with the remaining terminus of the other arm. To obtain the maximum yield of the desired concatamer, the ligation reaction should contain equimolar concentrations of the *ct* ends of each of the three species of DNA. Because each bacteriophage λ arm contains only a single *ct* end, whereas each potential insert contains two, the ratio of molecules in the reaction should be 2:1:2 (left arm:insert:right arm)—i.e., a 2:1 molar ratio of annealed arms to potential inserts.

The absolute concentration of DNA in the ligation mixture is also important. The concentration must be high enough to ensure that intermolecular ligation, which leads to formation of concatamers, is favored over self-ligation, which leads to cyclization of DNA molecules. A theoretical discussion of the effects of DNA concentration and size on the nature of the ligated products is presented by Dugaiczyk et al. (1975) (see also *Focus*, vol. 2, nos. 2 and 3, published by Bethesda Research Labs). Briefly, the ratio of cyclized to concatameric ligation products depends on two parameters, j and i. j is the effective concentration of one end of a DNA molecule in the neighborhood or volume of the other end of the same molecule. The value given to j is based on the assumption that duplex DNA behaves as a random coil; thus, j is inversely proportional to the length of the DNA molecule (as the DNA gets larger, the ends of a given molecule are less likely to interact).

The frequency with which the two ends of the molecule approach each other can then be calculated from the equation

$$j = \left(\frac{3}{2\pi lb}\right)^{3/2} \text{ ends/ml} \tag{1}$$

where l is the length of the DNA in centimeters (for bacteriophage λ DNA, $l = 13.2 \times 10^{-4}$ cm), and b is the length of a randomly coiled segment of DNA. The value of b is dependent on the ionic strength of the buffer, which affects the rigidity of the DNA. (For bacteriophage λ DNA dissolved in ligation buffer, $b = 7.7 \times 10^{-6}$ cm).

A more useful form of equation (1) is

$$j = j\lambda \left(\frac{MW\lambda}{MW}\right)^{3/2} \text{ends/ml} = \frac{5.5 \times 10^{22}}{MW^{3/2}} \text{ (ends/ml)} \qquad (2)$$

where $j\lambda = 3.22 \times 10^{11}$ ends/ml, and $MW\lambda = 30.8 \times 10^6$. Note that j is constant for a DNA molecule of given length and is independent of the DNA concentration.

Although j describes the effective concentration of the two ends of the same DNA molecule, i is a measure of the concentration of all complementary termini in the solution. For duplex linear DNA with self-complementary, cohesive ends

$$i = 2 \text{ N}_o \text{ M} \times 10^{-3} \text{ ends/ml} \qquad (3)$$

where N_o is Avogadro's number and M is the molar concentration of the DNA.

If the DNA molecule carries the termini that are not self-complementary, the concentration of each terminus is

$$i = \text{N}_o \text{ M} \times 10^{-3} \text{ ends/ml} \qquad (4)$$

Theoretically, then, when $j = i$, the end of a given DNA molecule is equally likely to make contact either with the other end of the same molecule or with the end of a different molecule. Thus, both circles and concatamers should be generated at equal rates in the ligation reaction. When $j > i$, circles are favored, and when $i > j$, concatamers are favored.

For DNA molecules with molecular weights in the range of 1×10^6 to 4×10^6 (1.5–6.0 kb), these predictions have been tested experimentally. It was found that concatamers were the predominant ligation product when $j = i$ and that circles were formed in significant amounts only when j was several times larger than i. One chief reason for the difference between the predicted and observed results is perhaps the fact that the equations describe an equilibrium situation. In a ligation reaction, however, both the concentration (i) and size (j) of the reactants change with time. Thus, although these values serve as a useful guide for setting up ligation reactions, empirical adjustment is still required.

Consider now the ligation of bacteriophage λ arms and potential inserts of eukaryotic DNA to form substrates for in vitro packaging. To generate long concatamers, we need to choose a DNA concentration such that $j \ll i$ for both the bacteriophage λ arms and the potential inserts. Recall also that we need to arrange the relative amounts of arms and inserts such that there is a 2:1 molar excess of arms to inserts in the ligation reaction. We can derive values of j for the arms and the potential inserts assuming that the size of the reannealed arms is 31 kb (1.9×10^7 daltons) and that the average size of the potential inserts is 20 kb (1.25×10^7 daltons).

By substitution in equation 2,

$$j_{arms} = 6.6 \times 10^{11} \text{ ends/ml}$$

and

$$j_{inserts} = 1.25 \times 10^{12} \text{ ends/ml}$$

These values differ from each other by a factor of 2. Because we want to ensure that $j \ll i$, we should use the larger of the two values in the subsequent calculation.

The value of i is a measure of the concentration of *all* cohesive (*ct*) termini in the reaction (i.e., potential inserts and arms). Let us choose i such that

$$\frac{i}{j_{inserts}} = 10$$

which should strongly favor the production of concatamers:

$$i = (10) \, j_{inserts} = (10) \, 1.25 \times 10^{12} \text{ ends/ml} = 1.25 \times 10^{13} \text{ ends/ml} \qquad (5)$$

Since $i = i_{inserts} + i_{arms}$,

$$i = (2N_o \, M_{inserts} + 2N_o \, M_{arms}) \times 10^{-3} \text{ ends/ml}$$

As stated above, we have chosen $M_{arms} = 2 \, M_{inserts}$; so

$$\begin{aligned} i &= (2N_o \, M_{inserts} + 2N_o \, 2 \, M_{inserts}) \times 10^{-3} \text{ ends/ml} \\ &= 6 \, N_o \, M_{inserts} \times 10^{-3} \text{ ends/ml} \end{aligned} \qquad (6)$$

Substituting values for i (equation 5) and N_o, we find that $M_{inserts} = 3.5 \times 10^{-9}$ M (43 μg/ml).

$$M_{arms} = 2 \, M_{inserts} = 7.0 \times 10^{-9} \text{ M (135 } \mu\text{g/ml)}$$

The reaction should therefore contain 43 μg/ml of inserts and 135 μg/ml of arms to achieve the optimal yield of highly concatamerized recombinant DNA molecules appropriate for packaging into particles of bacteriophage λ.

Test Ligation and Packaging

If enough arms, inserts, and packaging extracts are available, a series of test ligations and packagings should be performed to determine the ratio of arms to inserts that gives the greatest number of packageable molecules. Although this ratio is theoretically 2:1 (arms:inserts) (see page 286), some of the molecules may lack a cohesive terminus, so that the effective concentration of ends available for ligation may be less than the calculation suggests. For the test ligations, use molar ratios of 4:1, 2:1, 1:1, and 0.5:1 (arms:inserts).

1. Carry out the test ligations. Each ligation should contain a total of 2.0 μg of DNA in a volume of 10 μl. Also, set up a control containing only 1.5 μg of arms to estimate the background caused by contamination of the arms with stuffer fragment(s) or with intact bacteriophage λ DNA. The ligations should be carried out as follows. Mix:

   DNAs
   10× ligation buffer (see page 474)   1 μl
   H$_2$O   to a final volume of 10 μl

   Remove an aliquot (1 μl) from each reaction and add to 10 μl of TE (pH 7.9). Reserve for later analysis by gel electrophoresis.

2. Add bacteriophage T4 DNA ligase (20–200 Weiss units) to the remainder of the reaction and incubate for 12–16 hours at 12°C.

3. Remove another aliquot (1 μl) from each of the ligation reactions and add to 10 μl of TE (pH 7.9). Heat the aliquots, together with those set aside in step 1, for 5 minutes at 68°C in order to denature any unligated bacteriophage λ cohesive ends. Analyze the samples by electrophoresis through a 0.4% agarose gel, using as markers multimers of pBR322 (see page 474) and intact λ DNA.

4. If the ligation reactions have been successful, almost all of the DNA should be at least as large as intact bacteriophage λ DNA. Use 3 μl of each ligation reaction as the substrate for an in vitro packaging reaction (see Chapter 8). Be sure to carry out one mock packaging reaction, containing no DNA, to measure any background that might be contributed by the extracts. Also include a packaging reaction with your standardized preparation of bacteriophage λ DNA.

5. Measure the titer of bacteriophages in each of the packaging reactions as described on page 64.

6. Calculate the number of recombinant bacteriophage particles produced in each packaging reaction. From these results, determine the optimal ratio of arms to potential inserts in the ligation reaction. Use this ratio in the subsequent large-scale ligation/packaging reaction.

Notes

i. Save the test ligations and packaging reactions. Often they yield almost enough recombinant bacteriophages for a library.

ii. Some bacteriophage λ vectors contain as their stuffer fragment a segment of *E. coli* DNA coding for β-galactosidase. Such vectors form blue plaques when plated on lawns of *lac⁻* bacteria in the presence of the chromogenic substrate 5-bromo-4-chloro-3-indolyl-β-D-galactoside (Xgal). Recombinant bacteriophages in which the stuffer fragment has been replaced by a segment of foreign DNA give rise to conventional white plaques. It is therefore possible to use plaque color as a test to distinguish between recombinant bacteriophages and the parental vector.

   a. Use a bacteriophage λ sensitive, *lac⁻* strain of *E. coli* (e.g., CSH 18) to make a suspension of indicator bacteria (see page 63).

   b. Dissolve Xgal in dimethyl formamide at a concentration of 20 mg/ml. Store the stock solution at 4°C.

   c. Melt top agar or top agarose (in NZYCM or LB) in the normal way. Cool to 47°C. Add Xgal to a final concentration of 40 μg/ml (a 1:500 dilution of the stock solution).

   d. Plate the recombinant bacteriophage stock on the *lac⁻ E. coli* host in top agar or top agarose containing Xgal. The bottom agar does not need to contain Xgal. In parallel, titrate stocks of known *lac⁺* (e.g., Charon 4A) and *lac⁻* (e.g., Charon 21A) bacteriophages.

   e. Examine the plates after 9–15 hours of incubation at 37°C. *lac⁻* bacteriophages form colorless plaques; *lac⁺* bacteriophages form blue plaques.

Large-scale Ligation and Packaging

1. Set up a large-scale ligation reaction using the conditions determined in step 6, page 289. Again include a control of only vector arms. Take aliquots before and after ligation for analysis by gel electrophoresis. While the test gel is running, store the large-scale reaction at 4°C.

2. If the ligation is successful, package the ligated DNA, using as many aliquots of packaging extract as necessary. Be sure to work with only two or three packaging extracts at any one time so that the extracts do not thaw completely before the ligated DNA is mixed in. Carry out one mock packaging reaction, containing no DNA, to measure any background that may be contributed by the extracts. Also include a packaging reaction with your standardized preparation of bacteriophage λ DNA. Measure the titer of bacteriophages in each of the packaging reactions as described on page 64.

3. Calculate the total number of recombinant bacteriophages produced and their concentration in the packaging extract. If there are more than approximately 400,000 recombinant bacteriophages per milliliter, the library is sufficiently concentrated to be amplified directly (see pages 293ff). If need be, the recombinant bacteriophages can be purified and concentrated by centrifugation through cesium chloride step gradients, as described on the following pages.

Concentration of Recombinant Bacteriophages by Centrifugation through Cesium Chloride Step Gradients

1. Combine the packaging reactions and remove the cellular debris by centrifugation at 5000*g* for 5 minutes at 4°C. Measure the volume of the supernatant. Add 0.5 g of solid cesium chloride per milliliter of supernatant and adjust the volume to 7.5 ml or to a multiple of 7.5 ml, using SM containing 0.5 g of solid cesium chloride per milliliter.

2. In an ultracentrifuge tube (Beckman SW41 or equivalent), prepare a step gradient composed of three 1.5-ml steps (1.7, 1.5, and 1.45 g/ml CsCl in SM; see page 81). Mark the outside of the tube opposite the interface between the 1.5- and 1.45-g/ml steps.

3. Carefully load the bacteriophage suspension on top of the step gradient. Centrifuge at 32,000 rpm for 1.5 hours at 4°C. The bacteriophage particles will collect at the interface between the 1.5- and 1.45-g/ml CsCl solutions. However, the number of bacteriophage particles is too small to form a visible band. Therefore, puncture the tube near the bottom of the 1.5-g/ml step and collect 0.4-ml fractions. Locate the bacteriophage particles by spotting 5 μl of a 10^{-3} dilution of each fraction onto a lawn of indicator bacteria.

4. Pool the fractions containing bacteriophage particles and add sterile gelatin to a concentration of 0.02%. Dialyze overnight at 4°C against a solution of 0.1 M NaCl, 50 mM Tris · Cl (pH 7.5), and 10 mM $MgSO_4$ with one change of buffer. Use dialysis tubing that has been well rinsed with distilled water, autoclaved, and rinsed once with sterile dialysis buffer.

5. Recover the bacteriophage suspension from the dialysis bag and wash the bag with a small volume of SM. Measure the titer of the library, which is now ready for amplification.

AMPLIFICATION OF THE LIBRARY

The library of recombinant bacteriophages is amplified by growing a plate stock (see page 65) directly from the packaging mixture or from the concentrated bacteriophage suspension obtained by cesium chloride density gradient centrifugation (Maniatis et al. 1978).

1. Mix aliquots of the packaging mixture or the concentrated bacteriophage suspension containing 10,000–20,000 recombinant bacteriophages in a volume of 50 μl or less with 0.2 ml of plating bacteria (see page 63). Incubate at 37°C for 20 minutes.

2. Mix 6.5 ml of melted top agar or agarose with each aliquot of infected bacteria and spread onto a freshly poured, 150-mm plate of bottom agar.
   Alternatively, as many as 450,000 bacteriophages may be mixed with 14 ml of bacteria and plated in 75 ml of top agar or agarose onto 500 ml of bottom agar in a 23-cm × 33-cm glass baking dish.

3. Incubate at 37°C for a maximum of 8–10 hours. Do not allow the plaques to grow so large that they touch. This short period of growth minimizes (1) the chances for infection of bacteria with two different recombinants, thereby reducing the possibility of recombination between repetitive sequences carried by different recombinants with consequent "scrambling" of the library, and (2) the opportunity for changes in the bacteriophage population that may occur because of differences in the rate of growth of different recombinants.

4. Overlay the plates with 12 ml of SM (or 150 ml of SM if baking dishes are used). Store the plates at 4°C overnight in a level place.

5. Recover the bacteriophage suspension from each plate and transfer the suspension to a sterile polypropylene tube. Rinse the plate with 4 ml of SM. Add chloroform to 5%. Incubate for 15 minutes at room temperature with occasional shaking.

6. Remove cell and agar debris by centrifugation at 4000g for 5 minutes at 4°C.

7. Recover the supernatant to a glass tube or bottle. Add chloroform to 0.3% and store in aliquots at 4°C. The titer of the library will be stable for several years.

Note

Under some circumstances, it may be advisable to omit the amplification step and to screen the recombinant bacteriophages in the packaging mixture directly for those that contain the DNA sequences of interest. If, for example, the recombinant you seek grows poorly, it may be heavily underrepresented in the amplified library. Such recombinants, even those that form minute plaques, can sometimes be isolated if the amplification step is omitted. Packaging mixtures containing recombinant bacteriophages are plated out and the resulting plaques are screened directly by hybridization, as described in Chapter 10.

Construction of Genomic Libraries in Cosmid Vectors

Many of the procedures used to construct genomic DNA libraries in bacteriophage λ vectors also apply to cloning in cosmid vectors. In both systems, large segments of eukaryotic DNA, generated by quasirandom fragmentation, are ligated with vector DNA to form catenates that can be packaged into bacteriophage λ particles. As we have discussed, libraries constructed in bacteriophage λ vectors are stored and propagated in the form of such recombinant bacteriophage particles. In cosmid cloning, however, these particles merely serve as vehicles by which the recombinant DNA molecules are efficiently introduced into bacteria where they are propagated as large plasmids. Libraries constructed in cosmids, therefore, are maintained within populations of transformed bacteria. If each bacterium in the population carries a different recombinant cosmid, about 350,000 transformants must be generated and maintained in order to achieve a 99% probability that a particular, single-copy sequence of eukaryotic DNA will be represented in the library (see equation on page 271).

The current method of choice for constructing libraries of eukaryotic DNA in cosmid vectors is to clone DNA fragments obtained by partial digestion of high-molecular-weight DNA with restriction enzymes that recognize a 4-bp sequence and generate a cohesive end. In the two protocols given below, genomic DNA partially digested with *Mbo*I or *Sau*3A is cloned into the *Bam*HI site of the cosmid pJB8 (see Chapter 1). The two approaches differ in the method used to prevent self-concatamerization and packaging of the vector DNA. In one case (Meyerowitz et al. 1980; Grosveld et al. 1981), the linearized vector DNA is simply treated with phosphatase; in the other (Ish-Horowicz and Burke 1981), which involves more enzymatic steps, a modified form of directional cloning is used. Although both methods have been used to construct libraries of mammalian DNAs, the second is the more efficient.

CLONING IN PHOSPHATASE-TREATED COSMID VECTORS

The basic steps of this procedure are shown in Figure 9.5

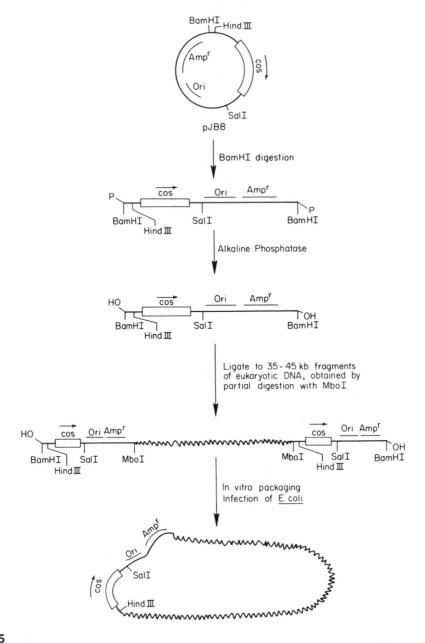

Figure 9.5

Cloning in phosphatase-treated vectors. DNA of the plasmid pJB8 is digested with *Bam*HI and dephosphorylated with alkaline phosphatase to yield a vector with protruding 5′ termini that can be ligated to 35–45-kb fragments of eukaryotic DNA generated by partial digestion with *Mbo*I or *Sau*3A. The resultant concatamers serve as substrates for in vitro packaging of bacteriophage λ particles. Following introduction into *E. coli*, the cosmid DNA recircularizes and replicates in the form of a large plasmid. The plasmid contains a β-lactamase gene that confers resistance to ampicillin on the host bacterium.

Preparation of Vector DNA

1. Digest 20 μg of closed circular pJB8 DNA, prepared according to methods given in Chapter 3, with a twofold to threefold excess of *Bam*HI for 1 hour. Stop the reaction by cooling to 0°C. Remove an aliquot (0.3 μg) and analyze by electrophoresis through a 0.8% gel, together with 0.3 μg of undigested pJB8 DNA. If digestion is incomplete, warm the reaction mixture to 37°C, add more restriction enzyme, and continue the incubation.

2. When digestion is complete, extract the sample with phenol/chloroform and precipitate the DNA with ethanol. Resuspend the DNA in 200 μl of 10 mM Tris·Cl (pH 8.0). Remove a 5-μl aliquot (0.5 μg) and store at -20°C.

3. Add 200 units (ATP hydrolysis units) of calf intestinal phosphatase (see pages 133ff) to the remainder of the sample. Incubate for 30 minutes at 37°C.

4. Add 1 μl of 0.5 M EDTA. Extract the DNA three times with phenol and once with chloroform and precipitate with ethanol.

5. Resuspend the DNA in 20 μl of TE (pH 8.0) and carry out a test ligation to determine the effectiveness of the phosphatase treatment. Set up three reactions containing:

   a. 0.5 μg of phosphatase-treated DNA;

   b. 0.5 μg of phosphatase-treated DNA plus 0.5 μg of bacteriophage λ DNA cleaved with *Bam*HI;

   c. 0.5 μg of pJB8 DNA cleaved with *Bam*HI but not treated with phosphatase (i.e., the aliquot set aside at step 2).

To each reaction add:

10× ligation buffer (see page 474)	1 μl
H$_2$O	to 10 μl

Remove an aliquot (2 μl) from each reaction and store at 4°C. Add 0.5 units of bacteriophage T4 DNA ligase to the remainder of the reactions and incubate for 4–8 hours at 12°C. Remove another 2-μl aliquot from each sample and assay all six aliquots by electrophoresis through a 0.7% agarose gel. Store the remainder of the reactions at -20°C.

The phosphatase-treated sample should show no evidence of ligation, whereas most of the untreated sample should have been converted into multimers. Because dephosphorylated, single-stranded, protruding ends of DNA can be ligated to complementary cohesive termini that carry a 5′ phosphate, at least some of the phosphatase-treated cosmid DNA should have become ligated to the fragments of bacteriophage λ/*Bam*HI DNA.

If the results of the test ligations are satisfactory, dispense the dephosphorylated DNA into aliquots and store at -20°C.

Partial Digestion of Eukaryotic DNA with *Mbo*I or *Sau*3A

Prepare the eukaryotic DNA for cloning by partial digestion with *Mbo*I or *Sau*3A by using the procedures described in detail on pages 282ff. Because large fragments of DNA are required for cloning in the cosmids selected, the eukaryotic DNA must be very large ($\geqslant$ 100 kb) before digestion. Following partial restriction endonuclease digestion, select DNA in the size range of 30–45 kb by electrophoresis through a 0.3–0.4% agarose gel. Intact bacteriophage λ DNAs of various sizes or pBR322 multimers should be used as markers.

Ligation and Packaging

To ensure that both ends of each potential insert become attached to cosmid sequences, ligation should be carried out with the phosphatase-treated vector DNA in 10-fold molar excess over the 30–45-kb fragments of eukaryotic DNA. To promote the formation of concatamers, the DNA concentration in the ligation reaction should be greater than 200 μg/ml (see page 286).

1. Mix:

phosphatase-treated vector DNA	1.5 μg
35–45-kb fragments of eukaryotic DNA	3 μg
10× ligation buffer (see page 474)	2 μl
H$_2$O	to 20 μl

The final concentration of DNA is 225 μg/ml. Remove an aliquot (1 μl) and store at 4°C. Add 20–200 Weiss units of bacteriophage T4 DNA ligase to the remainder of the reaction and incubate overnight at 12°C.

2. At the end of the ligation reaction, remove another 1-μl aliquot and analyze it by electrophoresis through a 0.4% agarose gel together with the aliquot set aside in step 1. If ligation was successful, some of the eukaryotic DNA should have been converted to high-molecular-weight concatamers. Most of the vector DNA should remain unligated.

3. Package the ligated DNA into bacteriophage λ particles, as described on pages 289–290, using no more than 0.5 μg of DNA per packaging reaction. As controls, package the samples set aside from step 5, page 297.

   For cloning in cosmids, packaging extracts should be prepared either according to protocol II or according to protocol I (see Chapter 8), using a buffer that contains spermidine but not putrescine. When putrescine is omitted from the packaging extract and the packaging reaction, the system exhibits selectivity in the size of DNA molecules that are packaged. Thus, DNA that is 80% of wild-type bacteriophage λ DNA in length is packaged 200-fold less efficiently than wild-type λ DNA itself. In the absence of putrescine, only cosmids that contain large inserts will be packaged.

4. After packaging is complete, add 0.5–1.0 ml of SM to each reaction. Store at 4°C. Mix a 10-μl aliquot of each reaction with 0.1 ml of SM and 0.2 ml of a fresh overnight culture of *E. coli* 1046 or DH1 grown in the presence of 0.2% maltose. 1046 and DH1 are less "leaky" *recA*⁻ strains than HB101 and may therefore reduce recombination between repetitive sequences of cloned eukaryotic DNA. Allow the bacteriophage particles to adsorb by incubating for 20 minutes at 37°C. Add 1 ml of L broth and continue the incubation for a further 45 minutes.

   Spread 0.5 ml and 0.1 ml of the bacterial culture onto agar plates containing ampicillin. After incubating the plates overnight at 37°C, count the number of bacterial colonies. Each microgram of ligated cosmid-eukaryotic DNA should yield 5×10^3 to 5×10^4 bacterial colonies.

5. Pick a number of individual colonies and grow small-scale (5-ml), overnight cultures. Isolate the plasmid DNA from 4–5 ml of each bacterial culture, using the alkaline lysis method (page 368). Digest the plasmid DNA with restriction enzyme(s) and analyze the size of the resulting fragments by gel electrophoresis.

CLONING IN COSMID VECTORS DIGESTED WITH TWO RESTRICTION ENZYMES AND TREATED WITH PHOSPHATASE

The principles of this procedure, which was developed by Ish-Horowicz and Burke (1981), were described in Chapter 1. The cosmid vector is digested in two separate aliquots with restriction enzymes (*Hind*III and *Sal*I) that cleave on either side of the *cos* site. The protruding, single-stranded termini of the resulting molecules are rendered incapable of ligation by any one of several enzymatic treatments (alkaline phosphatase is most commonly used, but nuclease S1 and the end-filling reaction of the Klenow fragment of *E. coli* DNA polymerase also work). The linear DNA molecules are then digested with *Bam*HI, and the fragments containing the *cos* sequence are isolated by gel electrophoresis. These fragments are ligated to fragments of eukaryotic DNA generated by partial cleavage with *Mbo*I or *Sau*3A to form simple concatamers that serve as substrates in the packaging reaction. Prior destruction of the protruding, single-stranded termini created by the first enzyme prevents the formation of more complex concatamers.

In their original procedure, Ish-Horowicz and Burke did not select DNA fragments of a given size to insert into the cosmid vector. Instead, they digested high-molecular-weight, eukaryotic DNA with *Mbo*I until the average size of the digestion products was 35–45 kb. The DNA was then dephosphorylated by treatment with alkaline phosphatase primarily to prevent the joining together of smaller fragments of DNA to form molecules of a size that could be packaged into bacteriophage λ particles. We have modified this procedure by including a sizing step and by omitting the alkaline phosphatase treatment. These changes improve the efficiency of the system and reduce the potential for production of cosmids carrying sequences of DNA that are noncontiguous in the original eukaryotic genome.

This modified approach is outlined in Figure 9.6 and described in detail on the following pages.

Figure 9.6

Cloning in cosmid vectors digested with two restriction enzymes. This procedure is a modification of the protocol of Ish-Horowicz and Burke (1981). Following digestion of pJB8 DNA with *Hind*III or *Sal*I and inactivation of the protruding termini (e.g., by alkaline phosphatase), the linear vector DNAs are cleaved with *Bam*HI. The two vector fragments containing the *cos* sequence are isolated and ligated to 35–45-kb fragments of eukaryotic DNA generated by partial digestion with *Mbo*I or *Sau*3A. Note that an entire complement of plasmid sequences is contained between the two *cos* sites. The concatamers are used as substrates for in vitro packaging of bacteriophage λ particles. Following introduction into *E. coli*, the cosmid DNA recircularizes and replicates in the form of a large plasmid. The plasmid contains a β-lactamase gene that confers resistance to ampicillin on the host bacterium.

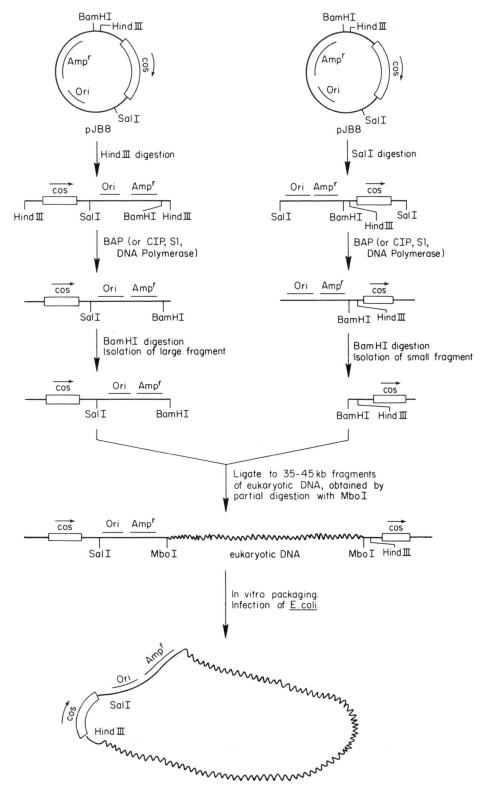

Figure 9.6

(*See legend on facing page.*)

Preparation of Vector DNA

1. Digest one 20-µg aliquot of closed circular pJB8 DNA with *Hin*dIII and a second 20-µg aliquot with *Sal*I. Use a twofold to threefold excess of the restriction enzymes for 1 hour at 37°C. Stop the reactions by cooling to 0°C. Remove a 0.3-µg aliquot from each reaction and analyze by gel electrophoresis through a 0.8% agarose gel together with 0.3 µg of undigested pJB8 DNA. If digestion is incomplete, warm the reactions to 37°C. Add more restriction enzyme(s) and continue the incubation.

2. When digestion is complete, extract the sample with phenol/chloroform and precipitate the DNA with ethanol. Resuspend each sample of DNA in 200 µl of 10 mM Tris · Cl (pH 8.0). Remove a 5-µl aliquot (0.5 µg) from each sample and store at −20°C.

3. Add 200 ATP hydrolysis units of alkaline phosphatase (see page 133) to the remainder of the samples. Incubate at 37°C (CIP) or 68°C (BAP).

   Note. Depending on the cosmid vector and the restriction enzymes used, other methods of inactivating the cohesive ends are possible (nuclease S1 or end-filling reactions with the Klenow fragment of *E. coli* DNA polymerase). For example, dephosphorylation of recessed 5'-phosphoryl termini is relatively inefficient, so that protruding 3' termini should be destroyed by digestion with nuclease S1 or mung-bean nuclease.

4. Add 1 µl of 0.5 M EDTA. Extract the sample three times with phenol and once with chloroform, and precipitate with ethanol.

5. Resuspend each DNA in 50 µl of TE (pH 8.0) and set up a series of three test ligations as described in step 5, page 297, to test the effectiveness of the phosphatase treatment. (Note that the bacteriophage λ DNA fragments in reaction C should be generated by digestion with *Hin*dIII or *Sal*I.) While carrying out the test ligations, store the remainder of the DNA samples at −20°C.

6. If the results of the test ligations are satisfactory, thaw the DNA samples and add to each:

10× restriction enzyme buffer	10 µl
H₂O	40 µl
twofold to threefold excess of *Bam*HI	

   Incubate at 37°C for 1 hour. Assay an aliquot of each reaction by electrophoresis through a 0.8% agarose gel.

7. If digestion is complete, separate the DNA fragments in the remainder of the samples by electrophoresis through a 0.8% agarose gel. Isolate the larger of the two *Bam*HI/*Hin*dIII fragments and the smaller of the two *Bam*HI/*Sal*I DNA fragments by electroelution (see pages 164ff).

8. Resuspend the recovered DNA fragments in 20 μl of TE (pH 7.9). Estimate the amount of DNA in the samples by ethidium bromide fluorescence (see page 468).

Partial Digestion of Eukaryotic DNA with MboI or Sau3A

Prepare the eukaryotic DNA for cloning by (1) partial digestion with MboI or Sau3A (see pages 282ff), and (2) isolation of 35–45-kb fragments by gel electrophoresis (see pages 164ff).

Ligation and Packaging

The total concentration of DNA in the ligation reaction should be greater than 120 μg/ml in order to favor the formation of mixed concatamers between the cosmid vectors and eukaryotic DNA (see the discussion on pages 286ff). Furthermore, the molar ratio of vector molecules to potential inserts should be 1:1:1 since the desired concatamer is vector 1 (BamHI/HindIII)–insert–vector 2 (BamHI/SalI).

1. Set up two ligation reactions as follows:

   a. 35–45-kb fragments of eukaryotic DNA   3 μg
       HindIII/BamHI fragment   0.15 μg
       SalI/BamHI fragment   0.15 μg
       10× ligation buffer (see page 474)   2 μl
       H$_2$O   to 20 μl

   b. HindIII/BamHI fragment   0.3 μg
       SalI/BamHI fragment   0.3 μg
       10× ligation buffer   2 μl
       H$_2$O   to 20 μl

   Remove an aliquot (2 μl) from each reaction and store at 4°C. Add 20–200 Weiss units of bacteriophage T4 DNA ligase to each reaction. Incubate at 12°C overnight.

2. At the end of the ligation reaction, remove another aliquot (2 μl) from each reaction and analyze them, together with the aliquots reserved in step 1 above, by electrophoresis through a 0.4% agarose gel. If ligation was successful, some of the eukaryotic DNA should have been converted to high-molecular-weight concatamers.

3. Proceed with packaging and amplification as described on pages 298–299 and 304–307.

AMPLIFICATION, STORAGE, AND SCREENING OF COSMID LIBRARIES

Following packaging into bacteriophage λ particles, cosmids are stable for long periods of time at 4°C. However, a permanent cosmid library can be established and maintained only in bacteria. The methods given below are intended to keep the library in a viable state while minimizing the opportunity for changes in the bacterial population that may occur because of differences in the rate of growth or viability of bacteria carrying different cosmids.

Replica Filters

In this method, bacterial colonies are grown on nitrocellulose filters laid on agar plates ($\sim 10^4$ colonies/150-mm filter; 30–50 filters/library) and frozen in situ essentially as described by Hanahan and Meselson (1980). Replicas are then prepared from the set of master filters for screening and propagation of the library.

1. Prepare agar plates with nitrocellulose filters (Millipore, Triton-free, HATF). For 85-mm diameter plates, 82-mm filters are used, and 127-mm filters are used for 150-mm plates. Dry filters are numbered with a soft pencil or ultra-ball-point pen. The filters are floated on the surface of a bath of water until they are wet, submerged, sandwiched between dry Whatman 3MM filters, wrapped in aluminum foil, and sterilized by autoclaving (30 minutes at 15 lb/in^2 on liquid cycle). The pack of sterilized filters is stored at room temperature in a sealed plastic bag. A sterile filter, picked up with sterile forceps, is laid on the surface of a day-old agar plate, peeled off, inverted, and replaced on the plate, numbered side up.

2. Mix an aliquot of the packaging mixture containing $\sim 2 \times 10^4$ packaged cosmids with 200 μl of a fresh overnight culture of *E. coli* 1046 or DH1 grown in the presence of 0.2% maltose.

   Incubate at 37°C for 20 minutes. If large amounts of packaging mixture inhibit attachment of the bacteriophage particles to the bacteria, or if the concentration of packaged cosmids is low, it may be necessary to carry out a preliminary step in which the bacteriophage particles are purified by centrifugation through cesium chloride step gradients (see page 292).

3. Add 1 ml of L broth to each infected culture. Continue incubation for a further 45 minutes at 37°C.

4. Apply 500 μl of the infected cell culture to the center of a numbered filter on the surface of an agar plate. Using a sterile glass spreader, disperse the fluid evenly over the surface of the filter. Leave a border 2–3 mm wide at the edge of the filter free of bacteria. Incubate the plate at 37°C until small colonies (0.1–0.2-mm diameter) appear (about 8–10 hours).

5. If desired, replica filters may be prepared at this stage (see step 7 below). Otherwise the filter should be transferred to a plate containing agar medium with 25% glycerol. Incubate the plate for 2 hours at 37°C.

6. Seal the plates well with parafilm, wrap them in a sealed plastic bag, and store them in an inverted position at −20°C. Replicas can be made after thawing the master plates at room temperature (still in an inverted position).

7. Replica filters are made as follows:

   a. Prepare in advance a stack of sterile Whatman 3MM filters (one for each filter plus a few spares).

   b. Peel the master nitrocellulose filter off the storage plate and lay it colony side up on a dampened pad of sterile, 3MM filter paper.

   c. Number a damp, sterile nitrocellulose filter and lay it on the master nitrocellulose filter.

   d. Press the two filters firmly together with a velvet replica-plating tool (see Fig. 9.7).

   e. Orient the two filters by placing a series of holes made in the pair of filters with an 18-gauge needle.

   f. Peel apart the filters. Lay the replica on a fresh agar plate and incubate at 37°C until colonies appear.

   g. Replace the master nitrocellulose filter on a fresh agar plate containing 25% glycerol. Incubate for 1 hour at 37°C and then freeze as described in step 6 above.

   Replica filters can be (1) used to replicate again, (2) stored at −20°C, or (3) used for screening colonies by in situ hybridization.

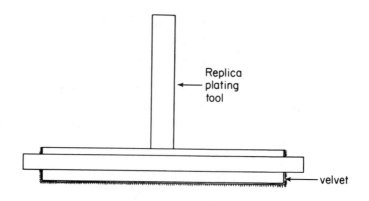

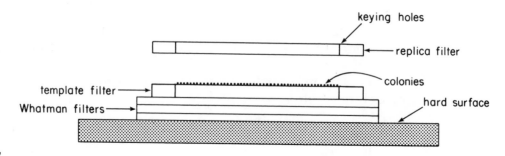

Figure 9.7

Amplification of Cosmid Libraries in Liquid Culture

Growing cosmid libraries in liquid-culture is simpler and more convenient than serial replication of bacterial colonies on nitrocellulose filters. However, because the bacteria are grown as a mixed population in liquid, there is a possibility that the composition of the library may change because of differences in the rate of growth of bacteria carrying different cosmids.

1. Infect 200-μl aliquots of an overnight culture of *E. coli* strain 1046 with approximately 2×10^4 of the packaged cosmids as described in step 2 (page 304).

2. Add 20 ml of L broth containing 25 μg/ml of ampicillin to each infected aliquot. Incubate with shaking at 37°C until the culture has reached mid-log phase.

3. Add sterile 100% glycerol to a final concentration of 15%, mix thoroughly, and dispense 1-ml aliquots of the cells to sterile vials. Store at -20°C. The cells are viable for more than a year when prepared in this manner.

4. To screen the library by hybridization, an aliquot from each stock is plated, either on nitrocellulose filters or on agar, as described in Chapter 10. pJB8 contains the ColE1 replicon, and replication is therefore relaxed. Nevertheless, the number of copies per cell of cosmids is quite small because of their huge size. Amplification with chloramphenicol is therefore advisable in order to obtain strong hybridization signals.

10

The Identification of
Recombinant Clones

Three procedures are commonly used to identify recombinant plasmid or λ bacteriophages that carry eukaryotic DNA sequences of interest.

The first, and by far the most general technique, is in situ hybridization of bacterial colonies or bacteriophage plaques to specific probes of ^{32}P-labeled DNA or RNA. The method is fast and can be carried out on a mass scale; as many as 5×10^5 to 1×10^6 plaques or colonies can be screened in a single experiment, and only 1–2 weeks are required to proceed from the initial step of plating the recombinants to the final stage of purification of the bacteriophage or colony of interest.

In the second method, cDNA clones carrying sequences complementary to specific mRNAs are identified by hybridization selection. The cloned DNAs are denatured, immobilized on a solid matrix, and hybridized to preparations of mRNA. The RNA·DNA duplex is heated to release the mRNA, which is then translated in cell-free, protein-synthesizing systems or in *Xenopus* oocytes. The translation products are identified by immunoprecipitation and/or SDS-polyacrylamide gel electrophoresis or, in rare cases, by biological assays.

Although this second procedure is often used to confirm the identification of cDNA clones isolated by other criteria, it was recently used to isolate directly, from a library of mouse cDNA, clones coding for β_2-microglobulin, a protein whose mRNA represents as little as 0.03% of the total cellular mRNA (Parnes et al. 1981). After enriching the mRNA 10-fold for β_2-microglobulin-specific sequences by sucrose density gradient centrifugation, a cDNA library of some 10^4 clones was constructed and divided into pools, each containing 14 independently derived clones. DNA prepared from four of these pools hybridized to RNA that could be translated in vitro into immunoprecipitable β_2-microglobulin. Individual β_2-microglobulin-specific clones were then isolated from the pools by subcloning. This entire procedure is labor-intensive and probably is worth attempting only with mRNAs that can be greatly enriched by physical fractionation. Arduous and limited though the method is, it is nevertheless the most generally applicable procedure for identifying cDNA clones corresponding to low-abundance mRNAs. The practicality of the method is somewhat enhanced by the fact that crude "minipreps" of plasmid DNAs can be used as hybridization probes.

Third, a rapid and sensitive method for screening bacteriophage λ libraries by recombination in *E. coli* has been developed (B. Seed, unpubl.). In this procedure, the DNA sequence that is to be used as a probe is cloned into a small plasmid (πVX) that carries as a selectable marker the tyrosine tRNA amber-suppressor. Bacteria are transformed by the recombinant plasmid and are then infected with a population of bacteriophages carrying a library of eukaryotic DNA sequences. The bacteriophage λ vector used in the construction of this library carries amber mutations in its arms. If the DNA cloned in the plasmid shares sequences with the DNA inserted into the bacteriophage genome, homologous recombination can occur and generate new bacteriophages that carry a copy of the tyrosine tRNA suppressor gene. Such bacteriophages are easily isolated by virtue of their ability to grow in strains of *E. coli* that lack an amber suppressor. By contrast to the first two techniques, the πVX system requires that a segment of DNA homologous to the desired gene be already available in cloned form. It is therefore useful only as a secondary screening technique.

In Situ Hybridization of Bacterial Colonies or Bacteriophage Plaques

Colony hybridization (Grunstein and Hogness 1975) is accomplished by transferring bacteria from a master plate to a nitrocellulose filter. The colonies on the filter are then lysed and the liberated DNA is fixed to the filter by baking. After hybridization to a ^{32}P-labeled probe, the filter is monitored by autoradiography. A colony whose DNA gives a positive autoradiographic result may then be recovered from the master plate.

In plaque hybridization (Benton and Davis 1977), the nitrocellulose filter is applied to the surface of a plate containing bacteriophage plaques so that there is direct contact between the plaques and the filter. Molecules of unpackaged bacteriophage DNA present in the plaques bind to the filter and are hybridized as described above.

SCREENING SMALL NUMBERS OF BACTERIAL COLONIES

This procedure (Grunstein and Hogness 1975) is used when it is necessary to screen small numbers (100–200) of colonies that are dispersed over several agar plates. The colonies are simultaneously consolidated onto a master agar plate and onto a nitrocellulose filter laid on the surface of a second agar plate. After a period of growth, the colonies on the nitrocellulose filter are lysed with alkali. After neutralization, the denatured plasmid DNA is fixed to the filter by baking. The master plate is stored at 4°C until the results of the screening procedure become available.

1. Place a nitrocellulose filter (Millipore HAWP) onto an agar plate containing the selective antibiotic. It is not necessary to use Triton-free or sterilized filters in this procedure; such filters need only be used when small numbers of bacteria are used as inocula. However, filters should be handled with gloved hands since finger oils prevent wetting of the filter and affect DNA transfer.

2. Using sterile toothpicks, transfer the individual bacterial colonies to be screened onto the filter and then onto a master agar plate that contains the selective antibiotic but no filter. Make small streaks (2–3 mm in length) or dots arranged in a grid pattern. Each colony should be streaked in an identical position on both plates. Up to 100 colonies can be streaked onto a single 85-mm plate. Finally, streak a colony containing a nonrecombinant plasmid (e.g., pBR322) onto both the filter and the master plate. This negative control is often useful and sometimes necessary to discriminate between specific annealing of the radioactive probe to a recombinant plasmid and nonspecific background hybridization.

3. Invert the plates and incubate at 37°C until the bacterial streaks have grown to 0.5–1.0 mm in width. At this stage, when the bacteria are still growing rapidly, the filter may be transferred to an agar plate containing chloramphenicol (10 μg/ml). Incubate for a further 12 hours at 37°C. This amplification step is necessary only when the copy number of the recombinant plasmids is expected to be low (e.g., if an unusually large segment [>10 kb] of foreign DNA has been inserted). Under normal circumstances, cloned DNA sequences can be detected very easily by hybridization without prior amplification of the recombinant plasmid.

4. Mark the nitrocellulose filter in three or more asymmetric locations by stabbing through it and into the agar beneath with a 18-gauge needle attached to a syringe containing waterproof black drawing ink (e.g., Higgins India Ink). Mark the master plate in approximately the same three locations.

5. Seal the master plate with parafilm and store it at 4°C in an inverted position until the results of the hybridization reaction are available.

6. Lyse the bacteria and bind the liberated DNA to the nitrocellulose filter by whichever of the two following procedures you find more convenient. Try to avoid (1) getting any of the solutions on the upper surface of the filter, and (2) trapping air bubbles under the filter.

Binding Liberated DNA—Procedure I

1. Cut four pieces of Whatman 3MM paper so that they fit neatly on the bottom of four Pyrex baking dishes (20 cm × 20 cm). Saturate one piece of paper with 10% SDS. Pour off any excess liquid.

2. Using blunt-ended forceps (e.g., Millipore forceps), peel the nitrocellulose filter from the plate and place it, colony side up, on the SDS-impregnated 3MM paper for 3 minutes. This treatment, which is optional, seems to result in a sharper hybridization signal. It probably works by limiting the diffusion of the plasmid DNA during denaturation and neutralization (E. F. Fritsch and P. Boyer, unpubl.).

3. Transfer the filter to the second sheet of 3MM paper that has been saturated with denaturing solution (0.5 M NaOH, 1.5 M NaCl). Leave the filter for 5 minutes. When transferring filters from one Pyrex tray to another, use the edge of the first tray as a scraper to remove as much fluid as possible from the underside of the filter.

4. Transfer the filter to the third sheet of 3MM paper that has been saturated with neutralizing solution (1.5 M NaCl, 0.5 M Tris·Cl [pH 8.0]). Leave the filter for 5 minutes.

5. Transfer the filter to the fourth sheet of 3MM paper that has been saturated with 2× SSPE. Leave the filter for 5 minutes.
   SSPE is usually made up as a 20× stock solution.

   20× SSPE

   3.6 M NaCl
   200 mM NaH_2PO_4 (pH 7.4)
   20 mM EDTA (pH 7.4)

6. Lay the filter, colony side up, on a sheet of dry 3MM paper. Allow to dry at room temperature for 30–60 minutes.

7. Sandwich the filter between two sheets of dry 3MM paper. Bake for 2 hours at 80°C in a vacuum oven.

8. Hybridize the filter to a ^{32}P-labeled probe as described on page 320.

Binding Liberated DNA—Procedure II[1]

1. Make a puddle (0.75 ml) of 0.5 M NaOH on a piece of Saran Wrap. Place the filter on the puddle, stretching the Saran Wrap so that the filter wets evenly. Leave for 2–3 minutes.

2. Blot the filter on a dry paper towel and repeat step 1, using a fresh piece of Saran Wrap and fresh 0.5 M NaOH.

3. Blot the filter and transfer to a fresh piece of Saran Wrap with 0.75 ml of 1 M Tris·Cl (pH 7.4). After 5 minutes, blot the filter dry and repeat.

4. Blot the filter and transfer to Saran Wrap with 0.75 ml of a solution of 1.5 M NaCl and 0.5 M Tris·Cl (pH 7.4). After 5 minutes, blot and transfer the filter to a piece of dry 3MM paper. Allow the filter to dry at room temperature for 30–60 minutes.

5. Sandwich the filter between two sheets of 3MM paper. Bake for 2 hours at 80°C in a vacuum oven.

6. Hybridize the filter to a ^{32}P-labeled probe as described on page 324.

Note

Any filters not used immediately in hybridization reactions should be wrapped loosely in aluminum foil and stored under vacuum at room temperature.

[1]D. Hanahan (unpubl.).

REPLICATING COLONIES ONTO NITROCELLULOSE FILTERS

Procedure I

This procedure (Hanahan and Meselson 1980) is used when you know ahead of time that you will need to screen large numbers of colonies. In this case, bacteria are plated directly from a transformation mixture onto detergent-free nitrocellulose filters; replica filters are prepared by filter-to-filter contact.

1. Number the dry nitrocellulose filters with a soft pencil or a ball-point pen and sterilize them as described on page 304. Prepare enough filters for a master and two replicas.

2. Using sterile, blunt-ended forceps, lay a sterile filter, numbered side down, on a day-old agar plate containing the appropriate antibiotics. Peel off the filter, invert it, and replace it on the plate, numbered side up.

3. Apply the bacteria in a small volume of liquid (< 0.2 ml, containing up to 50,000 bacteria for a 137-mm filter; < 0.1 ml, containing up to 15,000 bacteria for an 82-mm filter). Spread the liquid over the surface of the filter with a sterile, bent glass rod. Leave a border 2–3 mm wide at the edge of the filter free of bacteria. Let the plates stand at room temperature until all of the liquid has been absorbed.

4. Invert the plates and incubate at 37°C until very small colonies (0.1-mm diameter) appear (about 8–10 hours).

5. Prepare in advance a stack of Whatman 3MM papers sterilized by autoclaving (one for each filter plus a few spares)

6. Wet a numbered, sterile nitrocellulose filter (Millipore HAWP) by touching it to the surface of an agar plate containing the appropriate antibiotic. Leave the filter in contact with the surface of the agar, numbered side up. Try to arrange that the numbers on the replica filters correspond to those on the master filters.

7. Using sterile, blunt-ended forceps (e.g., Millipore forceps), gently remove the master filter from the first plate and place it on the stack of 3MM paper, colony side up.

8. Carefully place the second, wetted filter (numbered side down) on top of the master, being careful not to move filters once contact has been made. Press the filters together using sterile velvet stretched over a replica-plating tool (see Fig. 9.7, page 306).

9. Use an 18-gauge needle to make a characteristic pattern of keying holes in the filters while they are sandwiched together. Gently peel the filters apart and return the second filter to its plate, colony side up.

10. Prepare a second replica from the master in an identical manner. Key the replica to the existing holes in the master filter. Return both filters to their respective plates. If the master filter is to be used to make more than two replicas, it should be incubated for a few hours to allow the colonies to regenerate. Generally, it is best to make only two replicas from a single master in order to avoid problems caused by smearing of colonies.

11. Incubate the plates (master and replicas) at 37°C until colonies 1-2 mm in diameter have appeared. Colonies on the master plate reach the desired size more rapidly (6-8 hours).

   At this stage, while the bacteria are still growing rapidly, the replica filters may be transferred to agar plates containing chloramphenicol (10 μg/ml) and incubated for a further 12 hours at 37°C. As discussed on page 313, this amplification step is necessary only when the copy number of the recombinant plasmid is expected to be low.

12. Seal the master plates with parafilm and store at 4°C in an inverted position until the results of the hybridization reaction are available.

13. Lyse the bacteria and bind the liberated DNA to the replica filters using one of the two procedures described on pages 314-315.

Procedure II

This method is used to transfer many bacterial colonies simultaneously from the surface of agar plates to nitrocellulose filters. The method works with bacterial colonies of any size, but small colonies (0.1–0.2 mm) give the best results; they produce sharper hybridization signals and smear less than larger colonies. As many as 10^4 colonies per 150-mm plate can be screened using this technique.

1. After the bacterial colonies have grown to a diameter of 0.1–0.2 mm, remove the plate from the incubator and store it for 1–2 hours at 4°C in an inverted position.

2. Label a dry nitrocellulose filter (Millipore HAWP) with a soft pencil or ball-point pen and place it, numbered side down, on the surface of the agar medium in contact with the bacterial colonies until it is completely wet. Mark the filter and underlying agar in three or more asymmetric locations by stabbing through it with an 18-gauge needle attached to a syringe containing waterproof black drawing ink (e.g., Higgins India Ink).

   Although sterile filters are to be preferred, nonsterile filters can also be used as long as the master plate is not to be used again to make replicas.

3. Using blunt-ended forceps, peel off the filter.

4. At this stage several options are available:

   a. The bacteria adhering to the filter can be lysed immediately and the liberated DNA bound to the filter using one of the two procedures described on pages 314–315.

   b. The filter can be placed, colony side up, on the surface of a fresh agar plate containing the appropriate antibiotics. After incubation for a few hours, the large bacterial colonies 2–3 mm are lysed. This method is necessary only when transfer of the colonies to the filter is poor or uneven. This is not usually the case.

   c. The filter can be placed on a fresh agar plate containing chloramphenicol (10 μg/ml). As discussed earlier, this amplification is used only when the copy number of the recombinant plasmid is expected to be low.

   d. The filter can be used to prepare a second replica. The filter is placed, colony side up, on the surface of a fresh agar plate containing the appropriate antibiotics. A second, dry nitrocellulose filter is then laid carefully on top of the first and keyed to it. The filter sandwich is incubated for several hours at 37°C and the plasmids are amplified, if desired, by further incubation on an agar plate containing chloramphenicol. The filters are kept as a sandwich during the subsequent lysis and neutralization steps but are peeled apart before the final wash (Ish-Horowicz and Burke 1981).

5. Incubate the master plate for 10–12 hours at 37°C until the colonies have regenerated. Seal the plate with parafilm and store at 4°C in an inverted position.

SCREENING BACTERIOPHAGE λ PLAQUES BY HYBRIDIZATION[2]

To screen a library of mammalian DNA (genome complexity, 3×10^9 bp), a total of at least 300,000 recombinant plaques must be examined. Table 10.1 gives the maximum number of plaques that can be screened in culture dishes of different sizes.

In the baking-dish method, the DNA from the bacteriophage plaques is transferred to one large sheet of nitrocellulose. Handling a filter of this size is difficult and sometimes requires two people. Key the filter to the agarose in many spots, since accurate alignment of the plaques with the hybridization signal can be difficult.

In the following example, the volumes given are suitable for screening approximately 50,000 plaques in a 150-mm-diameter petri dish.

1. Mix aliquots of a packaging mixture or bacteriophage λ stock containing up to 50,000 bacteriophage particles in a volume of 50 μl or less with 0.3 ml of plating bacteria (see page 64). Incubate at 37°C for 20 minutes.

2. Add 6.5 ml of molten (50°C) top agarose (0.7%) and pour onto a 150-mm agar plate. The plates *must* be dry, otherwise the layer of top agarose peels off with the filter. Usually, 2-day-old plates that have been dried for several additional hours at 37°C with the lids slightly open work well. In humid weather, however, incubation for 1 or more days at 41°C may be necessary.

   Note. Be sure to use top agarose rather than top agar since the latter peels off even more easily than the former.

3. Incubate at 37°C until the plaques reach a diameter of approximately 1.5 mm and are just beginning to make contact with one another (10–12 hours). The plate should not show confluent lysis.

4. Chill the plates at 4°C for at least an hour to allow the top agarose to harden.

TABLE 10.1. NUMBERS OF PLAQUES IN CULTURE DISHES OF VARIOUS SIZES

Size of dish	Total area (cm²)	Volume of bottom agar (ml)	Volume of indicator bacteria (ml)	Volume of top agarose (ml)	Maximum number of plaques/dish
90-mm petri dish	63.9	30	0.1	2.5	15,000
150-mm petri dish	176.7	80	0.3	6.5	50,000
200-mm × 300-mm baking dish	600	300	1.2	25	200,000

[2]Benton and Davis (1977).

5. Number dry, nonsterile nitrocellulose filters (Millipore HAWP) with a soft pencil or a ball-point pen.

6. At room temperature, place a dry nitrocellulose circle neatly onto the surface of the top agarose so that it comes into direct contact with the plaques. Be careful not to trap air bubbles. The filter should be handled with gloved hands; finger oils prevent wetting of the filter and affect transfer of DNA. Mark the filter in three or more asymmetric locations by stabbing through it and into the agar beneath with an 18-gauge needle attached to a syringe containing waterproof black drawing ink.

   Once in contact with the top agarose, the filter wets very rapidly and transfer of bacteriophage DNA occurs quickly. Therefore, do not move the filter once contact with the plate is made. The easiest way of placing the filter on the plate is to hold it by its edges, bending it slightly so that the middle of the filter makes contact with the center of the plate. Let wetting action pull the rest of the filter onto the plate.

   Make certain that the keying marks are asymmetrically placed and that both the filter and the plate are marked. There must be enough ink on the plate to be easily visible when a second filter is in place. Large blotches of ink, however, are undesirable.

7. After 30–60 seconds, use blunt-ended forceps to peel off the first filter and immerse it, DNA side up, in a shallow tray of a denaturing solution (1.5 M NaCl, 0.5 M NaOH) for 30–60 seconds. Transfer the filter into neutralizing solution (1.5 M NaCl, 0.5 M Tris · Cl [pH 8.0]) for 5 minutes. Rinse the filter in 2× SSPE (see page 314) and place it on Whatman 3MM paper to dry.

8. Place a second, dry filter onto the same plate and mark it with ink at the same locations. After 1–2 minutes, peel the filter off the plate. Denature the DNA and neutralize as described in step 7 above.

   Generally, the first filter is left in contact with the plaques for 30–60 seconds and subsequent filters are left on about 30 seconds longer or until the filter is completely wet. As many as seven replicas from a single plate have been prepared (Benton and Davis 1977).

   If any top agarose peels off the plate with the filter, remove it by gently agitating the filter in the denaturing solution.

9. After all the filters are dry, wrap them between sheets of Whatman 3MM paper. Fix the DNA to the filter by baking for 2 hours at 80°C in a vacuum oven. Overbaking can cause the filters to become brittle.

10. Hybridize the filters to a ^{32}P-labeled probe (see page 324).

Note. Any filters not used immediately in hybridization reactions should be wrapped loosely in aluminum foil and stored under vacuum at room temperature.

SCREENING BY HYBRIDIZATION FOLLOWING IN SITU AMPLIFICATION OF BACTERIOPHAGE λ PLAQUES

Woo et al. (1978) and Woo (1979) have described a modification of the Benton and Davis screening method that involves amplification of the bacteriophages directly on the nitrocellulose filter prior to hybridization. Because more bacteriophage DNA becomes attached to the nitrocellulose filter, the autoradiographic signals from positive clones are stronger than those obtained by the original procedure. Amplification therefore leads to an improvement in the ratio of signal to noise and also allows the length of the autoradiographic exposure to be reduced.

1. Plate out the bacteriophages that are to be screened, as described on page 320.

2. Set up a fresh overnight culture of the host bacteria.

3. Prepare a set of numbered, sterile nitrocellulose filters (Millipore HAWP).

4. When the plaques are about 0.2 mm in diameter, remove the plates from the incubator and chill them at 4°C for at least an hour.

5. Dilute the overnight culture 10-fold in fresh medium. Submerge the sterile filters one at a time in the diluted cell suspension. Remove the filters and allow them to dry completely on a sterile sheet of Whatman 3MM paper in a laminar flow hood (usually 1-2 hours in the hood is necessary).

6. Gently place a nitrocellulose filter impregnated with bacteria onto the surface of one of the plates. Key the filter to the plate with marks made with India ink. Store the plate in the refrigerator for 5 minutes to allow transfer of the bacteriophages to take place.

7. Remove the filter from the plate and transfer it to a fresh agar plate. The side that was in contact with the bacteriophage plaques should face upward.

8. Repeat steps 5 and 6 with a second filter impregnated with bacteria.

9. Wrap the master plates in parafilm and store at 4°C in an inverted position until the results of the hybridization reaction are available.

10. Incubate the replica plates at 37°C for 6-12 hours. A bacterial lawn with plaques will grow up on the filter.

11. Saturate a piece of Whatman 3MM paper in a baking or petri dish with 0.5 M NaOH and 1.5 M NaCl. Pour off the excess liquid. Remove the filter from the plate and place it on the sheet (plaque side up) for 5 minutes.

12. Transfer the filter to a piece of 3MM paper saturated with a solution of 0.5 M Tris · Cl (pH 8.0) and 1.5 M NaCl, and then to a piece of 3MM paper saturated with 2× SSPE.

13. Dry the filters in air for an hour and then bake them for 2 hours at 80°C under vacuum.

14. Hybridize the filters to ^{32}P-labeled probe as described on page 324.

HYBRIDIZATION OF DNA OR RNA IMMOBILIZED ON FILTERS TO RADIOACTIVE PROBES

There are many methods available to hybridize radioactive probes in solution to DNA or RNA immobilized on nitrocellulose filters. These methods differ in the following aspects:

the solvent and temperature used (68°C in aqueous solution or 42°C in 50% formamide);

the volume of solvent and the length of hybridization (large volumes for periods as long as 3 days or minimal volumes for times as short as 4 hours);

the degree and method of agitation (continuous shaking or stationary);

the concentration of the labeled probe and its specific activity;

the use of compounds, such as dextran sulfate, that increase the rate of reassociation of nucleic acids;

the stringency of washing following the hybridization.

Although the choice depends to a large extent on personal preference, we would like to offer the following guidelines.

1. Hybridization reactions in 50% formamide at 42°C are easier to set up, present less of an evaporation problem, and are less harsh on the filters than is hybridization at 68°C in an aqueous solution. The kinetics of the hybridization reaction in 80% formamide are approximately three to four times slower than in an aqueous solution (Casey and Davidson 1977). Assuming a linear relationship between the rate of hybridization and formamide concentration, the rate in 50% formamide should be two times slower than in an aqueous solution.

2. The smaller the volume of hybridization solvent, the better. The kinetics of nucleic acid reassociation are faster, and the amount of probe needed may be reduced so that the DNA on the filter acts as the driver for the reaction. All these are important parameters when detecting clones of low-abundance mRNAs. However, it is essential that sufficient liquid be present for the filters to remain at all times covered by a film of the hybridization solution.

3. Continual movement of the probe solution across the filter is unnecessary, even for a reaction driven by DNA immobilized on the filter. However, if a large number of filters are hybridized simultaneously, agitation is advisable in order to prevent the filters from adhering to each other.

4. The kinetics of the hybridization reaction are difficult to predict from theoretical considerations, partly because the exact concentration of the immobilized nucleic acid and its availability for hybridization are unknown.

When using probes made by nick translation of double-stranded DNA, the following rule of thumb is useful: Allow the hybridization to proceed for a time sufficient to enable the probe in solution to achieve 1–$3 \times C_{o}t_{1/2}$. In 10 ml of hybridization solution, 1 μg of a probe of 5-kb complexity will reach $C_{o}t_{1/2}$ in 2 hours. To determine the time of half-renaturation for any other probe, simply enter the appropriate values into the following equation:

$$\frac{1}{X} \times \frac{Y}{5} \times \frac{Z}{10} \times 2 = \text{number of hours to achieve } C_{o}t_{1/2}$$

where,

X = the weight of probe added (in μg)
Y = its complexity (for most probes, complexity is proportional to the length of the probe in kb)
Z = the volume of the reaction (in ml)

After hybridization for $3 \times C_{o}t_{1/2}$ has been reached, the amount of the probe available for additional hybridization to the filter is negligible. For single-stranded cDNA probes, the hybridization time may be shortened since the lack of a competing DNA strand in solution favors hybridization to DNA bound to the filter.

5. In the presence of dextran sulfate, the rate of association of nucleic acids is accelerated because the nucleic acids are excluded from the volume of the solution occupied by the polymer. Their effective concentration is therefore increased. The rate of association reportedly increases 10-fold in the presence of 10% dextran sulfate (Wahl et al. 1979).

   Although dextran sulfate is useful in circumstances where the rate of hybridization is the limiting factor in detecting sequences of interest, it is unnecessary for most purposes. It is also difficult to handle because of its viscosity and sometimes can lead to high backgrounds.

6. In general, the washing conditions should be as stringent as possible; i.e., a combination of temperature and salt concentration should be chosen that is slightly (5°C) below the T_{m} of the hybrid under study. The temperature and salt conditions can often be determined empirically in preliminary experiments where Southern blots (see pages 382ff) of genomic DNA are hybridized to the probe of interest and then washed under conditions of different stringency.

HYBRIDIZATION TO NITROCELLULOSE FILTERS CONTAINING REPLICAS OF BACTERIOPHAGE PLAQUES OR BACTERIAL COLONIES

The following protocol is designed for (a) two 20-cm × 30-cm nitrocellulose filters or (b) 30 circular, 82-mm-diameter filters. Appropriate adjustments should be made to the volumes when carrying out hybridization reactions with different numbers or sizes of filters.

1. Float the baked filters on the surface of a tray of 6× SSC until they have become thoroughly wetted from beneath. Submerge the filters for 5 minutes.

2. Transfer the filters (a) to a rectangular, flat-bottomed plastic box (22 cm × 32 cm) or (b) to a circular, glass crystallizing dish. Stack the filters on top of one another.

3. Add (a) 300 ml or (b) 100 ml of prewashing solution. Incubate at 42°C for 1–2 hours.

    In this and all subsequent steps, the circular filters in the crystallizing dish should be agitated on a rotating platform so that they do not stick to one another. The large, rectangular filters may be stationary.

    The prewashing solution removes from the filters any absorbed medium, fragments of agarose, or loose bacterial debris.

    Prewashing solution

    50 mM Tris · Cl (pH 8.0)
    1 M NaCl
    1 mM EDTA
    0.1% SDS

4. Pour off the prewashing solution. Incubate the filters for 4–6 hours at 42°C in (a) 100–150 ml or (b) 60 ml of prehybridization solution.

    The filters should be completely covered by the prehybridization solution. During prehybridization, sites on the nitrocellulose filter that bind single- or double-stranded DNA nonspecifically become saturated by unlabeled, salmon sperm DNA, SDS, or components in the Denhardt's solution. When using ^{32}P-labeled cDNA or RNA as a probe, poly(A) should be included in the prehybridization solution and hybridization solutions at a concentration of 1 μg/ml to prevent the probe from binding to T-rich sequences that are found fairly commonly in eukaryotic DNA.

    Prehybridization solution

    50% formamide
    5× Denhardt's solution
    5× SSPE
    0.1% SDS
    100 μg/ml denatured, salmon sperm DNA

After all the components have dissolved, centrifuge the prehybridization solution at 1000g at 15°C for 15 minutes or filter it through Whatman 1MM paper using a Buchner funnel. Sterilize the solution by filtration through disposable Nalgene filters. Store frozen at $-20°C$ in 25-ml aliquots.

Formamide. Many batches of reagent-grade formamide are sufficiently pure to be used without further treatment. However, if any yellow color is present, the formamide should be deionized by stirring on a magnetic stirrer with Dowex XG8 mixed-bed resin for 1 hour and filtering twice through Whatman 1MM paper. Deionized formamide should be stored in small aliquots under nitrogen at $-70°C$.

Denhardt's solution (50×)

Ficoll	5 g
polyvinylpyrrolidone	5 g
BSA (Pentax Fraction V)	5 g
H$_2$O	to 500 ml

20× SSPE. See page 314.

Denatured, salmon sperm DNA. This is prepared as follows: Dissolve the DNA (Sigma Type-III sodium salt) in water at a concentration of 10 mg/ml. If necessary, stir the solution on a magnetic stirrer for 2–4 hours at room temperature to help the DNA to dissolve. Shear the DNA by passing it several times through an 18-gauge hypodermic needle. Boil the DNA for 10 minutes and store at $-20°C$ in small aliquots. Just before use, heat the DNA for 5 minutes in a boiling-water bath. Chill it quickly in ice water.

5. Denature the ^{32}P-labeled probe DNA by heating for 5 minutes to 100°C. Add the denatured probe to the prehybridization solution covering the filters. Incubate at 42°C until 1–3× $C_0 t_{1/2}$ is achieved (see page 325). During the hybridization, the containers holding the filters should be tightly closed to prevent loss of fluid by evaporation.

6. After the hybridization is completed, discard the hybridization solution. Wash the filters 3–4 times, for 5–10 minutes each wash, in a large volume (300–500 ml) of 2× SSC and 0.1% SDS at room temperature. Invert the filters at least once during washing. At no stage during the washing procedure should the filters be allowed to dry.

7. Wash the filters twice for 1–1.5 hours in (a) 500 ml or (b) 300 ml of a solution of 1× SSC and 0.1% SDS at 68°C. At this point, the background is usually low enough to put the filters on film. If the background is still high or if the experiment demands washing at higher stringencies, immerse the filters for 60 minutes in (a) 500 ml or (b) 300 ml of a solution of 0.2× SSC and 0.1% SDS at 68°C.

8. Dry the filters in air on a sheet of Whatman 3MM paper at room temperature. Tape the filters (numbered side up) onto sheets of 3MM paper and place pieces of tape marked with radioactive ink at several locations on the 3MM paper. These markers serve to align the autoradiograph with the filters.

   Radioactive ink is made by mixing a small amount of ^{32}P with a water-proof black ink. We find it convenient to make the ink in three grades: very hot (>2000 cps on a minimonitor); hot (>500 cps on a minimonitor); and cool (>50 cps on a minimonitor). Use a fiber-tipped pen to apply ink of the desired hotness to the pieces of sticky tape.

9. Cover the Whatman 3MM paper and filters in Saran Wrap. Apply to X-ray film (Kodak XR or equivalent) and expose overnight at $-70°C$ with an intensifying screen (see pages 470ff).

10. After development, align the film with the filters using the marks left by the radioactive ink. Use a fiber-tipped pen to mark the film with the position of the asymmetrically located dots on the numbered filters. Tape a piece of tracing paper to the film. Mark the position of positive hybridization signals onto the tracing paper. Also mark (in a different color) the positions of the asymmetrically located dots. Remove the tracing paper from the film. Identify the positive colony or plaque by aligning the dots on the tracing paper with those on the agar plate.

   Some batches of nitrocellulose filters swell and distort during hybridization so that it becomes difficult to align the two sets of dots. This problem can be alleviated to some extent by autoclaving the filters before use (see pages 304–305).

11. Each positive plaque should be picked as described in Chapter 2 and placed into 1 ml of SM containing a droplet of chloroform. Often, the alignment of the filters with the plate does not permit identification of a single hybridizing plaque. In this case, an agar plug containing several plaques is picked. An aliquot (usually 50 μl of a 10^{-2} dilution) of the bacteriophages that elute from the agarose plug should be replated so as to obtain approximately 500 plaques on an 85-mm plate. These plaques should then be screened a second time by hybridization. A single, well-isolated, positive plaque should be picked and used to make a plate stock (see page 64).

   Each positive bacterial colony should be picked with a sterile toothpick into 2 ml of medium containing the appropriate antibiotics. The bacteria are then replated so as to obtain approximately 300 colonies on an 85-mm plate. If the original colony was picked from an uncrowded area of the original plate, a small number of the secondary colonies should be picked and grown overnight in 2-ml cultures. The plasmids in these bacterial cultures should be isolated and analyzed by one of the methods described in Chapter 11. If the original colony was picked from a very crowded area of the original plate, it may be worthwhile screening the secondary colonies by hybridization before isolating and analyzing plasmid DNA.

Identification of cDNA Clones by Hybridization Selection

Specific mRNAs can be purified from total cellular mRNA by hybridization to viral or cloned DNA that has been denatured and immobilized on nitrocellulose filters (Prives et al. 1974; Paterson et al. 1977; Harpold et al. 1978; Ricciardi et al. 1979; Parnes et al. 1981) or activated cellulose paper (Goldberg et al. 1979; Seed 1982). Although nitrocellulose filters are simple to prepare and are able to bind large quantities of DNA, they become brittle and gradually release DNA during the hybridization step. Activated cellulose papers, on the other hand, are more difficult to prepare, but they can be reused many times because the DNA is irreversibly bound.

HYBRIDIZATION SELECTION USING DNA BOUND TO NITROCELLULOSE

When setting up to screen clones by hybridization selection, the following points should be borne in mind.

1. A minimum of 0.10 ng of a single species of mRNA is needed to detect a specific protein by in vitro translation followed by immunoprecipitation and/or SDS-polyacrylamide gel electrophoresis.

2. The efficiency of the hybridization selection procedure is quite low, varying between 10% and 30% for different mRNAs. Thus, only 0.1–0.3 μg of every microgram of mRNA included in the hybridization reaction can be recovered in a functional form.

3. If the average length of the inserts in a cDNA library carried in pBR322 is 800 bp, then about 10% of the weight of the recombinant plasmid DNA bound to the filter is complementary to mRNA. In theory, therefore, as little as 10 ng (0.1 ng × 10 × 10) of DNA of a recombinant cDNA plasmid bound to a nitrocellulose filter can be identified by hybridization selection. In practice, the rate of the hybridization reaction is too slow to be useful when such small quantities of DNA are employed. To drive the hybridization reaction at a reasonable rate, about 1.0 μg of each recombinant plasmid is bound to the filter. Because this quantity of DNA occupies only $^1/_{10}$–$^1/_{20}$ of the filter's capacity for DNA, it is possible to screen up to 20 plasmids simultaneously by applying approximately 20 μg of a mixture of different DNAs to the filter.

Binding DNA to Nitrocellulose[3]

1. Dissolve DNA in water at a concentration of 500 μg/ml.

2. Heat to 100°C for 10 minutes.

3. Chill the sample quickly on ice. Add an equal volume of 1 M NaOH and incubate at room temperature for 20 minutes.

4. Using a sterile scalpel and wearing gloves, cut a sheet of nitrocellulose filter (Millipore HAWP) into 3-mm squares. Place the cut filters on the virgin side of a piece of parafilm.

5. Neutralize the DNA sample by adding 0.5 volumes of a solution of 1 M NaCl, 0.3 M sodium citrate, 0.5 M Tris·Cl (pH 8.0), and 1 M HCl. Mix well and immediately chill the DNA sample in ice.

6. Using an automatic micropipette, spot 5 μl of the DNA solution onto each of the filters. Allow it to absorb and then spot another 5 μl. Repeat the process until each filter has been loaded with approximately 20 μg of DNA.

7. Allow the filters to dry in air for an hour.

8. Place the dried filters into a sterile, 50-ml, screw-capped, conical tube. Wash the filters twice with 50 ml of 6× SSC at room temperature. Redistribute the filters onto a fresh sheet of parafilm.

9. Blot the filters dry with Kimwipes. Allow the filters to dry in air for an hour.

10. Place the dried filters into a sterile, glass test tube fitted with a loose metal cap and bake them for 2 hours at 80°C in a vacuum oven. Store the filters at room temperature under vacuum.

[3]Parnes et al. (1981).

Hybridization and Elution of RNA

1. Place the filters containing the DNA to be hybridized in a small siliconized glass vial or a sterile, 1.5-ml Eppendorf tube.

2. If the filters are new and have not been hybridized previously, add 1 ml of H_2O and heat in a boiling-water bath for 1 minute. Cool in ice and remove the H_2O by aspiration. Add 1 ml of H_2O and mix by vortexing briefly. Remove the H_2O by aspiration.

   This procedure causes any DNA that is loosely bound to the filter to be eluted.

   If the filters have been used previously in hybridization selection experiments, they should be washed successively with:

   a. 1 ml of a solution of 2× SSC and 0.1 N NaOH (30 minutes at room temperature);

   b. five 1-ml aliquots of 2× SSC (vortex for 10 seconds; remove 2× SSC by aspiration);

   c. 1 ml of H_2O.

   This treatment is designed to remove any RNA that did not elute from the DNA bound to the filter during a previous hybridization selection.

3. Prepare the hybridization solution. The amount of hybridization solution used should be kept to a minimum.

   Hybridization solution
   100–500 μg/ml poly(A)$^+$ mRNA
   65% (v/v) deionized formamide (see page 327)
   20 mM 1,4-piperazinediethanesulfonic acid (PIPES; pH 6.4)
   0.2% SDS
   0.4 M NaCl
   100 μg/ml calf liver tRNA (Boehringer-Mannheim)

4. Heat the hybridization solution at 70°C for 10 minutes.

5. Add the hybridization solution to the filters and incubate at 50°C for 3 hours.

6. Following hybridization, remove the hybridization solution by aspiration and add 1 ml of wash solution (10 mM Tris [pH 7.6], 0.15 M NaCl, 1 mM EDTA, and 0.5% SDS) to the filters. The wash solution should be preheated to 65°C and maintained at 65°C during the washing.

7. Vortex the sample for several seconds and remove the wash buffer by aspiration.

8. Repeat the washing (steps 6 and 7) nine more times.

9. Wash twice with the same buffer without SDS.

10. Transfer the filters to a suitably sized, siliconized polypropylene centrifuge tube.

11. Add 300 μl of H_2O and 30 μg of calf liver tRNA per 0.5 cm^2 of filter-surface area.

12. Place the tubes in a boiling-water bath for 1 minute and then snap-freeze the samples in a dry-ice/ethanol bath.

13. Thaw the samples in ice and remove the filters. Extract the eluted RNA with an equal volume of phenol/chloroform. Precipitate the RNA by adding 60 μl of 2 M sodium acetate (pH 5.2) and 1 ml of ethanol. Store the tube in dry ice/ethanol for 20 minutes. Recover the RNA by centrifugation for 10 minutes in an Eppendorf centrifuge.

14. Wash the precipitate twice with 70% ethanol.

15. Dry the precipitate in a vacuum desiccator and resuspend in 5 μl of H_2O for translation in vitro.

16. Dry the filters in air and store them under vacuum at room temperature until they are to be reused.

Notes

i. Many different mRNAs have been selected by using the hybridization conditions given above. However, the thermal stability of DNA · RNA hybrids can vary over quite a large range, depending on the G + C content of the hybrid. Thus, the conditions given are not universal, and some adjustments to the temperature of hybridization, concentration of formamide, and temperature of washing may be required in order to select a particular mRNA with maximal efficiency.

ii. Nitrocellulose filters can be used up to three times in hybridization-selection experiments. However, their efficiency declines with every use, probably because of loss of DNA. When maximum sensitivity is required, fresh filters should be used.

Preparation of Pools of Plasmid DNAs

1. Grow individual recombinant cDNA clones overnight in 1 ml of medium containing the appropriate antibiotic.

2. Inoculate 0.1 ml of each of 10 saturated cultures into a 125-ml flask containing 25 ml of rich medium (LB or χ1776 medium). Incubate the mixed culture at 37°C until the $OD_{600} \simeq 0.7$ (~ 2 hours).

3. Add chloramphenicol to a final concentration of 170 μg/ml. Continue the incubation for a further 12–16 hours.

4. Prepare DNA from the culture by the alkaline lysis procedure described on page 368.

Preparation of Diazobenzyloxymethyl Paper

By contrast to nitrocellulose, to which DNA becomes bound by hydrophobic interactions, cellulose derivatized with diazobenzyloxymethyl (DBM) groups binds DNA and RNA covalently (Noyes and Stark 1975; Alwine et al. 1977).

As shown in Figure 10.1, Whatman 50 paper is reacted with m-nitrobenzyloxymethyl pyridinium chloride (NBM) to produce m-aminobenzyloxymethyl (ABM) paper. The ABM paper is then activated with acid and sodium nitrite to produce diazobenzyloxymethyl (DBM) paper. The diazonium salt can then covalently bind single-stranded DNA. (See page 336 for the steps in preparing DBM paper.)

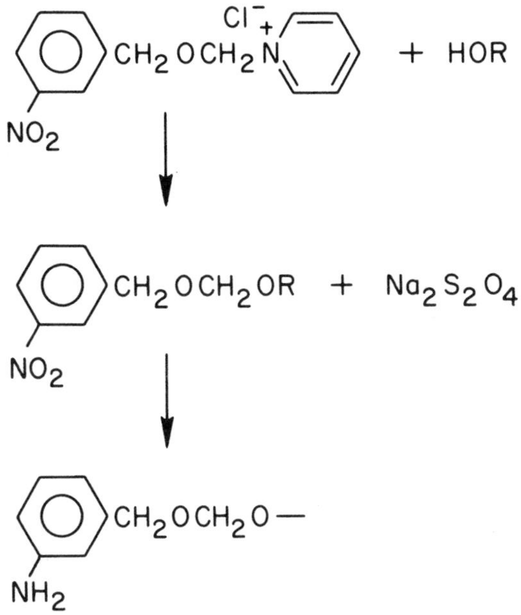

Figure 10.1

Chemical synthesis of m-aminobenzyloxymethyl (ABM) paper. (*HOR*) Cellulose hydroxyl group.

Caution

Steps 2 through 5 should be carried out in a well-ventilated fume hood.

1. Cut Whatman 50 paper so that it fits snugly in the bottom of a square 20-cm × 20-cm Pyrex baking dish.

2. For each square centimeter of paper, prepare a solution consisting of:

NBM (1-*m*-nitrobenzyloxymethyl pyridinium chloride)	2.3 mg
sodium acetate · 3H$_2$O	0.7 mg
H$_2$O	28.5 μl

Note. NBM is available from Gallard-Schlesinger Chemical Mfg. Corp. (584 Mineola Ave., Carle Place, NY 11514).

3. In a fume hood, pour the solution evenly over the paper, remove bubbles with a gloved hand, and gently rub the solution into the paper until the paper is dry. (The paper will change from a translucent appearance to opaque white as it dries.)

4. Incubate the paper at 60°C for 10 minutes. Be certain the paper is completely dry before proceeding.

5. Bake the paper at 130–135°C for 30–40 minutes. **Beware of fumes!**

6. Wash the paper several times with distilled water for a total of approximately 20 minutes.

7. Wash the paper three times with acetone for 20 minutes and allow it to dry in the air.

8. The paper is stable for prolonged periods when stored in Saran Wrap at 4°C.

Activation of ABM Paper

1. Place a sheet of paper in a glass dish at 60°C in a **fume hood**. Add 0.4 ml of a 20% aqueous solution of sodium dithionite for each square centimeter of paper. Incubate for 30 minutes.

2. Wash the paper with a large volume (500–1000 ml) of distilled water.

3. Wash the paper once with 500 ml of a 30% (v/v) solution of acetic acid.

4. Wash the paper with several changes of distilled water for 5 minutes.

5. Transfer the paper to an ice-cold solution of 1.2 M HCl (0.3 ml/cm^2 paper).

6. For every 100 ml of HCl solution, add (with mixing) 2.7 ml of a solution of $NaNO_2$ (10 mg/ml in H_2O, prepared immediately before use). Leave the glass dish on ice for 30 minutes.

7. The paper is now activated and may be wrapped in Saran Wrap and stored at −70°C for several months without detectable loss of activity.

Application of DNA to DBM Paper

1. DBM paper can bind 15–25 μg of single-stranded DNA/cm^2 (i.e., its capacity is much less than that of nitrocellulose).

2. Dissolve 10 μg of DNA in 100 μl of TE, add 50 μl of 1 M NaOH, and boil for 5 minutes.

3. Add (in this order):

2 M sodium acetate (pH 5.2)	30 μl
1 M HCl	40 μl
ethanol	500 μl

   Store at −70°C for 15 minutes.

4. Recover the DNA by centrifugation for 5 minutes in an Eppendorf centrifuge. Dissolve the DNA in 25 μl of H$_2$O and heat the solution to 100°C for 1 minute. Snap-freeze the solution by immersing it in a bath of dry ice/ethanol.

5. Using a sterile razor blade or hole punch, cut the DBM paper that has been treated with HCl (see step 7, page 337) into 5-mm squares or small circles.

6. Wash the pieces of DBM paper twice with ice-cold, distilled water and twice more with 20 mM sodium acetate (pH 4.0).

7. Remove any excess liquid from the DBM paper by blotting with Whatman 3MM paper.

8. Spot the 25 μl of DNA (10 μg) directly onto a square of DBM paper. Allow the paper to dry in air and then store for 12–16 hours in a closed Eppendorf tube.

9. Wash the paper three times with water, four times with 0.4 N NaOH, and four times with water.

10. Rinse the paper with prehybridization buffer preheated to 42°C.

    Prehybridization buffer

    50% deionized formamide (see page 327)
    0.1% SDS
    0.6 M NaCl
    80 mM Tris·Cl (pH 7.8)
    4 mM EDTA

11. Incubate the paper at 65°C for 1 hour in elution buffer.

   Elution buffer
      99% (v/v) deionized formamide (see page 327)
      10 mM Tris · Cl (pH 7.8)

12. Wash the paper in prehybridization buffer (200 μl/filter).

RNA Hybridization and Elution

1. Prepare the hybridization solution:

   50–100 μg/ml poly(A)$^+$ RNA
   50% (v/v) deionized formamide (see page 327)
   0.1% SDS
   0.6 M NaCl
   80 mM Tris·Cl (pH 7.8)
   4 mM EDTA

   The amount of the hybridization solution should be kept to a minimum. Usually 1.5-ml Eppendorf tubes or small, siliconized glass vials are used.

2. Heat the hybridization solution to 70°C for 10 minutes.

3. Add the filters to the hybridization solution and incubate at 37°C for 18 hours.

4. After the hybridization reaction is completed, transfer the pieces of DBM paper to a fresh tube.

5. Wash the paper by vortexing 10 times at 37°C with 1 ml of wash buffer. Remove each wash by aspiration with a sterile pasteur pipette.

   Wash buffer
   50% (v/v) deionized formamide
   0.2× SET
   0.1% SDS

   20× SET
   3 M NaCl
   0.4 M Tris·Cl (pH 7.8)
   20 mM EDTA

6. Elute the RNA from the DNA·RNA hybrid by incubating the filter in 200 μl of elution buffer for 5 minutes at 65°C.

   Elution buffer
   99% (v/v) deionized formamide
   10 mM Tris·Cl (pH 7.8)

7. Remove the eluate to a fresh tube. Add:

H_2O	200 μl
2 M sodium acetate (pH 5.2)	40 μl
calf liver tRNA (Boehringer-Mannheim)	15 μg
ethanol	1100 μl

Store the tube in dry ice/ethanol for 20 minutes.

8. Recover the RNA by centrifugation (10 minutes in an Eppendorf centrifuge). Wash the pellet twice with 70% ethanol.

9. Dissolve the RNA in 5 μl of H_2O and translate in a cell-free system (see page 344).

10. Wash the filters twice with water. Allow them to dry in air and store them under vacuum at room temperature. The filters may be reused for hybridization selection after washing them in:

   a. 0.1 N NaOH for 20 minutes at room temperature;

   b. water (six times);

   c. hybridization buffer.

   If the filters are regenerated in this way, there is little or no loss of efficiency even after several cycles of hybridization/selection.

Preparation of 2-Aminophenylthioether Paper

Recently, a new arylamine paper derivative for diazotization and coupling to nucleic acids was developed by B. Seed (1982). 1,4 Butanediol diglycidyl ether (BDG) reacts with Whatman 50 paper to produce oxirane cellulose, which is then coupled with o-aminothiolphenol in alkaline solution to give o-aminophenylthioether (APT) cellulose (Fig. 10.2). The APT paper is easier to prepare and more stable than ABM paper, and the diazotized form (DPT paper) is also more stable than diazotized ABM (DBM) paper.

Caution

All of the following operations (up to the final water wash) must be carried out in a well-ventilated fume hood. Some form of skin protection is essential; neoprene gloves are best for this purpose.

  1,4 Butanediol diglycidyl ether (BDG) can be purchased from Aldrich (12, 419-2). It is probably a carcinogen and should be handled with care. 2-Aminothiophenol (Aldrich 12, 313-7) is also toxic.

1. Cut sheets of Whatman 50 filter paper to the desired size and place approximately 20 g in a heat-sealable polyester bag (e.g., Sears' Seal-N-Save).

2. Add 70 ml of 0.5 M NaOH to the bag, followed by 30 ml of BDG. Seal the bag, leaving 100–200 ml of air to allow good mixing. Rotate the bag end-over-end (axis of rotation in the plane of the bag) for 12–16 hours at 30 rpm.

3. Pour off the excess reagent into a waste container and allow it to inactivate for at least 2 days before disposal.

4. Remove the papers one by one into 500 ml of a solution formed by dissolving 2-aminothiophenol to 2% (v/v) in ethanol and then adding an equal volume of 0.5 M NaOH. Leave the papers for 2 hours in the 2-aminothiophenol/ethanol/NaOH solution.

5. Wash the coupled papers by sequential immersion and agitation in ethanol followed by 0.1 M HCl for a total of three repetitions of the cycle. Each wash should be approximately 15 minutes long.

6. Rinse the papers in water for 1 to 2 hours.

7. Wash the papers in ethanol.

8. Dry the papers in air. The dried papers should be stored at −20°. The application of DNA to APT paper is carried out using the procedure described on page 338.

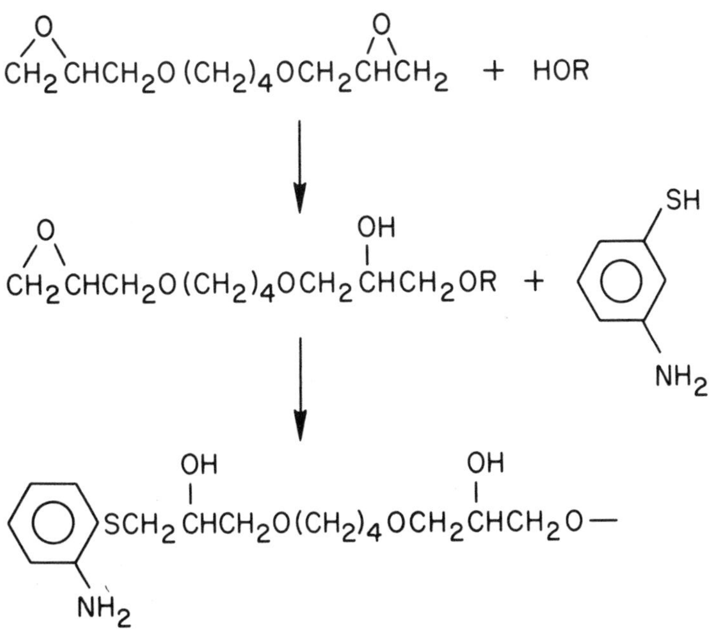

Figure 10.2

Chemical synthesis of o-aminophenylthioether (APT) paper. (HOR) Cellulose hydroxyl group.

Note

An equally effective and simpler procedure for coupling the paper to 2-aminothiophenol is to:

i. Carry out steps 1 and 2 (page 342).

ii. Add 10 ml of 2-aminothiophenol to 40 ml of ethanol.

iii. Add the mixture to the contents of the plastic bag (following step 2).

iv. Reseal and rotate the bag for an additional 10 hours; then proceed to step 6.

TRANSLATION OF HYBRIDIZATION-SELECTED RNA IN RETICULOCYTE LYSATES

RNA recovered from the filter hybridization reactions may be translated in cell-free extracts or in frog oocytes to produce a protein that can be immunoprecipitated or tested for biological activity. In vitro translation kits prepared from reticulocyte lysates or extracts of wheat germ can be obtained from New England Nuclear (NEN) or Bethesda Research Labs (BRL); however, we find that the efficiency of the commercial translation kits is much lower than that of "homemade" in vitro translation systems. We have therefore included a brief description of a procedure for preparing rabbit reticulocyte lysate, provided by B. Roberts. We have also included a protocol for injection of frog oocytes, provided by D. Melton.

Preparation of Rabbit Reticulocyte Lysate

1. To obtain anemic rabbit blood, subcutaneously inject six rabbits, each weighing 2–3 kg, with acetylphenylhydrazine (a 1.2% solution neutralized to pH 7.5 with 1 M HEPES, pH 7.0) according to the following schedule:

   1.0 ml on day 1
   1.6 ml on day 2
   1.2 ml on day 3
   1.6 ml on day 4
   2.0 ml on day 5

2. On days 7 and 8, bleed the rabbits in the following way: Swab one ear with xylene-saturated cotton and make a single incision with a new razor blade in the posterior ear vein midway along the length of the ear. Each rabbit should yield 50–60 ml of blood, which is collected into 50 ml of chilled, normal saline containing 0.001% heparin.

3. Filter the blood through cheesecloth and then centrifuge at 2000g for 5 minutes. Wash the packed cells three times with normal saline. The last centrifugation step should be carried out at 5000g.

4. Measure the volume of the packed cells and lyse at 0°C by adding an equal volume of cold water. After 1 minute, centrifuge the lysate at 20,000g for 20 minutes.

5. Divide the supernatant into 0.5-ml aliquots and freeze at −70°C. The lysate is stable for several months at this temperature.

Treatment of Reticulocyte Lysate with Micrococcal Nuclease

The purpose of this treatment is to destroy endogenous mRNAs so that the translation activity of the reticulocyte lysate becomes completely dependent on exogenously added mRNAs. Micrococcal nuclease is active only in the presence of calcium ions, and the digestion is terminated by the addition of EGTA. The subsequent translation reaction is not affected by the presence of EGTA nor by the presence of inactive micrococcal nuclease.

1. Prepare the following stock solutions:

    a. 50 mM $CaCl_2$;
    b. 100 mM EGTA (pH 7.0);
    c. micrococcal nuclease; 150,000 units/ml, stored in 50 mM glycine (pH 9.2) and 5 mM $CaCl_2$;
    d. creatine phosphokinase type I; 40 units/ml in 50% glycerol;
    e. hemin stock; 4 mg/ml in ethylene glycol.

Hemin stock solution is made up as follows:

    a. Dissolve 20 mg of hemin in 400 μl of 0.2 M KOH. Add the hemin slowly and vortex between each addition.

    b. Add:

H_2O	600 μl
1 M Tris·Cl (pH 7.8)	100 μl

    Adjust the pH to pH 7.8 by adding 0.1 M HCl, if necessary.

    c. Add 4 ml of ethylene glycol. Mix by vortexing.

    d. Centrifuge at 5000g for 5 minutes. Discard the insoluble pellet.

2. To 100 parts of reticulocyte lysate, add:

    1 part hemin solution
    2 parts $CaCl_2$ solution

    Mix.

3. Add 0.05 parts of micrococcal nuclease (150,000 units/ml) and mix. Incubate for 15 minutes at 20°C.

4. Add 2 parts of 100 mM EGTA and mix.

5. Add 0.4 parts of creatine phosphokinase (40 units/ml). Mix. Dispense the treated lysate in 250–500-μl aliquots and store at −70°C.

Note

Hemin is included in the reticulocyte lysate because it is a powerful suppressor of an inhibitor of the initiation factor E1f2a. In the absence of hemin, protein synthesis in reticulocyte lysates ceases after a short period of incubation.

Translation in Reticulocyte Lysates

1. Make up the following translation cocktail.

100 mM spermidine	200 μl
800 mM creatine phosphate	400 μl
5 mM amino acids (− methionine)	200 μl
1 M dithiothreitol	80 μl
500 mM HEPES (pH 7.4)	1600 μl
H$_2$O	710 μl
	3200 μl

Dispense the translation cocktail in 50–100-μl aliquots and store at $-70°$C.

5 mM amino acids (− methionine) is a solution containing all amino acids except methionine, at a concentration of 5 mM.

2. The standard reaction mixture for translation contains:

translation cocktail	2 μl
1 M KCl	2 μl
32.5 mM magnesium acetate	0.5 μl
[^{35}S]methionine (100 μCi)	10 μl
micrococcal-nuclease-treated reticulocyte lysate	10 μl
H$_2$O	0.5 μl
RNA	1 μl

Incubate for 1 hour at 37°C.

Notes.

i. The optimal concentration of potassium chloride in the translation reactions varies from preparation to preparation of reticulocyte lysate and from mRNA to mRNA. The amount of potassium chloride required to achieve the maximum efficiency of translation should be determined for every new batch of reticulocyte lysate using a standard preparation of mRNA. Likewise, the effect of different concentrations of potassium chloride on the efficiency of translation of the particular mRNA of interest should be ascertained.

ii. Up to 10 μg of total cytoplasmic RNA or 0.2 μg of poly(A)$^+$ RNA can be added to the reaction before saturation is reached. Usually, cleaner results are obtained with poly(A)$^+$ RNA than with total cytoplasmic RNA.

iii. The addition of saturating amounts of mRNA to the system should stimulate the incorporation of [^{35}S]methionine into acid-precipitable material some 10-fold to 25-fold. Under optimal conditions, about 3×10^6 to 5×10^6 cpm of [^{35}S]methionine should be incorporated per 25 μl of translation reaction.

iv. If necessary, the amount of radioactivity in the reaction can be increased by using evaporation to concentrate the solution of [^{35}S]methionine supplied by the manufacturer.

3. The products of in vitro translation reactions may be analyzed by immunoprecipitation and/or SDS-polyacrylamide gel electrophoresis. A good description of relevant immunoprecipitation techniques may be found in an article by Kessler (1981). Detailed and standardized procedures for electrophoresis in polyacrylamide-SDS gels are also available (Laemmli 1970; Maizel 1971).

Table 10.2 and the formula for the stacking gel below give the amounts of the different ingredients needed to make polyacrylamide gels of different porosities.

Stacking gel

30% acrylamide	1.7 ml
2% bisacrylamide	0.7 ml
1 M Tris·Cl (pH 6.9)	1.25 ml
20% SDS	50 μl
H$_2$O	6.35 ml
TEMED	25 μl
10% ammonium persulfate	50 μl

TABLE 10.2. SDS-POLYACRYLAMIDE SEPARATING GELS

Ingredients	Percent acrylamide:mg bisacrylamide/ml						
	5%:2.6	7.5%:1.95	10%:1.3	12.5%:1.0	15.0%:0.86	17.5%:0.73	20%:0.65
30% Acrylamide (ml)	5.0	7.5	10.0	12.5	15.0	17.5	20.0
2% Bisacrylamide (ml)	3.9	2.9	2.0	1.5	1.3	1.1	1.0
1 M Tris·Cl (pH 8.7) (ml)	11.2	11.2	11.2	11.2	11.2	11.2	11.2
20% SDS (ml)	0.15	0.15	0.15	0.15	0.15	0.15	0.15
H$_2$O (ml)	9.45	8.25	6.65	4.65	2.35	—	—
TEMED (μl)	25	25	25	25	25	25	25
10% Ammonium persulfate (μl)	100	100	100	100	100	100	100

Notes

i. Ammonium persulfate is unstable in aqueous solution. A new solution should be made up every few days and stored at 0°C.

ii. Acrylamide is **toxic** and can be absorbed through the skin. **Wear gloves** when working with the chemical, both dry or in solution.

iii. Highly purified acrylamide is expensive. Cheaper grades can be easily purified as follows:

a. Dissolve acrylamide in H_2O (30% w/v)

b. Add about 20 g of mixed-bed resin (MB-1, Mallinckrodt) per liter of solution. Stir at room temperature overnight.

c. Filter the solution through Whatman 1MM filter paper. Store in tightly sealed bottles.

TRANSLATION OF MESSENGER RNA INJECTED INTO FROG OOCYTES

During oogenesis in the frog *(Xenopus)*, immature eggs or oocytes accumulate a vast amount of major enzymes, organelles, and precursors. For example, an oocyte contains about 200,000 more ribosomes than a somatic *Xenopus* cell and has an excess of 10,000 tRNA molecules compared with somatic cells (reviewed by Laskey 1980). These maternal stockpiles are normally used early during frog development in the formation of the tadpole. However, these stored materials are biologically active and provide a sensitive test system for the translation of exogenous mRNA.

Gurdon and his colleagues first demonstrated the usefulness of the oocyte translation system using rabbit globin mRNA (Gurdon et al. 1971). Their studies showed that oocytes will synthesize hemoglobin when injected with reticulocyte 9S RNA and hemin. Subsequent studies have shown that all eukaryotic mRNAs so far tested, but not prokaryotic mRNAs, will direct the synthesis of the appropriate protein when the RNA is injected into living oocytes (reviewed by Lane and Knowland 1975).

In addition to translating injected mRNAs into proteins, the oocyte's enzymes correctly modify or mature the newly synthesized proteins. These modifications can include cleavage of precursor proteins, phosphorylation, and glycosylation (Berridge and Lane 1976; Colman et al. 1981; Mathews et al. 1981). Moreover, in cases where the mRNAs code for secretory proteins such as interferon, the newly synthesized interferon is secreted from the injected oocyte (Colman and Morser 1979; Colman et al. 1981). Thus, unlike most in vitro translation systems, injected oocytes provide a complete system for the study of all events associated with protein synthesis and secretion.

Messenger RNAs injected into oocytes are stable and efficiently translated. For example, the half-life of globin mRNA in injected oocytes is estimated to be more than two weeks (Gurdon et al.1973). Consequently, injected oocytes can be cultured for many days, during which time they will continue to synthesize proteins from the injected mRNAs. It is of interest to note that oocytes translate injected mRNAs much more efficiently than do cell-free systems. Rabbit globin mRNA is translated about 100–1000 times more efficiently in frog oocytes than in a reticulocyte lysate. Moreover, the rate of translation in injected oocytes compares favorably with that in normal, unmanipulated reticulocytes: Oocytes are about one quarter as efficient as normal reticulocytes (Gurdon et al. 1971).

Methods

Animals

Adult male and female *Xenopus laevis* can be obtained from a variety of animal suppliers, including K. Evans (716 Northside, Ann Arbor, MI 48105). The frogs can be kept in any type of water tank, without aeration, at 18–22°C. Regular feeding (twice per week) with fragments of beef liver and beef heart will help maintain a healthy colony.

Preparation of Oocytes

Oocytes should be obtained from healthy, adult, female *Xenopus*. This is easily accomplished by anesthetizing the frog in a 1:1000 (w/v) solution of ethyl *m*-aminobenzoate in water for 10–30 minutes. A small incision (1 cm) on the posterior ventral side gives ready access to the frog's ovary. After removing a segment of the ovary, the incision can be sutured, and the frog will quickly recover in water. Typically, one female has enough oocytes for 3–5 separate experiments.

The ovary should be placed immediately in modified Barth saline (MBS-H) and separated into small groups of 5–20 large oocytes by using watchmaker's forceps.

Oocytes can be stored at 19°C in MBS-H for days and will remain viable for injection experiments.

MBS-H

88 mM NaCl
1 mM KCl
0.33 mM $Ca(NO_3)_2$
0.41 mM $CaCl_2$
0.82 mM $MgSO_4$
2.4 mM $NaHCO_3$
10 mM HEPES (pH 7.4)

Benzyl penicillin (10 mg/l) and streptomycin sulfate (10 mg/l) are often added to suppress the growth of bacterial contaminants.

Preparation of RNA

RNA can be isolated from cells by virtually any standard procedure. To increase the concentration of mRNA in the injection solution, it is helpful, but not necessary, to isolate poly(A)$^+$ RNA. Normally, RNA is prepared by phenol/chloroform (1:1) extraction, followed by precipitation with ethanol. The RNA should be dissolved in distilled water, MBS-H, or a low-salt solution for injection. Detergents must be avoided because they are highly toxic to oocytes.

Injection of RNA

Large, fully grown oocytes (stages V and VI, according to Dumont 1972) are injected using a fine micropipette, a micrometer syringe, and any standard dissecting stereomicroscope. The construction of a suitable injection pipette is described by Gurdon (1974). Oocytes are placed on a microscope slide, blotted dry with a paper towel, and the slide is then placed on an ice box on the microscope stage. Before and during the insertion of the pipette, the oocytes can be positioned with blunt watchmaker's forceps.

After the pipette has penetrated the oocyte, a 30–50-nl aliquot of the RNA solution (up to 500 ng) is delivered using a micrometer syringe. The injection volumes can be measured by marking the shaft of the pipette with ink and following the movement of the meniscus. Immediately after the injection, the oocytes are transferred to MBS-H and incubated at 19°C, usually for 24–48 hours.

Radioactive Labeling

Oocytes will take up exogenous isotopes from the MBS-H medium. Radioactive amino acids such as [^{3}H]histidine, [^{3}H]leucine, [^{3}H]proline, [^{3}H]valine, or [^{35}S]methionine can be used to label newly synthesized proteins. The radioactive isotope is usually dried in vacuo and redissolved in MBS-H at 0.5–2.0 mCi/ml. Damaged oocytes should be discarded before the labeling period is begun. Injected oocytes are most conveniently labeled by incubation in small wells of microtiter plates containing 2–10 μl of the radioactive precursor solution per oocyte.

Extraction of Labeled Proteins

Proteins can be extracted from injected oocytes by homogenizing five oocytes in 200 μl of 75 mM Tris · Cl (pH 6.8), 1% β-mercaptoethanol, 1% SDS, and 1 mM phenylmethylsulfonyl fluoride. After centrifugation for 5 minutes at room temperature in a microfuge at 10,000g, there will be two distinct phases: an upper phase, which contains the labeled proteins, and a lower phase of yolk and pigment granules. The supernatant can be diluted with sample buffer and analyzed by electrophoresis in polyacrylamide gels (Laemmli 1970).

Proteins secreted from injected oocytes can be recovered from the incubation media by precipitation, with 0.2 volumes of ice-cold trichloroacetic acid (50% w/v) in the presence of carrier protein (BSA, 2 μg). The precipitates are collected by centrifugation, washed with ice-cold acetone, and dried before being resuspended in sample buffer (Colman and Morser 1979) for gel electrophoresis. Methods for the subcellular fractionation of oocytes (into membrane-bound and cytosol fractions) have been described by Zehavi-Willner and Lane (1977) and Colman et al. (1981).

Screening Bacteriophage λ Libraries for Specific DNA Sequences by Recombination in *Escherichia coli*

THE PRINCIPLE FOR SELECTION BY RECOMBINATION

In this method (B. Seed, unpubl.), "probe" sequences are inserted into a very small plasmid vector, πVX, and introduced into recombination-proficient bacterial cells. Genomic bacteriophage libraries are then propagated in the cells, allowing those bacteriophages that bear sequences homologous to the probe to acquire an integrated copy of the plasmid by reciprocal recombination. Bacteriophages bearing integrated plasmids can be purified from the larger pool of bacteriophages lacking integrated plasmids by growth under the appropriate selective conditions.

The DNA sequence and a restriction map of πVX are shown in Figures 10.3 and 10.4. The πVX plasmid has three functional segments: an origin of replication ($\sim$ 600 bp) derived from the pMB1 replicon; a tyrosine tRNA amber-suppressor gene (synthetic *supF* gene); and a "polylinker" or short sequence bearing multiple restriction sites useful for insertion of cloned fragments. The πVX plasmid is maintained in *E. coli* by selecting for amber-

Figure 10.3

The complete nucleotide sequence of πVX. The sequence begins at the *Eco*RI site marked with an arrow in the lower panel of Fig. 10.4 and proceeds around the circular molecule in a clockwise direction.

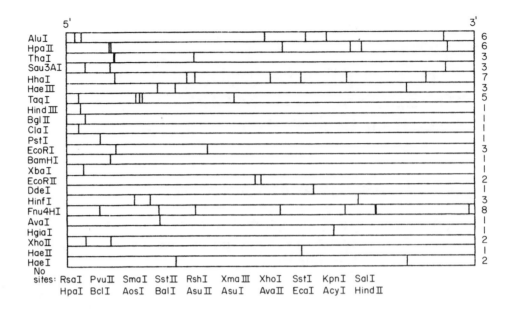

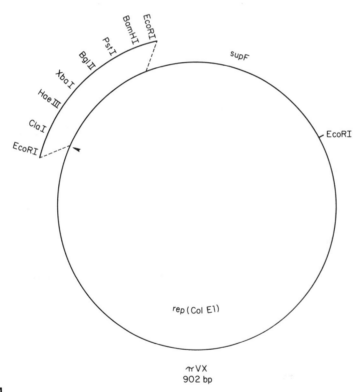

π VX
902 bp

Figure 10.4

A complete restriction map of the microplasmid πVX. The top panel shows the location of all known restriction endonuclease cleavage sites, as determined by a computer scan of the DNA sequence of πVX. The circular πVX plasmid is represented in linear form by breaking the molecule at the EcoRI site marked with an arrow between the ColEl origin and the polylinker (see lower panel). The map proceeds in a clockwise direction around the plasmid DNA. The number of sites recognized by various restriction enzymes is shown at the right-hand end of the linear map.

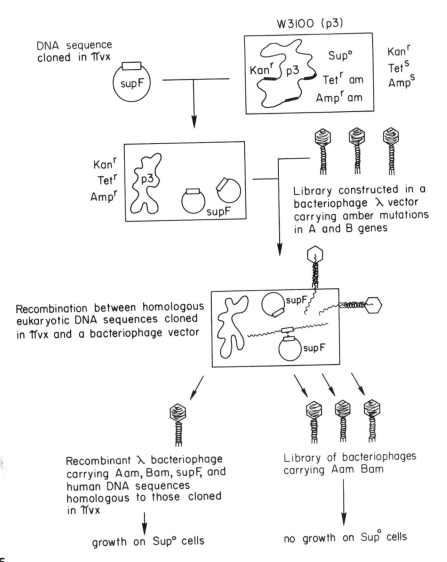

Figure 10.5

A restriction fragment containing a probe sequence is cloned into πVX. The hybrid plasmid that contains a functional *supF* gene transforms *E. coli* strain W3110 r⁻ m⁺ (p3) to tetracycline and ampicillin resistance by suppressing the amber mutations in the *tet* and *bla* genes of the resident plasmid p3. The population of transformed cells is then infected with a library of bacteriophage λ recombinants that carry mammalian DNA fragments inserted into a vector bearing amber mutations in its *A* and *B* genes. If the mammalian DNA sequence cloned into πVX is homologous to the sequence cloned into the infecting bacteriophage, recombination may occur and generate a bacteriophage genome that carries a functional *supF* gene derived from πVX. This bacteriophage can grow in Su⁻ cells.

suppressor function in cells that also contain the 60-kb plasmid p3 (a derivative of RP1), which bears amber mutations in both the ampicillin- and tetracycline-resistance elements (see Fig. 10.5). Transformation of cells already carrying p3 with πVX confers upon them simultaneous resistance to ampicillin and tetracycline. In the absence of πVX, the p3 plasmid is maintained in the bacterial population by selecting for kanamycin resistance.

After the probe DNA has been cloned into πVX, bacteria containing the recombinant microplasmid are infected with a library of bacteriophage λ recombinants constructed with a vector that bears at least two amber mutations in its essential genes. If the eukaryotic DNA sequence cloned into πVX is homologous to a sequence present in the infecting bacteriophage, reciprocal recombination results in the insertion of the entire microplasmid into the bacteriophage. This bacteriophage, by contrast to its parent, can grow in Su$^-$ cells because of the amber suppressor contributed by the microplasmid. Bacteriophages in which the two amber mutations in the vector arms have simultaneously reverted can also grow in Su$^-$ cells. However, the frequency of spontaneous double reversion is very low, and in the absence of recombination with πVX, the proportion of the progeny that can grow in Su$^-$ cells is approximately 10^{-10}–10^{-12}. The frequency of recombinants obtained from cells carrying πVX depends in part on the length of the common segment of DNA that is shared by the recombinant plasmid and bacteriophage genomes. For a homologous segment 500 bp in length, the frequency of recombination is at least 10^{-3}. Even though only one bacteriophage in 10^5 in a mammalian library may contain the desired homologous sequence, the net frequency of recombination (10^{-8}) is well above the frequency for reversion.

STRAINS USED IN πVX SCREENING

The three bacterial strains used in πVX screening are:

W3110 r$^-$m$^+$, Su$^-$, a spontaneous *thy*$^+$ revertant of W3110 r$^-$m$^+$, *thy*$^-$.

W3110 r$^-$m$^+$ (p3), which carries a *tra*$^-$ derivative of pLM2. This plasmid, which bears two amber mutations, is itself a derivative of RP1 (Mindich et al. 1978).

W3110 r$^-$m$^+$ (p3) (πVX).

Strains carrying πVX or its derivatives should be grown in the presence of tetracycline (7.5 μg/ml) and ampicillin (12.5 μg/ml).

PREPARATION OF πVX

πVX may be prepared from W3110 r$^-$m$^+$ (p3) (πVX) by any of the procedures described in Chapter 3. The yield is usually about 50 μg/liter.

The πVX vector is separated from the p3 plasmid by digesting the mixed plasmid DNA preparation with one or two restriction enzymes that yield the termini of interest and by isolating the πVX fragment by electrophoresis through a 1% agarose gel.

TRANSFORMATION OF W3110 r⁻m⁺ (p3)

The best transformation protocol for W3110 r⁻m⁺ (p3) seems to be that described by Dagert and Ehrlich (1979). Using this method, W3110 r⁻m⁺ (p3) yields as many as 10^7 transformants per microgram of supercoiled pBR322. The selective medium should contain 15 μg/ml of tetracycline and 25 μg/ml of ampicillin. Occasionally, the competent cells will contain bacteria with chromosomal suppressors; this problem can be minimized by using 24-hour-old competent cells in the transformation protocol.

PLATING BACTERIOPHAGE λ LIBRARIES ON W3110 r⁻m⁺ (p3) (πVX)

As many as 10^6 recombinant bacteriophages can be used to infect 0.25 ml of an overnight culture of miniplasmid-bearing cells, which can then be plated on a 85-mm plate containing 7.5 μg/ml of tetracycline and 12.5 μg/ml of ampicillin.

The bacteriophages harvested from this plate (see Chapter 3) are then plated on Su⁻ cells. Up to 5×10^9 bacteriophages can be used to infect 0.25 ml of an overnight culture of W3110 r⁻m⁺, Su⁻.

GENETIC TESTS FOR BACTERIOPHAGE CONTAINING A SUPPRESSOR GENE

Two bacterial strains are used to demonstrate the presence of a suppressor gene on the bacteriophages recovered from Su⁻ cells. The strain NK5486 contains an amber mutation in the β-galactosidase gene of the *lac* operon (*lacZ*am). When bacteriophages containing a suppressor gene are grown on NK5486 in the presence of the chromogenic substrate 5-bromo-4-chloro-3-indolyl-β-D-galactoside (Xgal), the *lacZ* amber mutation is suppressed, resulting in a blue plaque. This test, which can be conveniently carried out by spot titration, should include both negative and positive controls using Su⁻ bacteriophages and bacteriophages known to contain a suppressor.

The other strain used to test for suppressor activity is AB1157, which contains an amber mutation in the *recA* gene of *E. coli*. The *recA*⁻ cells will not support the growth of *fec*⁻ recombinant bacteriophages (see Chapter 1) unless those bacteriophages also contain a suppressor gene. Thus, *fec*⁻ bacteriophages containing a suppressor gene will form plaques, whereas *fec*⁻ bacteriophages without a suppressor will not form plaques. Again, this test can be conveniently carried out by spot titration.

TESTING THE πVX SYSTEM

To test the πVX system, a reconstruction experiment is carried out (see Fig. 10.6). The bacteriophage strains used include λ Charon 4A (which contains Aam and Bam mutations), a library of human genomic DNA prepared in the λ Charon 4A vector, and a recombinant clone (λHβG1) containing the human β-globin-gene region. λHβG1 was isolated from the library prepared in λ Charon 4A. The plasmid strains used include the microplasmid vectors πVX and πβΔP, which contains a 600-bp sequence located 2 kb to the 5′ side of the human β-globin gene. Each of the bacteriophage strains is grown on bacteria harboring either πVX or πβΔP, and the lysates are then tested for bacteriophages that carry the suppressor gene by plating on a Su⁻ bacterial host. Growth of the bacteriophages in cells carrying πVX should yield no progeny capable of growth on Su⁻ cells since no homology exists between the πVX vector and the genomes of any of the bacteriophages. Similarly, growth of λ Charon 4A on πβΔP should result in no Su⁺ phage. Growth of λHβG1 and the total library on πβΔP should produce bacteriophages capable of growth on a Su⁻ host at frequencies of 10^{-3} and 10^{-8}, respectively. These bacteriophages are then tested for the presence of the suppressor gene and the DNA fragment from the β-globin-gene region.

1. Prepare overnight cultures of two bacterial strains, W3110 r⁻m⁺ (p3) (πVX) and, W3110 r⁻m⁺ (p3) (πβΔP).

2. Obtain high titer stocks of each of the following bacteriophages:

   a. a library of human DNA fragments cloned in bacteriophage λ Charon 4A.

   b. the recombinant bacteriophage λHβG1, which contains a segment of DNA that includes the sequence cloned in πβΔP.

   c. bacteriophage λ Charon 4A.

3. Prepare one set of plate stocks on W3110 r⁻m⁺ (p3) (πVX) and another on W3110 r⁻m⁺ (p3) (πβΔP) as follows:

   a. Add 0.2 ml of the overnight culture of W3110 r⁻m⁺ (p3) (πVX) to each of seven tubes (1–7).

   b. Add 0.2 ml of the overnight culture of W3110 r⁻m⁺ (p3) (πβΔP) to each of seven tubes (8–14).

Then add the bacteriophages as follows:

Tube number		
1	4×10^5 pfu	$\lambda H\beta G1$
2	4×10^5 pfu	λ Charon 4A
3–7	1×10^6 pfu	human DNA library
8	4×10^5 pfu	$\lambda H\beta G1$
9	4×10^5 pfu	λ Charon 4A
10–14	1×10^6 pfu	human DNA library

Incubate for 20 minutes at 37°C. Add 3 ml of molten agar (47°C) and plate the contents of each tube on a fresh, 85-mm plate of bottom agar containing 7.5 μg/ml of tetracycline and 12.5 μg/ml of ampicillin. Incubate at 37°C.

4. Start overnight cultures of K802 (*su*II) and W3110 r⁻m⁺ (p3) (Su⁻).

5. Harvest the plate stocks (see page 65) and titrate the bacteriophages on K802 (*su*II) and W3110 r⁻m⁺ (p3) (Su⁻). The expected titers (pfu/ml) are given in Table 10.3.

6. Start overnight cultures of AB1157 and K5624. Prepare plates containing the chromogenic substrate Xgal (see page 290).

7. Pick 1 plaque from the $\lambda H\beta G1$/Su⁻ plate and 5 plaques from the library/Su⁻ plate and elute the bacteriophages in SM (see page 65). About 10^6 bacteriophages should be recovered from each plaque. Spot 100–200 of each bacteriophage onto lawns of AB1157 and NK5486 (on an Xgal plate) to verify the presence of the suppressor gene (see page 290).

8. Prepare minilysates of the bacteriophages grown on W3110 r⁻m⁺ (p3) (Su⁻) using the procedure described in Chapter 11. Prepare small quantities of the bacteriophage DNAs.

TABLE 10.3. BACTERIOPHAGE TITERS

		Plated on	
Initial host	Bacteriophage	K802 (*su*II)	W3110 r⁻m⁺ (p3) (Su⁻)
W3110 r⁻m⁺ (p3) ($\pi\beta\Delta P$)	$\lambda H\beta G1$	10^{10}	10^7
	λ Charon 4A	10^{10}	0
	library	10^{10}	10^2
W3110 r⁻m⁺ (p3) (πVX)	$\lambda H\beta G1$	10^{10}	<10
	λ Charon 4A	10^{10}	<10
	library	10^{10}	<10

9. Digest each DNA sample with *Eco*RI and compare the fragments with those obtained from λHβG1 DNA, by electrophoresis through a 1% agarose minigel. λHβG1 DNA should yield bands of 0.5, 3.1, 2.25, 1.75, 5.2, and 3.1 kb in addition to the left (20 kb) and right (10 kb) arms of λ Charon 4A. Recombinants between λHβG1 and πβΔP should lack the 5.2-kb fragment but should yield four new fragments 3.6 kb, 2.7 kb, 0.6 kb, and 0.2 kb in length.

Note

The 0.6-kb and 0.2-kb fragments are too small to be easily detected.

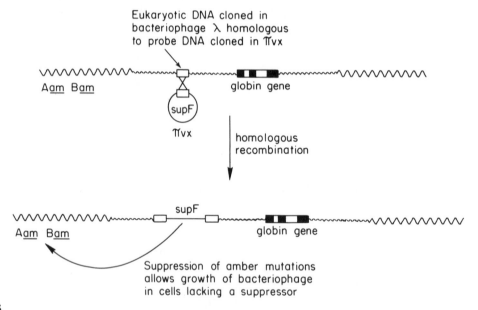

Figure 10.6

The upper line shows in diagrammatic form the recombination event between a πVX derivative containing a sequence that maps near to a globin gene. The recombination results in (1) a duplication of the probe sequence and (2) insertion of the *supF* gene into the bacteriophage λ chromosome. The integrated suppressor, by suppressing the amber mutations in the λ *A* and *B* genes, allows the recombinant bacteriophage to grow on Su⁻ cells.

USING THE πVX SYSTEM

The πVX system provides a potentially powerful method for rapid isolation of bacteriophages that contain sequences homologous to a cloned DNA fragment. It should be particularly useful when multiple mutants or divergent homologs of an existing cloned gene are to be isolated. However, it can also be used to retrieve from a library bacteriophages that contain the desired sequences but grow poorly or that for other reasons are underrepresented in a genomic library. For example, when the πVX system was applied to an existing library of human DNA, several novel bacteriophages containing β-globin genes were recovered that had been missed during multiple rounds of conventional screening by hybridization.

There are two disadvantages to the system. First, the DNA fragment to be used as a probe must be cloned into πVX before the library can be screened, and second, the recombination event results in the unnatural interruption of eukaryotic DNA sequences. The effects of this interruption can be minimized either by choosing a probe fragment that lies to one side of and does not overlap the sequences of interest or by using more than one fragment if the exact location of the sequences of interest within a segment of DNA is not known. Once a bacteriophage has been recovered, it is a fairly simple matter to reclone into πVX one or more segments of the eukaryotic insert. In this way, bacteriophages can be recovered in a second round of screening that do not contain an interruption in the sequence of interest.

Special care must be taken when using the πVX system to avoid contamination of libraries with nonamber bacteriophages. Unnecessary characterization of false recombinants can be avoided by plating the putative recombinant bacteriophages on *lacZ*am or *recA*am bacterial hosts, as described above.

11

Analysis of Recombinant DNA Clones

In this chapter we describe several methods that are used routinely to analyze clones of recombinant DNA. These include protocols for the rapid isolation of small amounts of bacteriophage λ or plasmid DNAs, for the construction of maps of restriction sites in and around the sequences of interest, and for subcloning fragments of DNA into plasmid vectors.

Rapid Isolation of Plasmid or Bacteriophage λ DNA

A number of rapid procedures are available for isolating DNA of bacteriophages or plasmids grown from individual plaques or bacterial colonies. These procedures yield DNA in sufficient quantity and purity for analysis by restriction endonuclease digestion, gel electrophoresis, Southern blotting, or DNA sequencing. Thus, plaques or colonies that are identified by their ability to hybridize a given probe can be rapidly characterized in more detail and prepared for subsequent cloning procedures.

RAPID, SMALL-SCALE ISOLATION OF PLASMID DNA

Over the years many methods have been developed to isolate plasmid DNA from small cultures inoculated with single colonies of bacteria. Some of these are unnecessarily tedious, some are irreproducible, and a few only work with certain strains of *Escherichia coli*. The two methods that follow are rapid, can be carried out with many samples simultaneously, work with all commonly used strains of *E. coli*, and give high yields of plasmids. Typically, 1.5 ml of an unamplified, overnight culture carrying a plasmid such as pXf3 or pAT153 yields 2-3 μg of plasmid DNA.

Boiling Method[1]

1. Inoculate 5 ml of medium containing the appropriate antibiotic with a single bacterial colony. Incubate at 37°C overnight with vigorous agitation.

2. Pour 1.5 ml of the culture into an Eppendorf tube. Centrifuge for 1 minute in an Eppendorf centrifuge. Store the remainder of the overnight culture at 4°C.

3. Remove the medium by aspiration, leaving the bacterial pellet as dry as possible.

4. Resuspend the cell pellet in 0.35 ml of:

   8% sucrose
   0.5% Triton X-100
   50 mM EDTA (pH 8.0)
   10 mM Tris·Cl (pH 8.0)

5. Add 25 μl of a freshly prepared solution of lysozyme (10 mg/ml in 10 mM Tris·Cl, pH 8.0). Mix by vortexing for 3 seconds.

6. Place the tube in a boiling-water bath for 40 seconds.

7. Centrifuge immediately for 10 minutes at room temperature in an Eppendorf centrifuge.

8. Remove the pellet from the Eppendorf tube with a toothpick.

9. To the supernatant, add 40 μl of 2.5 M sodium acetate and 420 μl of isopropanol. Mix by vortexing and store for 15 minutes in a dry-ice/ethanol bath.

[1]Holmes and Quigley (1981).

10. Centrifuge for 15 minutes at 4°C in an Eppendorf centrifuge.

11. Dry the pellet and resuspend it in 50 μl of TE (pH 8.0) containing DNase-free RNase (50 μg/ml). Incubate for 10 minutes at 37°C. This treatment eliminates RNA that can mask small fragments of DNA in agarose gels.

12. Remove 10 μl of the solution to a fresh Eppendorf tube. Add 1.2 μl of the appropriate buffer and 1 unit of the desired restriction enzyme. Incubate for 1–2 hours at the appropriate temperature. Store the remainder of the minipreparation at −20°C.

13. Analyze the DNA fragments in the restriction digest by gel electrophoresis.

Alkaline Lysis Method[2]

1. Inoculate 5 ml of medium containing the appropriate antibiotic with a single bacterial colony. Incubate at 37°C overnight with vigorous shaking.

   7.5ml

2. Pour 1.5 ml of the culture into an Eppendorf tube. Centrifuge for 1 minute in an Eppendorf centrifuge. Store the remainder of the overnight culture at 4°C.

   45ml

3. Remove the medium by aspiration, leaving the bacterial pellet as dry as possible.

4. Resuspend the pellet by vortexing in 100 μl of an ice-cold solution of:

   1ml / 500ul

   3ml

   50 mM glucose
   10 mM EDTA
   25 mM Tris · Cl (pH 8.0)
   4 mg/ml lysozyme

   Add powdered lysozyme to the solution just before use.

5. Store for 5 minutes at room temperature. The top of the tube need not be closed during this period.

   Make 2ml → 6ml
   for 1ml
   2. 4.2ml 1N NaOH
   1. 250ul 20% SDS
   9.75ml H₂O

   2ml

6. Add 200 μl of a freshly prepared, ice-cold solution of:

   0.2 N NaOH
   1% SDS

   Close the top of the tube and mix the contents by inverting the tube rapidly two or three times. Do not vortex. Store the tube on ice for 5 minutes.

   3ml   *4.5 ml*   *750ul*

7. Add 150 μl of an ice-cold solution of potassium acetate (~ pH 4.8) made up as follows: To 60 ml of 5 M potassium acetate, add 11.5 ml of glacial acetic acid and 28.5 ml of H_2O. The resulting solution is 3 M with respect to potassium and 5 M with respect to acetate.

   Close the cap of the tube and vortex it gently in an inverted position for 10 seconds. Store on ice for 5 mintues.

   RC; 10' at 8.5K

8. Centrifuge for 5 minutes in an Eppendorf centrifuge at 4°C.

9. Transfer the supernatant to a fresh tube.

---

[2]This protocol, contributed by D. Ish-Horowicz, is a modification of the method of Birnboim and Doly (1979).

1 min

2ml
5' at 8.5k

10. Add an equal volume of phenol/chloroform. Mix by vortexing. After centrifuging for 2 minutes in an Eppendorf centrifuge, transfer the supernatant to a fresh tube.

into clean
10ml
Oakridge

11. Add two volumes of ethanol at room temperature. Mix by vortexing. Stand at room temperature for 2 minutes.

10min at 8.5k

12. Centrifuge for 5 minutes in an Eppendorf centrifuge at room temperature.

13. Remove the supernatant. Stand the tube in an inverted position on a paper towel to allow all of the fluid to drain away.

14. Add 1 ml of 70% ethanol. Vortex briefly and then recentrifuge. *5' at 8.5k*

ultra

15. Again remove all of the supernatant. Dry the pellet briefly in a vacuum desiccator.

Resuspend
in 300 µl
TE

16. Add 50 µl of TE (pH 8.0) containing DNase-free pancreatic RNase *OMIT* (20 µg/ml). Vortex briefly.

17. Remove 10 µl of the solution to a new Eppendorf tube. Add 1.2 µl of the appropriate buffer and 1 unit of the desired restriction enzyme. Incubate for 1–2 hours at the appropriate temperature. Store the remainder of the preparation at −20°C.

18. Analyze the DNA fragments in the restriction digest by gel electrophoresis.

Rapid Disruption of Colonies to Test for Inserts in Plasmids

Frequently, it is possible to analyze the size of plasmid DNA without restriction endonuclease digestion to ascertain whether or not an insert is present. The protocol described below yields enough DNA to load on a single lane of an agarose gel (Barnes 1977).

1. Grow bacterial colonies to a large size (2–3 mm) on agar medium containing the appropriate antibiotic.

2. Using a sterile toothpick, transfer a small quantity of the colony to a master plate. Transfer the remainder of the colony to an Eppendorf tube or to a well of a microtiter dish containing 25 μl of 50 mM NaOH, 0.5% SDS, 5 mM EDTA, and 0.025% bromocresol green (cracking buffer). Dispense the colony by stirring the solution gently with the toothpick.

3. Cap the tube or seal the microtiter well with electrical tape and incubate at 68°C for 45–60 minutes.

4. Add 2.5 μl of 25% Ficoll and load the contents onto a 0.7% agarose gel without ethidium bromide.

5. After electrophoresis, stain the gel by soaking for 45 minutes in a solution of ethidium bromide (0.5 μg/ml H_2O).

   The rate of migration of superhelical DNA in the absence of ethidium bromide more faithfully reflects its molecular weight than in the presence of ethidium bromide.

RAPID, SMALL-SCALE ISOLATION OF BACTERIOPHAGE λ DNA

Plate Lysate Method[3]

1. Using a pasteur pipette, pick a single, well-isolated plaque (see page 64) into 1 ml of SM containing a drop of chloroform. Store at 4°C for 4-6 hours to allow the bacteriophage particles to diffuse out of the top agarose.

2. Mix 50-100 μl of the bacteriophage suspension ($\sim 10^5$ pfu) with 100 μl of indicator bacteria (see page 63). Incubate for 20 minutes at 37°C. Add 2.5 ml of molten, 0.7% top *agarose* and spread on the surface of a freshly made, 85-mm plate containing 30 ml of NZYCM plus 1.5% *agarose*.

   Note. Do not use agar since most batches of agar contain potent inhibitors of restriction endonucleases.

3. Invert the plate and incubate at 37°C for 9-14 hours until the plaques cover almost the entire surface of the plate.

4. Add 5 ml of SM directly onto the plate and allow the bacteriophages to elute by storing for at least 1-2 hours at room temperature with constant, gentle shaking.

5. Transfer the SM to a 12-ml polypropylene centrifuge tube, and remove the bacterial debris by centrifugation at 8000g for 10 minutes at 4°C.

6. Recover the supernatant and add RNase A and DNase I, each to a final concentration of 1 μg/ml. Incubate for 30 minutes at 37°C.

7. Add an equal volume of a solution containing 20% (w/v) polyethylene glycol and 2 M NaCl in SM and incubate for 1 hour at 0°C (ice water). This solution may be prepared in advance and stored at 4°C.

8. Recover the precipitated bacteriophage particles by centrifugation at 10,000g for 20 minutes at 4°C.

9. Remove the supernatant by aspiration. Stand the tube in an inverted position on a paper towel to allow all of the fluid to drain away. *transfer to eppendorf - extract w/ CHCl₃*

10. Add 0.5 ml of SM and resuspend the bacteriophage particles by vortexing.

11. Centrifuge at 8000g for 2 minutes at 4°C to remove debris.

[3] E. F. Fritsch (unpubl.).

12. Transfer the supernatant to a fresh Eppendorf tube. Add 5 μl of 10% SDS and 5 μl of 0.5 M EDTA (pH 8.0). Incubate at 68°C for 15 minutes.

13. Extract once with phenol, once with phenol/chloroform (1:1), and once with chloroform. Transfer the aqueous phase to a fresh Eppendorf tube between each extraction.

14. To the final aqueous phase add an equal volume of isopropanol. Store at −70°C for 20 minutes. Thaw and centrifuge in an Eppendorf centrifuge for 15 minutes at 4°C.

15. Wash the pellet with 70% ethanol. Dry the pellet and resuspend it in 50 μl of TE (pH 8.0).

16. Remove 10 μl of the solution to a fresh Eppendorf tube. Add 1.2 μl of the appropriate buffer and 1–2 units of the desired restriction enzyme. Incubate for 1–2 hours at the appropriate temperature. Store the remainder of the preparation at −20°C.

17. Analyze the DNA fragments in the restriction digest by gel electrophoresis.

Notes

i. The addition of DNase-free pancreatic RNase (20 μg/ml) to the restriction enzyme buffer sometimes improves the digestion.

ii. The infectivity of DNA prepared in this way and packaged in vitro is identical to that of more highly purified, bacteriophage λ DNA.

Liquid Culture[4]

1. Pick a single, well-isolated plaque (see page 63) and put it into 1 ml of SM containing a drop of chloroform. Store at 4°C for 4–6 hours to allow the bacteriophage particles to diffuse out of the top agarose.

2. In a 25-ml tube, mix 0.5 ml of the bacteriophage suspension (approximately 3×10^6 bacteriophages) with 1.6×10^8 bacterial cells in stationary phase. Incubate for 15 minutes at 37°C.

   Both the absolute amounts of bacteriophages and cells and the ratio between them are important. It is best to use bacteriophage stocks or plaque eluates of known titer.

3. Add 4 ml of NZCYM medium. Incubate at 37°C with agitation for approximately 9 hours. The culture should clear but very little debris should be evident.

4. Add 0.1 ml of chloroform to the culture and shake it for a further 15 minutes at 37°C. Transfer the lysate to a 5-ml polypropylene centrifuge tube. Centrifuge at 8000g for 10 minutes at 4°C.

5. Continue with the plate lysate method at step 6 (see page 65).

[4]Leder et al. (1977).

Constructing Maps of Sites Cleaved by Restriction Endonucleases

A number of strategies are used to construct maps of the sites at which restriction enzymes cleave DNA; it is usually necessary to employ more than one of them to obtain maps that are sufficiently accurate and detailed to be useful. The techniques most commonly used are:

simultaneous digestion with combinations of restriction enzymes;
sequential digestion of an isolated DNA fragment with a second restriction enzyme;
partial digestion, either of unlabeled DNA or DNA labeled specifically at only one terminus;
partial exonucleolytic digestion of the DNA followed by digestion with a restriction enzyme.

Which of these techniques to employ depends in part on the size of the DNA mapped and in part on the level of detail to be required. For example, to map a piece of DNA more than 2–3 kb in length, it is usual to begin by analyzing the fragments produced by simultaneous digestion with pairs of restriction enzymes that cleave the DNA infrequently (e.g., enzymes that recognize hexanucleotide sequences).

From the sizes of the DNA fragments, it is usually possible to deduce the relative locations of at least some of the cleavage sites. As the number of pairwise combinations of enzymes increases, the number of defined sites increases until, eventually, no ambiguities remain. The resolution of maps constructed in this way depends on the accuracy with which the sizes of the DNA fragments can be determined relative to those of the markers. However, even in the best of circumstances, it is difficult to produce a large-scale map that is accurate to less than 100–200 bp. To construct fine-structure maps, it is usually necessary to isolate a particular fragment of DNA, to cleave it with combinations of enzymes that recognize tetranucleotide sequences, and to measure the size of the fragments by electrophoresis through polyacrylamide gels.

At all stages of mapping, it is useful to work from a fixed point (e.g., one of the termini of linear DNA). Two procedures have been developed for this purpose. The first involves partial digestion of an end-labeled fragment of linear DNA (Smith and Birnstiel 1976). By adjusting the conditions of digestion so that on average only one cleavage occurs per molecule, a ladder of discrete, labeled DNA fragments is generated. The sizes of these fragments reflect the distance between the labeled end of the DNA and a given restriction site. The difference in size between two adjacent fragments on the gel defines the distance between neighboring restriction sites. In the second procedure, a DNA fragment is digested to different extents with a progressive, double-strand exonuclease such as *Bal*31. DNA isolated at each time point is then digested with one or more restriction enzymes. Fragments disappear from the digest in the order in which restriction sites occur in the DNA (Legerski et al. 1978; see pages 137ff).

MAPPING BY SINGLE AND MULTIPLE DIGESTIONS

1. Digest the cloned DNA with restriction enzymes that could be expected to cleave only a small number of sites in the insert and for which the target sites in the vector are known. Digest individual aliquots (5 μg) of DNA with a different enzyme and analyze 0.5-μg samples by electrophoresis through a 0.9% agarose gel together with markers of the appropriate size. Meanwhile, store the remainder of the samples at $-20°C$. Count the number of DNA bands seen in each digest, measure the distance each of them has traveled, and calculate their molecular weights.

2. Cleave aliquots (0.5 μg) of each primary digest with a series of additional restriction enzymes, choosing enzymes that cut at only a few sites within the insert.

   It may be necessary

   a. to purify the DNA from the primary digest by extraction with phenol/chloroform and precipitation with ethanol, or

   b. to dilute the DNA if the second restriction enzyme will not cut efficiently in the buffer used for the primary digest (see pages 102ff).

   Again, analyze the size of the digestion products by gel electrophoresis, this time including not only markers of known size, but also samples of the primary digest.

   By using this procedure, you can be sure that the first restriction enzyme digest has gone to completion before proceeding with the second. If the pairs of primary and secondary digests are loaded into adjacent slots on the gel, it is easy to detect small differences in migration.

   For a linear DNA, the number of double-digestion fragments equals the number of fragments generated by the first enzyme plus the number of fragments generated by the second enzyme minus 1.

   For a circular DNA, the number of double-digestion fragments equals the number of fragments generated by the first enzyme plus the number of fragments generated by the second enzyme.

   If the number of observed fragments is less than the expected number, search for very small fragments or doublets (two fragments that migrate at the same rate through agarose or polyacrylamide gels.)

   Maps are built up from these data by a process that is part trial and error and part an exercise in basic logic (see Lawn et al. 1978). A complete map is obtained when all cleavage sites are assigned to unambiguous locations that are internally consistent with one another.

SEQUENTIAL DIGESTION

Several techniques are available for the isolation of fragments of DNA from gels (see Chapter 5); all of them yield preparations that can be cleaved by restriction enzymes. However, for the purposes of sequential digestion, the isolation of DNA from gels cast with low-melting-temperature (LMT) agarose has a major advantage: A gel slice containing a DNA fragment can be dissolved at 65°C without denaturing the DNA, cooled to 37°C without gelling of the agarose, and incubated with a second restriction enzyme. Thus, sequential digestion of DNA can be performed rapidly and efficiently (Parker and Seed 1980).

1. Dissolve LMT agarose in electrophoresis buffer (TBE or TAE; see page 454) and cast the gel in the usual fashion. The strength of solidified LMT gels is less than that of gels cast with standard agarose, and it is best to pour and run them in the cold, where they gain some rigidity. LMT agarose gels may be poured with or without ethidium bromide.

   Note. Avoid using electrophoresis buffers containing phosphate; such buffers alter the melting characteristics of LMT agarose.

2. Load the DNA sample and carry out electrophoresis in the usual way.

3. After electrophoresis, stain the DNA, if necessary, by soaking the gel in a solution of ethidium bromide (0.5 μg/ml in water) for 30 minutes. Destain by immersing the gel for 10 minutes in water and examine the DNA fragments using a long-wave UV light. Using a razor blade, cut out the appropriate band in as thin a slice of gel as possible. Trim any excess agarose from the DNA.

   After cutting out the band of DNA, photograph the gel to obtain a permanent record.

4. Place the slice of gel in a small tube and add at least 10 volumes of cold water. Soak the gel slice for 1–2 hours at 4°C. During this time, most of the electrophoresis buffer and the free ethidium bromide diffuses out of the gel.

5. Remove the excess liquid and store the gel slice at 4°C in a tightly sealed tube until needed.

6. When required, place the gel slice at 65°C until the agarose completely melts (2–5 minutes) and then either transfer the entire sample directly to a water bath at 37°C or dispense aliquots into tubes prewarmed to 37°C. Add 0.1 volume of the appropriate 10× restriction buffer (warmed to 37°C).

7. Add nuclease-free bovine serum albumin (BSA) from a 10% solution in water to a final concentration of 0.1% (the BSA may be mixed with the 10× restriction buffer and added with it, if desired).

8. Add the restriction enzyme. Usually, two to three times more enzyme is required for complete digestion of DNA in the presence of LMT agarose than under normal conditions. Incubate for the appropriate length of time.

9. Heat the reaction mixture to 65°C for 5 minutes, add gel-loading buffer I (see page 455), and load in a slot of a horizontal gel cast with conventional agarose. The LMT agarose should be allowed to harden in the wells before electrophoresis is begun. Before loading, samples containing marker DNAs should be adjusted so that they contain approximately the same concentration of LMT agarose and BSA as the samples under study.

10. Electrophoresis is carried out, and the gel is stained and photographed as usual.

Notes

i. Because the DNA that is melted out of the gel cannot be further concentrated, the amount of DNA that can be applied to the second gel is dependent on the volume of the gel slice and the amount of DNA applied to the LMT agarose gel. For optimal efficiency, as much DNA as possible should be loaded in a small volume into a narrow slot of a thin, LMT agarose gel, and the resulting slice should be trimmed of any excess agarose.

In many cases, it is advisable to concentrate the DNA after the first digestion by extraction with phenol and precipitation with ethanol.

ii. DNA can be digested to completion in the presence of as much as 1.2% LMT agarose.

PARTIAL DIGESTION

Two types of partial digestion have been used to map restriction sites in DNA. The aim of the first method is to compare the sizes of partial- and complete-digestion products and to deduce which fragments might be adjacent to one another in the original DNA.

Unfortunately, the method is of limited usefulness because the number of possible partial-digestion products quickly becomes unmanageable as the number of restriction sites increases.

For example, in a linear DNA molecule:

number of restriction sites	1	2	3	4	5	6
number of possible partial-digestion products		0	2	5	9	14	20

In general, the number of partial-digestion products (F) of a linear DNA molecule that contains N restriction sites is given by the formula

$$F_{N+1} = \frac{N^2 + 3N}{2}$$

The picture is much simplified if the fragment of DNA is labeled at one end and the products of partial digestion are monitored by autoradiography (Smith and Birnstiel 1976). In this case:

i. The number of labeled partial-digestion products is equal to the number of restriction sites within the DNA.

ii. The labeled fragments form a simple overlapping series with a common labeled terminus.

iii. The ascending order of fragments in the gel corresponds directly to the order of restriction sites along the DNA.

iv. The products of partial digestion by several different enzymes can be analyzed simultaneously on the same gel, so that the relative positions of restriction sites can be obtained from a single autoradiograph.

Methods for labeling isolated DNA fragments at their 5′ termini with [^{32}P]phosphoryl groups using polynucleotide kinase and [γ-^{32}P]ATP or their recessed 3′ termini using [α-^{32}P]dNTPs with the Klenow fragment of *E. coli* DNA polymerase I are given in Chapter 4. The labeled DNA is then cleaved asymmetrically with a suitable restriction enzyme into two fragments that are isolated by gel electrophoresis and separately mapped by partial digestion.

Below we describe a method to end-label crude, small-scale plasmid DNA preparations that have been cleaved with a restriction enzyme that creates recessed 3′ termini. This labeled DNA can then be mapped by the techniques described above.

A Rapid Method to Label Recessed 3′ Termini in Small-scale Preparations of Plasmid DNA[5]

1. Prepare plasmid DNA from 1.5 ml of an overnight culture of bacteria using the alkaline lysis procedure (see page 368).

2. Set up a digestion with a restriction enzyme that creates recessed 3′ termini at a suitable position in the plasmid DNA. Mix:

DNA preparation	10	μl
10× restriction enzyme buffer	1.2	μl
desired restriction enzyme	1	unit

   Incubate for 1–2 hours at the appropriate temperature.

3. Add 10 μCi (~1 μl) of the appropriate [α-^{32}P]dNTP (400 Ci/mmole) in stabilized aqueous form. The nucleotide should be complementary to the first unpaired nucleotide of the protruding 5′ terminus; e.g., when labeling a terminus created by *Bam*HI, [α-^{32}P]dGTP is added to the reaction:

4. Add 0.2 μl (~1–2 units) of the Klenow fragment of *E. coli* DNA polymerase I. Incubate at 20°C for 30 minutes.

5. Separate the labeled DNA from unincorporated nucleotides by passage through a column of Sephadex G-75 or by spun-column chromatography using Sephadex G-50 (see Appendix A). The labeled DNA may then be mapped by partial digestion with a second restriction enzyme.

Notes

i. The end-labeling reaction works well in all restriction enzyme buffers.

ii. This method of end-labeling is also used when it is necessary to identify small (< 200 bp) fragments in digests of small-scale preparations of plasmid DNA. In this case, load about $^1/_{10}$ of the end-labeling reaction directly onto the slot of an agarose or an acrylamide gel. After electrophoresis, cover the gel with Saran Wrap and expose it to X-ray film to obtain an autoradiographic image (see page 470).

[5]Drouin (1980).

small; a specified amount of "carrier" DNA (usually plasmid or bacteriophage λ DNA) is therefore added in order to more easily control the rate at which the restriction enzyme digests the labeled DNA (Smith and Birnstiel [1976], adapted by D. Engel [pers. comm.]).

1. Mix:

   > ~ 10^4 cpm of DNA labeled at its 3′ or 5′ end
   >   (dissolved in no more than 8 μl of H_2O)
   > carrier DNA   1 μg
   > 10× restriction buffer   1 μl
   > H_2O to 10 μl
   > restriction enzyme   1–2 units

2. Incubate at 37°C, withdrawing 1.8-μl aliquots at 2, 5, 10, 15, and 30 minutes into 1 μl of 0.5 M EDTA.

   At this stage, all the aliquots should be combined in order to enhance the detection of poorly represented, partial-digestion products.

3. Add 2 μl of gel-loading dye I, and load 5 μl of the combined reaction onto a single lane of a 5% polyacrylamide gel and 5 μl onto a single lane of a 1.4% agarose gel. As markers, use 10^4 cpm of a mixture of end-labeled fragments of pBR322 DNA.

4. Carry out electrophoresis until the bromophenol blue has migrated two thirds of the length of each gel.

5. Cover the gels with Saran Wrap. Expose them to X-ray film at −70°C to obtain autoradiographic images (see pages 470ff).

Note

Unambiguous results can be easily obtained for a purified DNA fragment containing heterologous termini (e.g., BamHI and EcoRI) that can be labeled in separate reactions with different dNTPs—[α-^{32}P]dGTP in the case of a BamHI terminus and [α-^{32}P]dATP in the case of an EcoRI terminus. Since only one of the two ends of the DNA become labeled in each reaction, partial digestion can be carried out without further treatment of the DNA fragment.

Mapping by Digestion of Exonuclease-treated DNA

The construction of maps by restriction enzyme digestion of DNA that has been degraded to various extents by Bal31 is described on pages 137ff.

Southern Transfer

The techniques described on the previous pages yield a map that displays the order of restriction sites and the sizes of restriction fragments. Localization of particular sequences of DNA within these fragments is usually accomplished by the transfer technique described by Southern (1975). DNA fragments that have been separated according to size by electrophoresis through an agarose gel are denatured, transferred to a nitrocellulose filter, and immobilized. The relative positions of the DNA fragments in the gel are preserved during their transfer to the filter. The DNA attached to the filter is then hybridized to ^{32}P-labeled DNA or RNA, and autoradiography is used to locate the position of any bands complementary to the radioactive probe. This technique can be used not only to locate specific sequences in cloned DNA, but also to identify sequences within digests of total eukaryotic DNA (Botchan et al. 1976; Jeffreys and Flavell 1977). The techniques described on the following pages are applicable to both genomic and cloned DNA, with only minor modifications.

TRANSFER OF DNA FROM AGAROSE GELS TO NITROCELLULOSE PAPER

1. After electrophoresis is completed, stain the DNA with ethidium bromide and photograph the gel. It is sometimes useful to place a ruler alongside the gel so that the distance that any given band of DNA has migrated can be read directly from the photographic image.

   Half a microgram of the DNA of a recombinant λ bacteriophage or 0.2 μg of a recombinant plasmid DNA is more than sufficient to allow inserted DNA sequences to be easily detected by Southern hybridization. However, 10 μg of total mammalian DNA (haploid genome = 3×10^9 bp) must be applied to a single gel-slot in order to be able to detect sequences that occur only at the single-copy level.

2. Transfer the gel to a glass baking dish and trim away any unused areas of the gel with a razor blade.

3. Denature the DNA by soaking the gel in several volumes of 1.5 M NaCl and 0.5 M NaOH for 1 hour at room temperature with constant stirring or shaking.

   Note. After electrophoresis, some investigators prefer to hydrolyze the DNA partially by acid depurination (by soaking the gel twice for 15 minutes in 0.25 M HCl at room temperature) prior to alkali denaturation (Wahl et al. 1979). This acid-induced cleavage aids in the transfer of large DNA fragments. However, it is important not to let the hydrolysis reaction proceed too far; otherwise the DNA is cleaved into fragments that are too short to bind efficiently to the filter (< 300 bp). In most cases, sufficient breakage occurs during the time that the ethidium-bromide-stained DNA is exposed to UV irradiation to allow efficient transfer of DNA up to 20 kb in length.

4. Neutralize the gel by soaking in several volumes of a solution of 1 M Tris · Cl (pH 8.0) and 1.5 M NaCl for 1 hour at room temperature with constant shaking or stirring.

5. Wrap a piece of Whatman 3MM paper around a piece of plexiglass or a stack of glass plates. Place the wrapped support inside a large baking dish. The support should be longer and wider than the gel. Fill the dish with 10× SSC (see page 447) almost to the top of the support and smooth out all air bubbles in the 3MM paper with a smooth glass rod.

   Note. 20× SSPE can also be used as the transfer buffer.

6. Invert the gel so that its original underside is now uppermost. Place the gel on the damp 3MM paper. Make sure there are no air bubbles between the 3MM paper and the gel.

7. Using a fresh scalpel or a paper cutter, cut a piece of nitrocellulose filter (Schleicher & Schuell BA 85 or Millipore HAHY) about 1–2 mm larger than the gel in both dimensions. Use gloves and Millipore forceps to handle the nitrocellulose.

8. Float the nitrocellulose filter on the surface of a solution of 2× SSC until it wets completely from beneath. Then immerse the filter in the 2× SSC for 2–3 minutes.

    The rate at which different batches of nitrocellulose wet varies enormously. If the filter is not saturated after several minutes, it should be replaced with a new piece of nitrocellulose since the transfer of DNA to unevenly wet nitrocellulose is unreliable.

    Note. Nitrocellulose that has been touched by greasy hands will never wet!

9. Place the wet nitrocellulose filter on top of the gel, so that one edge extends just over the line of slots at the top of the gel. Be careful to remove all air bubbles that are trapped between the gel and the filter.

10. Wet two pieces of Whatman 3MM paper, cut to exactly the same size as the gel, in 2× SSC and place them on top of the nitrocellulose filter. Again remove all air bubbles.

11. Cut a stack of paper towels (5–8 cm high) just smaller than the 3MM paper. Place the towels on the 3MM paper. Put a glass plate on top of the stack and weigh it down with a 500-g weight (see Fig. 11.1). The objective is to set up a flow of liquid from the reservoir through the gel and the nitrocellulose paper, so that DNA fragments are eluted from the gel and are deposited onto the nitrocellulose paper. To prevent short circuiting of fluid between the paper towels and the 3MM paper under the gel, many workers surround the gel with a water-tight border of Saran Wrap.

12. Allow transfer of DNA to proceed for about 12–24 hours. As the towels become wet, they should be replaced.

    The rate of transfer of DNA depends on the size of the DNA fragment and the porosity of the gel. Small fragments of DNA (< 1 kb) transfer from an 0.8% agarose gel within an hour or 2 while transfer of DNA greater than 15 kb takes 15 hours or more.

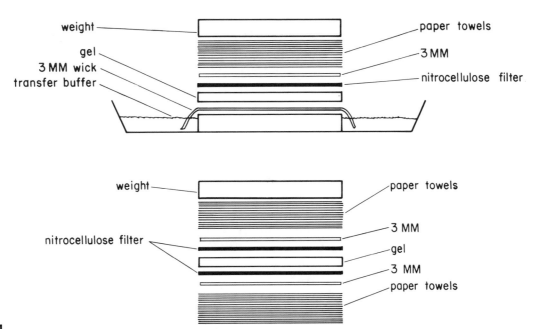

Figure 11.1

Methods for transfer of DNA from agarose gels to nitrocellulose filter paper. *(Top)* The most common system for transfer of DNA (see text for explanation). *(Bottom)* A system for preparing duplicate nitrocellulose filters from a single gel; the transfer buffer is supplied only by the liquid in the agarose gel itself.

13. Remove the towels and the 3MM filters above the gel. Turn over the dehydrated gel and filter and lay them, gel side up, on a dry sheet of 3MM paper. Mark the positions of the gel slots on the filter with a very soft pencil or a ball-point pen.

14. Peel off and discard the gel. Soak the filter in 6× SSC at room temperature for 5 minutes.

15. Allow excess fluid to drain from the filter and set the filter to dry at room temperature on a sheet of 3MM paper.

16. Place the dried filter between two sheets of 3MM paper. Bake for 2 hours at 80°C under vacuum.

    If the filter is not to be used immediately in hybridization experiments, it should be stored at room temperature under vacuum between sheets of 3MM paper.

Notes

i. The procedure given above and illustrated in Figure 11.1 (top) is the most widely used of many different methods that are employed to transfer DNA from gels to filters. Of the alternative set-ups, perhaps the best is that shown in Figure 11.1 (bottom) in which DNA is transferred from a single gel to two nitrocellulose filters simultaneously. Transfer of small DNAs (< 5 kb) occurs extremely rapidly in this system and is essentially complete in 3–4 hours. However, the only source of transfer buffer is the liquid trapped in the gel itself, and transfer of high-molecular-weight DNA fragments (> 10 kb) is therefore somewhat inefficient.

ii. Marker DNAs that will hybridize to the radioactive probe can serve simultaneously to orient the filter and to provide size markers directly on the autoradiograph. The amount of DNA in the marker lane that will hybridize to the probe should be equal to or only slightly more than the amount of probe-specific DNA in the sample lanes.

HYBRIDIZATION OF SOUTHERN FILTERS

1. Float the baked filter on the surface of 6× SSC until it wets from beneath. Immerse the filter in the 6× SSC for 2 minutes.

2. Slip the wet filter into a heat-sealable plastic bag (e.g., Sears' Seal-n-Save).

3. Add 0.2 ml of prehybridization fluid warmed to 68°C for each square centimeter of nitrocellulose filter.

   Prehybridization fluid
   6× SSC
   0.5% SDS
   5× Denhardt's solution (see page 448)
   100 μg/ml denatured, salmon sperm DNA (see page 327)

4. Squeeze as much air as possible from the bag. Seal the open end of the bag with the heat sealer. Incubate the bag for 2–4 hours submerged in a water bath at 68°C.
   Often, small bubbles of air form on the surface of the filter as the temperature of the prehybridization solution rises to 68°C. It is important that these bubbles be removed by occasionally agitating the fluid in the bag; otherwise the components of the prehybridization fluid will not be able to coat the filter evenly.

5. Remove the bag from the water bath. Open the bag by cutting off one corner with scissors. Squeeze out as much prehybridization solution as possible.

6. Using a pasteur pipette, add the hybridization solution to the bag. Use just enough solution to keep the filter wet (50 μl/cm^2 of filter).

   Hybridization solution
   6× SSC
   0.01 M EDTA
   ^{32}P-labeled denatured probe DNA
   5× Denhardt's solution
   0.5% SDS
   100 μg/ml denatured, salmon sperm DNA

   Typical hybridization conditions for Southern filters are given in Table 11.1.

7. Squeeze as much air as possible from the bag. Seal the cut edge with the heat sealer so that as few air bubbles as possible are trapped in the bag.

TABLE 11.1 HYBRIDIZATION CONDITIONS FOR SOUTHERN FILTERS

DNA on filter	Sp. act. of probe DNA (cpm/μg)	Amount of probe added	Time of hybridization (hr)
Fragments of cloned DNA ($\sim$ 100 ng/fragment)	10^7	10^5–10^6 cpm (0.01–0.1 μg)	3–4
Total eukaryotic DNA (10 μg)	10^8	1×10^7 cpm -5×10^7 (0.1–0.5 μg)	12–16

8. Incubate the bag submerged in a water bath at 68°C for the required hybridization period.

9. Remove the bag from the water bath and quickly cut along the length of three sides. Using gloves, remove the filter and immediately submerge it in a tray containing a solution of 2× SSC and 0.5% SDS at room temperature.

   Note. Do not allow the filter to dry out at any stage during the washing procedure.

10. After 5 minutes, transfer the filter to a fresh tray containing a solution of 2× SSC and 0.1% SDS and incubate for 15 minutes at room temperature with occasional gentle agitation.

11. Transfer the filter to a flat-bottomed plastic box containing a solution of 0.1× SSC and 0.5% SDS. Incubate at 68°C for 2 hours with gentle agitation. Change the buffer and continue incubating for a further 30 minutes.

    Note. If the homology between the probe and the DNA bound to the filter is inexact, the washing should be carried out under less stringent conditions. In general, washing should be carried out at $T_m = -12$°C.
    The following relationships are useful:

    a. $T_m = 69.3 + 0.41 \cdot (G + C)\%$ (Marmur and Doty 1962)

    b. The T_m of a duplex DNA decreases by 1°C with every increase of 1% in the number of mismatched base pairs (Bonner et al. 1973).

    c. $(T_m)\mu_2 - (T_m)\mu_1 = 18.5 \log_{10} \frac{\mu_2}{\mu_1}$

    where μ_1 and μ_2 are the ionic strengths of two solutions (Dove and Davidson 1962).

12. Dry the filter at room temperature on a sheet of Whatman 3MM paper.

13. Wrap the filter in Saran Wrap and apply to X-ray film to obtain an autoradiographic image (see page 470).

Notes

Hybridization may also be carried out in:

a. flat-bottomed plastic boxes.

b. buffers containing formamide. Each increase of 1% in the formamide concentration lowers the T_m of a DNA duplex by 0.7°C (McConaughy et al. 1969; Casey and Davidson 1977).

Subcloning Small DNA Fragments into Plasmid Vectors

Subcloning fragments of DNA from recombinants originally constructed in bacteriophage λ or plasmid vectors is one of the most frequently used procedures in the molecular cloning of DNA. The procedure is used to obtain well-defined hybridization probes, to simplify the task of constructing fine-structure maps of restriction sites, to obtain large amounts of specific fragments of DNA for sequencing, and to construct recombinant plasmids that contain novel combinations of foreign DNA fragments.

As mentioned previously, the major difficulty in cloning in plasmids is to distinguish between the desired recombinants and the recircularized vector DNA. However, this problem can be minimized by (1) choosing ratios of vector to insert DNA that favor intermolecular ligation at the expense of intramolecular ligation and (2) cloning by insertional inactivation or by directional insertion. When these procedures are combined with screening by the Grunstein-Hogness (1975) technique, subcloning of DNA fragments into plasmids usually presents few problems.

SUBCLONING DNA FRAGMENTS WITH COHESIVE ENDS

1. Mix together 200 ng of linearized plasmid DNA (phosphatase treated, if desired) and a threefold molar excess of the fragment to be subcloned. Precipitate the mixed DNAs with ethanol and wash the pellet.

2. Dissolve the DNA in 8 µl of TE (pH 8.0). Add 1 µl of 10× ligation buffer (see page 474). Reserve an aliquot (1 µl) for later analysis by gel electrophoresis. Add 10 units of T4 ligase and mix by vortexing briefly. Incubate for 8 hours at 12°C (for cohesive ends created by *Eco*RI) or 16–20°C (for all other cohesive ends).

3. After ligation, remove a second aliquot (1 µl). Analyze both aliquots by electrophoresis through an agarose gel to check for successful ligation.

4. Use 2.5 µl of the remaining sample to transform bacteria to antibiotic resistance and screen the resulting colonies by hybridization, insertional inactivation, or by analyzing the structure of plasmids isolated by one of the rapid methods described on pages 361ff.

Note

There are times when it is necessary to subclone a restriction fragment that contains one protruding and one blunt terminus. For example, consider the problem of subcloning a fragment with *Eco*RI/*Hae*III termini into pBR322. A suitable vector may be prepared by digesting the plasmid with *Hin*dIII and filling in the resulting protruding ends by the Klenow fragment of DNA polymerase I. The DNA then is digested with *Eco*RI and purified by electrophoresis through an agarose gel or by chromatography through Sepharose CL-4B (see page 465). The eluted vector DNA, containing one blunt end and one *Eco*RI sticky end, is ligated to the *Eco*RI/*Hae*III fragment. Because the ends of the vector DNA cannot ligate to one another, the majority of transformed cells carry plasmids containing the desired insert.

THE USE OF SYNTHETIC DNA LINKERS IN SUBCLONING

There are circumstances in which it is convenient to use synthetic DNA linkers to clone fragments of DNA—for example, during cDNA cloning or when subcloning a blunt-ended fragment. Although such fragments can be cloned by direct ligation to blunt-ended, linear plasmid DNA, the efficiency of this process is poor for three reasons. First, the K_m for the activity of T4 ligase on blunt-ended DNA is nearly 100 times higher than its K_m on DNA with cohesive ends. Thus, ligation of blunt-ended DNA requires a high concentration of enzyme and a high concentration of DNA ends (greater than 1 μM). Very large amounts of the restriction fragment to be cloned are therefore needed. Second, during blunt-end ligation, a fraction of the plasmid vector will recircularize. And third, because of the high concentration of the fragments to be cloned, many recombinant plasmids will contain more than one insert of foreign DNA.

A better way to clone a blunt-ended DNA fragment is to use synthetic, duplex DNA linkers (Bahl et al. 1976; Scheller et al. 1977). These blunt-ended, synthetic DNAs contain one or more restriction sites that generate a cohesive end after they are cleaved by the appropriate restriction enzyme (see Fig. 7.2). Synthetic linkers containing recognition sites for many different restriction enzymes are available from several commercial manufacturers.

Cloning blunt-ended DNA fragments with synthetic linkers involves two ligation reactions. In the first, the linkers (which are themselves blunt-ended) are attached to the DNA fragment. Because the synthetic linkers are very small (8–12 bp), it is relatively easy to achieve the high concentration of ends required for blunt-end ligation. The ligation reaction is "driven" by a high concentration of the linkers (20–50 μg/ml), and a relatively low concentration of the fragment to be cloned is required.

After the synthetic linkers are attached, the DNA is cleaved by the appropriate restriction enzyme in order to generate cohesive ends that can be joined to a linear, plasmid DNA fragment with compatible termini. By dephosphorylating the 5' termini of the plasmid, efficient utilization of the foreign DNA fragment can be achieved.

An alternative to synthetic linkers is synthetic "adaptors," which are preformed restriction sites that do not require cleavage with a restriction site in order to create a cohesive end (Bahl and Wu 1978). For example, an *Eco*RI adaptor consists of the sequence

$$^{5'}_{OH}\text{A-A-T-T-C-C-C-G-G-G}^{3'}_{OH}$$

whereas a *Sma*I adaptor consists of the sequence

$$^{5'}_{OH}\text{C-C-C-G-G-G}^{3'}_{OH}$$

To create an *Eco*RI site, the *Sma*I adaptor is first phosphorylated so it can be ligated to the blunt-ended fragment to be cloned and then annealed to the *Eco*RI adaptor to form the duplex

```
5' OH                            3'
   └A-A-T-T-C-C-C-G-G-G
            ┌G-G-G-C-C-C
       3' OH              5'
```

The blunt end of this adaptor can be ligated, but the protruding terminus cannot because of the 5′-hydroxyl group. After blunt-end ligation has been carried out, the protruding 5′-hydroxyl terminus is phosphorylated using polynucleotide kinase. This *Eco*RI "adapted" restriction fragment can now be efficiently ligated to plasmid DNA that has been linearized by *Eco*RI and treated with phosphatase.

Conversion of Fragments with Protruding 5′ Ends to Blunt Ends[6]

Protruding 5′ ends are filled using the DNA polymerizing activity of the Klenow fragment of *E. coli* DNA polymerase I.

1. Set up the following reaction:

   restriction fragment (up to 1 μg of DNA in 10 μl)
   2 mM solution of all four dNTPs   1 μl
   10× nick-translation buffer (see page 112)   2.5 μl
   H_2O to 25 μl

2. Add 2 units of the Klenow fragment of DNA polymerase I. Mix and incubate for 15–30 minutes at 22°. Heat to 70°C for 5 minutes to inactivate the enzyme.

3. Ligate the blunt-ended DNA fragment to synthetic, phosphorylated linkers as described on page 396.

   There is no need to purify the blunt-ended DNA since the end-filling and ligation reactions can be carried out sequentially in the same reaction mixture.

[6]Wartell and Reznikoff (1980).

Conversion of Fragments with Protruding 3' Ends to Blunt Ends

Restriction fragments with protruding 3' ends are made blunt-ended using the 3' exonuclease activity of bacteriophage T4 DNA polymerase. In the presence of all four dNTPs, T4 DNA polymerase will remove unpaired 3' tails from restriction fragments and will stop when it reaches the first paired base if the appropriate complementary dNTP is present (see pages 117ff). Although *E. coli* DNA polymerase is capable of carrying out the same reaction, T4 DNA polymerase is preferred because its 3' exonuclease is one thousand times more active than that of the *E. coli* enzyme.

1. Set up the following reaction:

   restriction fragment (up to 1 μg of DNA in 10 μl)
   10× T4 polymerase buffer   2 μl
   H$_2$O   to 19 μl
   2 mM solution of all four dNTPs   1 μl
   T4 DNA polymerase   1 μl ($\sim$ 2.5 units)

   10× T4 polymerase buffer
   0.33 M Tris acetate (pH 7.9)
   0.66 M potassium acetate
   0.10 M magnesium acetate
   5.0 mM dithiothreitol
   1 mg/ml bovine serum albumin (BSA Pentax Fraction V)

2. Incubate for 5 minutes at 37°C.

3. Add 1 μl of 0.5 M EDTA. Extract once with phenol/chloroform. Precipitate the DNA with ethanol.

4. Collect the DNA by centrifugation. Wash the pellet with 70% ethanol and recentrifuge.

5. Dry the pellet and then dissolve the DNA in 20 μl of TE (pH 7.6).

6. Ligate the blunt-ended DNA fragment to synthetic, phosphorylated linkers as described below.

Attachment of Synthetic Linkers

Most linkers are supplied by the manufacturer with 5′-hydroxyl ends. Because T4 DNA ligase requires 5′-phosphate ends, it is necessary to phosphorylate the linkers before they can be joined to DNA. However, the kinase and ligase reactions can be carried out sequentially in the same reaction mix (Maniatis et al. 1978). As mentioned above, the critical factors in blunt-end ligation are the concentration of ends (1 μM; 50 μg/ml linkers is adequate) and the concentration of ligase. Thus, the reaction is carried out in as small a volume as possible.

Set up the reaction mix as follows.

1. Mix:

10× buffer	1 μl
linkers (1.0–2.0 μg)	1 μl
H$_2$O	6 μl

    10× Linker-kinase buffer
    0.66 M Tris·Cl (pH 7.6)
    10 mM ATP
    10 mM spermidine
    0.1 M MgCl$_2$
    150 mM DTT
    2 mg/ml gelatin or BSA

2. Add 2 units of T4 DNA kinase and incubate for 1 hour at 37°C.

3. Add this reaction mixture directly to 10 μl of the same buffer containing 0.4 μg of the blunt-ended restriction fragment.

4. Add 2 μl (1 unit) of T4 DNA ligase. Incubate at 22°C for 6 hours.

5. Add 1 μl of 0.5 M EDTA. Extract once with phenol/chloroform. Precipitate the DNA with ethanol. Recover the DNA by centrifugation. Wash the pellet with 70% ethanol and recentrifuge.

6. Dry the pellet and then dissolve the DNA in 90 μl of TE.

7. The next step is to digest the "linkered" DNA with the appropriate enzyme to generate cohesive ends. This seems like a straight-forward task, but there are two problems. First, because the molar amount of the linkers is so large, a large amount of restriction enzyme is required to achieve a complete digest. Second, the digested linkers can join to the vector, which can lead to a high background of "nonrecombinant" plasmids. For best results, the DNA fragment should be separated from the linkers before and after digestion. This can be accomplished by gel filtration chromatography (Sepharose CL-4B) or gel electrophoresis. Because it is not essential to cleave all of the linkers that are polymerized to the DNA fragment, the restriction digest is therefore usually carried out directly without making any attempt to remove unligated or self-ligated linkers. The DNA fragment to be cloned (which often contains more than one linker at its ends) is then separated from the digested linkers by preparative gel electrophoresis or chromatography through Sepharose CL-4B.

8. Add 10 μl of 10× restriction enzyme buffer and 20 units of restriction enzyme. Incubate for several hours at the appropriate temperature.

9. Add 2 μl of 0.5 M EDTA. Extract once with phenol/chloroform.

10. Recover the supernatant and add 6.0 μl of 5 M NaCl. Apply the supernatant to a 5-ml column of Sepharose CL-4B equilibrated in TE (pH 7.6) containing 0.3 M NaCl. Collect twelve 0.3-ml fractions and analyze 50 μl of each fraction by agarose gel electrophoresis. Pool the fractions containing the fragments of DNA. Precipitate the DNA with ethanol.

11. Ligate the fragment to plasmid DNA and transform bacteria as described in Chapter 8.

CREATING RESTRICTION SITES BY LIGATION OF BLUNT-ENDED DNA MOLECULES

If protruding termini created by restriction enzymes are filled and the resulting blunt-ended DNA molecules are ligated together, it frequently happens that one or both of the original restriction sites may be regenerated. In a few cases, entirely new sites may be created.

An example of a regenerated site is:

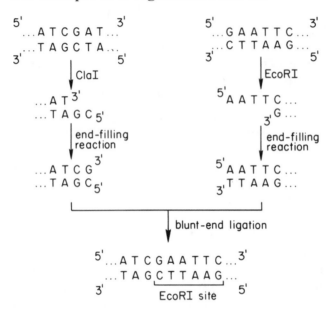

An example of a newly created site is:

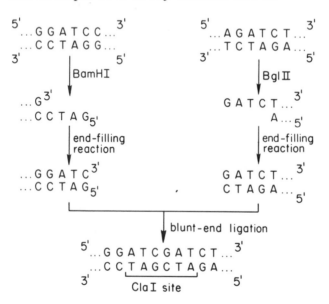

The creation and regeneration of sites is extremely useful since it provides a simple way to screen recombinant clones for those that are of interest. Furthermore, the presence of a restriction site at the junction of the fragments means that they can be easily separated from one another at any time in the future.

The following tables may be used to find out whether a restriction site is created or regenerated when two blunt-ended fragments are ligated together. Table 11.2 shows which base is required to regenerate or create a site when a protruding terminus has been filled (e.g., a filled *Bcl*I terminus needs to be ligated to a DNA fragment with an A residue at its 5' end in order to regenerate a *Bcl*I site).

Table 11.3 shows the 5' bases at the termini that are created by cleavage of double-stranded DNA with restriction enzymes and subsequent repair.

**TABLE 11.2. 5' TERMINAL BASE
REQUIRED TO COMPLETE
A RESTRICTION SITE**

Terminus created by	Nucleotide required
*Bcl*I	A
*Taq*I	A
*Xba*I	A
*Bgl*I	T
*Hind*III	T
*Dde*I	G
*Hpa*II	G
*Xho*I	G
*Xma*I	G
*Xma*III	G
*Ava*II	C
*Bam*HI	C
*Bst*EII	C
*Eco*RI	C
*Hinf*I	C
*Sal*I	C
*Sau*96I	C

Termini generated by *Bcl*I, *Bgl*II, *Bam*HI, *Mbo*I, or *Sau*3A can be filled and ligated together to create a *Cla*I site.

TABLE 11.3. 5′ BASE AFTER REPAIR OF TERMINUS

Nucleotide	Restriction endonuclease
A	*Eco*RI, *Hind*III, *Hinf*I, *Hpa*I[1], *Rsa*I[1]
T	*Dde*I, *Mst*II, *Sal*I, *Xho*I
G	*Ava*II, *Bam*HI, *Bcl*I, *Bgl*II, *Bst*EII, *Hae*II[1], *Mbo*I, *Mst*I[1], *Pst*I[2], *Sac*II[2], *Sau*3A, *Sau*96I, *Sma*I[1], *Xma*II
C	*Alu*I[1], *Bal*I[1], *Cla*I, *Fnu*DII[1], *Hgi*AI[2], *Hha*I[2], *Hpa*II, *Kpn*I[2], *Nar*I, *Pru*I[1], *Pvu*II[1], *Sac*I[2], *Sph*I[2], *Stu*I[1], *Taq*I, *Xba*I, *Sma*I

[1]The enzyme leaves a blunt end; no repair is necessary.
[2]The enzyme leaves an overhanging 3′ terminus, which needs to be removed by treatment with nuclease S1, mung-bean nuclease, or T4 DNA polymerase.

CARRYING OUT SEQUENTIAL CLONING STEPS RAPIDLY

Often, when constructing plasmids, it is necessary to carry out several enzymatic reactions in sequence. It is not always necessary to purify the DNA by extraction with phenol/chloroform and precipitation with ethanol between each of these reactions. The following guidelines may be useful.

1. Many enzymes are inactivated by heat. For example exposure for 15 minutes to 70°C inactivates all polymerases, ligases, and almost all commonly used restriction enzymes except *Bam*HI, *Bcl*I, *Bst*NI, *Hin*dIII, *Sal*I, *Taq*I, and *Tha*I. Because most enzymes are more heat-labile in the absence of divalent cations, just sufficient EDTA should be added to the reaction mixture before heating to chelate all divalent cations.

   Note. Bacterial alkaline phosphate *cannot* be inactivated by heating to 70°C (see page 133).

2. Most restriction enzymes may be inactivated by adding diethylpyrocarbonate (DEPC) to a final concentration of 0.1% (i.e., add 0.01 volumes of a 10% solution of DEPC in ethanol) and heating for 20 minutes at 37°C or 10 minutes at 56°C. To counteract the drop in pH resulting from the carbon dioxide released during breakdown of DEPC, the solution should be cooled in ice, and 5 μl of 1 M Tris · Cl (pH 8.0) should be added for every 1 μl of DEPC solution used.

3. DNA can be rapidly precipitated from enzyme reactions by adding (a) sufficient EDTA to chelate any divalent cations that may be present, (b) 0.4 volumes of 5 M ammonium acetate, and (c) 2 volumes of isopropanol. After 10 minutes at room temperature, the DNA can be recovered by centrifugation. Proteins are not precipitated by isopropanol in the presence of 2 M ammonium acetate.

   Note. This procedure should not be used before kinasing DNA, since T4 polynucleotide kinase is strongly inhibited by ammonium ions.

12

Vectors That Express Cloned DNA in *Escherichia coli*

Expression vectors contain sequences of DNA that are required for the transcription of cloned copies of genes and the translation of their mRNAs in *Escherichia coli*. Such vectors have been used both to express eukaryotic genes in *E. coli* and to increase production of prokaryotic gene products.

The three major requirements for expression of a cloned gene in *E. coli* are:

1. The coding region of the gene must not be interrupted by intervening sequences.

2. The gene must be placed under the control of an *E. coli* promoter that is efficiently recognized by *E. coli* RNA polymerase.

3. The mRNA must be relatively stable and efficiently translated. In addition, to be recovered, the foreign protein produced in *E. coli* must not be rapidly degraded by bacterial proteases.

PROMOTERS

A promoter is a DNA sequence that directs RNA polymerase to bind to DNA and to initiate RNA synthesis. Different promoters work with different efficiencies: Strong promoters cause mRNAs to be initiated at high frequency; weak promoters direct the synthesis of rarer transcripts. Comparison of the sequences of a number of different promoters reveals two highly conserved regions, one located about 10 bp (−10 region or Pribnow box; Pribnow 1975) and the other about 35 bp (−35 region) upstream from the point at which transcription starts (for a review, see Rosenberg and Court 1979). These two regions are thought to be important in determining promoter strength because mutations that decrease the frequency of transcription usually decrease the amount of homology with the conserved sequences. However, other, more moderately conserved regions of promoters may also contribute to promoter strength. Furthermore, the number of nucleotides that separate the conserved sequences is important for efficient promoter function. For example, 16 to 19 nucleotides separate the −10 region from the −35 region; mutations altering the spacing between these two conserved regions in a *lac* promoter (*lacP*ˢ; de Crombrugghe et al. 1971; Stefano and Gralla 1982) and in the β-lactamase promoter (Jaurin et al. 1981) change the "strength" of the promoter. These results indicate either that there is an optimal spacing common to all promoters or that there is an optimal spacing for any individual promoter that is dependent on its particular DNA sequence.

The only true test of the efficiency of a promoter is to measure the frequency with which the synthesis of the appropriate mRNA is initiated. Because this value is difficult to obtain from in vivo studies, the efficiency of a promoter is frequently deduced indirectly from the level at which the relevant protein product is expressed. However, a major difficulty in comparing the strengths of promoters by this method is that different mRNAs contain different untranslated leader sequences at their 5′ termini that may affect the efficiency of translation of the mRNA. The level of expression of the protein product may therefore be due to the strength of the promoter, to the composition and length of the untranslated leader sequence, or to a combination of the two.

Despite these uncertainties, it is well-established that many *E. coli* genes are controlled by relatively weak promoters. Thus the expression of such genes can be increased by placing them downstream from an efficient promoter (e.g., *lac*uv5, *trp*, *tac*, *trp-lac*uv5 hybrid promoter, λ*p*ʟ, *ompF*, *bla*). Eukaryotic promoters function extremely poorly, if at all, in *E. coli*, and efficient expression of eukaryotic proteins has been achieved only when the coding sequence is placed under the control of a strong *E. coli* promoter.

The most useful promoters for expressing foreign genes in *E. coli* are those that are both strong and also regulated. Obviously if the product of the cloned gene is toxic to *E. coli* cells, then coupling the gene to a strong, unregulated promoter is not desirable. In addition, a constitutively high level transcription may interfere with plasmid DNA replication and lead to plasmid instability (Remaut et al. 1981). The presence of efficient terminators of transcription placed downstream from the promoter may circumvent this problem. In fact, strong promoters (e.g., certain bacteriophage T5 promoters) require the presence of a strong downsteam termination signal if they are to be stably maintained on a plasmid (Gentz et al. 1981).

Vectors Using the p_L Promoter of Bacteriophage λ

The p_L promoter of bacteriophage λ is a strong, well-regulated promoter that has been used in several expression vectors (Hedgpeth et al. 1978; Bernard et al. 1979; Remaut et al. 1981; Shimatake and Rosenberg 1981). A gene encoding a temperature-sensitive λ repressor (e.g., λ*c*I*ts*857) may either be included in the cloning vector or may be provided by a prophage resident in the bacterial chromosome (Bernard et al. 1979). At low temperature (31°C), the p_L promoter is maintained in a repressed state by the *c*I-gene product. After the activity of the repressor is destroyed by raising the temperature of the culture, the p_L promoter directs the synthesis of large quantities of mRNA. Several vectors utilizing the λp_L promoter are described below.

pPLa2311

Insertion of DNA into the *Pst*I site of pPLa2311 (Remaut et al. 1981; see Fig. 12.1) inactivates the plasmid's ampicillin-resistance (*bla*) gene. Transformed cells that are resistant to kanamycin and sensitive to ampicillin are therefore likely to have insertions of foreign DNA at the *Pst*I site. The λp_L promoter directs the synthesis of large amounts of mRNA encoding the gene product to be expressed when the foreign DNA is inserted in the correct orientation.

pPLa8

pPLa8 was constructed by inserting a *Bam*HI linker in the *Pst*I site of pPLa2311 (Remaut et al. 1981). Bacteria carrying pPLa8 are therefore sensitive to ampicillin, so that insertional inactivation cannot be used to screen for recombinant plasmids that carry foreign DNA at the *Bam*HI site.

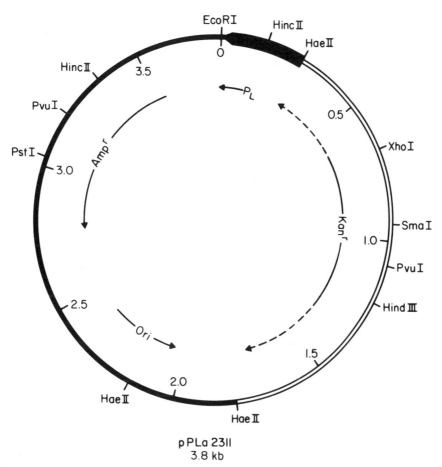

Figure 12.1

pPLa2311, a plasmid 3.8 kb in length with unique *Eco*RI and *Pst*I sites, that carries the p_L promoter of bacteriophage λ. The heavy arrow indicates the location of the p_L region and the direction of transcription. The solid lines represent DNA derived from pBR322. The open line shows the region carrying the gene that confers kanamycin resistance (whose reading direction is unknown). Insertion of foreign DNA into the *Pst*I site inactivates the ampicillin-resistance gene. Genes inserted into either the *Pst*I site or the *Eco*RI site can be regulated by introducing the recombinant plasmid into a temperature-sensitive λ lysogen (*c*Its857). The cells carrying this plasmid are grown to late log phase at 32°C and are then shifted to 42°C to inactivate the *c*I-gene product and to turn on the p_L promoter. Plasmid pPLa8 has the same structure as pPLa2311, except that a *Bam*HI linker has been inserted into the *Pst*I site (Remaut et al. 1981).

pKC30

The plasmid pKC30 (Shimatake and Rosenberg 1981) has been shown to direct the production of large amounts of a toxic protein (in this case, the bacteriophage λ *c*II-gene product; see Fig. 12.2) which under normal circumstances is produced by bacteriophage λ in low amounts and is rapidly degraded when present in small amounts.

The plasmid contains an *Hpa*I site located 321 bp downstream from the bacteriophage λ p_L transcription start site (see Fig. 12.2). When the bacteriophage λ *c*II gene was cloned into this site in the correct orientation, a lysogen (in which the resident prophage was making λ repressor) could be transformed at high efficiency, while a nonlysogen could not be transformed (Shimatake and Rosenberg 1981). In the nonlysogen, there was no *c*I repressor to reduce synthesis of the *c*II-gene product, and the constitutive synthesis of large amounts of the protein proved to be lethal for *E. coli*. The fact that a lysogen was able to tolerate the plasmid indicates that an integrated copy of bacteriophage λ makes enough repressor to reduce the expression of the *c*II gene to a nonlethal level.

Large amounts of the *c*II-gene product (4% of total cell protein) could be produced from this plasmid by raising the temperature of a lysogen in which the defective prophage carried a temperature-sensitive mutation in the repressor gene (λ*c*I*ts*857). The bacteriophage λ *N*-gene product was also required for high levels of production of *c*II protein; presumably the *N* protein blocked termination of transcription at t_{R1}, a rho-dependent transcription termination site located upstream from the *c*II gene (see Chapter 1).

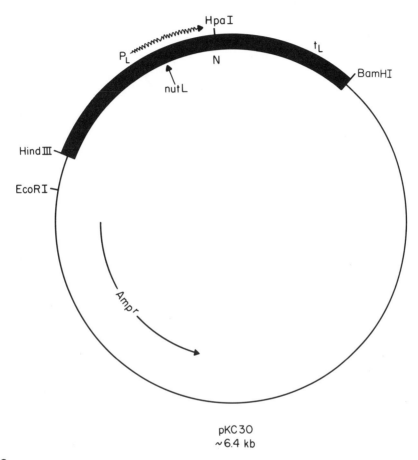

Figure 12.2
pKC30, a plasmid ~ 6.4 kb in length that carries the p_L promoter of bacteriophage λ and an *Hpa*I recognition site located 321 nucleotides downstream from the p_L transcriptional start site. The plasmid is a derivative of pBR322 and contains a *Hind*III-*Bam*HI fragment (black box) derived from bacteriophage λ inserted between the *Hind*III and *Bam*HI sites within the tetracycline-resistance gene. The insertion contains the promoter signal, p_L, a site recognized by the *N*-gene product (*nut*L), the *N* gene itself, and the strong rho-dependent transcription-termination signal t_L. The *Hpa*I recognition site lies within the coding region of the *N*-gene coding region. Sequences inserted into the *Hpa*I site can be regulated by introducing the recombinant plasmid into a temperature-sensitive λ lysogen (*cI*ts857). The cells are grown to late log phase at 32°C and then shifted to 42°C to inactivate the *cI*-gene product and to turn on the p_L promoter. This vector has been used to express the λ*c*II protein at a level such that the protein comprises 4% of the total protein of the cell (Shimatake and Rosenberg 1981).

Other Promoters

Expression of cloned genes in bacteria can also be controlled by other promoters. For example:

1. The *ompR* gene codes for a positive regulatory protein that controls the expression of *ompF*, a gene coding for a major outer membrane protein of *E. coli*. A cold-sensitive mutant mapping in the *ompR* gene has been isolated. Therefore, the transcription from the *ompF* promoter can be activated by raising the temperature of the culture (T. Silhavy, pers. comm.).

2. The *trp* promoter is regulated by *trp* repressor and can be induced by the addition of 3-indolylacetic acid to the medium (Morse et al. 1970) or by tryptophan starvation (see Miller and Reznikoff 1978).

3. The *lac* promoter is regulated by *lac* repressor and can therefore be induced by the addition of the inducer isopropyl-β-D-thiogalactoside (IPTG) to the bacterial culture (see Miller and Reznikoff 1978).

4. Finally, a hybrid promoter consisting of the $trp-35$ region fused to the *lac* -10 region and the *lac* operator has been constructed by de Boer et al. (1982). This strong *trp-lac* hybrid promoter (or *tac* promoter) is regulated by the *lac* repressor.

As discussed above, increased expression of an *E. coli* protein encoded by a gene that does not have a strong promoter may be achieved by placing the gene downstream from one of these or other strong *E. coli* promoters (Selker et al. 1977; Backman and Ptashne 1978).

RIBOSOME-BINDING SITES

To achieve high levels of gene expression in *E. coli*, it is necessary to use not only strong promoters to generate large quantities of mRNA, but also ribosome-binding sites to ensure that the mRNA is efficiently translated. In *E. coli*, the ribosome-binding site includes an initiation codon (AUG) and a sequence 3–9 nucleotides long located 3–11 nucleotides upstream from the initiation codon (Shine and Dalgarno 1975; Steitz 1979). This sequence, which is called the Shine-Dalgarno (SD) sequence, is complementary to the 3' end of *E. coli* 16S rRNA. Binding of the ribosome to mRNA is thought to be promoted by base pairing between the SD sequence in the mRNA and the sequence at the 3' end of the 16S rRNA (Steitz 1979).

The efficiency of translation of an mRNA could be affected by several factors:

1. The degree of complementarity between the SD sequence and the 3' end of the 16S rRNA.

2. The spacing and possibly the DNA sequence lying between the SD sequence and the AUG (Roberts et al. 1979a,b; Guarente et al. 1980a,b). The level of expression of genes has been measured in plasmids in which this was systematically altered; an optimal spacing between the *trpL* SD sequence and the ATG of two genes was determined (D. Goeddel, unpubl.). Comparison of different mRNAs shows that there are statistically preferred sequences from positions −20 to +13 (where the A of the AUG is position 0) (Gold et al. 1981). Leader sequences have been shown to influence translation dramatically (Roberts et al. 1979a,b).

3. The nucleotide following the AUG, which affects ribosome binding (Taniguchi and Weissman 1978).

EXPRESSION OF EUKARYOTIC GENES

Many of the vectors described below are designed so that genes can be placed downstream from a promoter. However, for efficient expression of a eukaryotic gene, a bacterial ribosome-binding site also must be provided. This can be accomplished in the following two ways.

Vectors That Express Unfused Eukaryotic Proteins

The first method of providing a bacterial ribosome-binding site involves the synthesis of a protein that initiates at the AUG of the eukaryotic mRNA and that contains no bacterial sequences at its amino terminus. One way to accomplish this is, first, to introduce a restriction site immediately upstream from the ATG of the gene to be expressed and then to clone the gene directly into a compatible restriction site immediately downstream from an SD sequence (Goeddel et al. 1979a, 1980b; Edman et al. 1981). A "hybrid" ribosome-binding site (Backman and Ptashne 1978) is thus constructed and is composed of a bacterial SD sequence and the ATG of the gene to be expressed. The protein is synthesized, unfused to any other protein. If the DNA sequence to be expressed lacks an ATG codon, then one must be provided by chemical DNA synthesis (e.g., Goeddel et al. 1979a; Davis et al. 1981).

For example, the "mature" form of human growth hormone (HGH) does not begin with a methionine, because normally it is synthesized as a precursor with an aminoterminal signal peptide that is cleaved during secretion. To construct a plasmid expressing the mature form of HGH in *E. coli*, chemically synthesized DNA containing an *Eco*RI site, an ATG codon, and the first 24 codons of mature HGH was inserted, along with cDNA encoding the rest of HGH, into a vector with an *Eco*RI site just downsteam from the *lac* promoter and SD sequence. The SD sequence was separated from the chemically synthesized ATG by 11 bp of DNA generated by ligation of the *Eco*RI ends. This plasmid directed synthesis of HGH under control of the *lac* promoter.

An alternative means of introducing an ATG codon into a cloned gene and of maintaining the bacterial ribosome-binding site is to use the vector pAS1, described below. With this vector, the λp_L promoter, the λcII SD sequence, and ATG codon are fused directly to the aminoterminal coding portion of the eukaryotic gene to be expressed.

It may be critical to optimize the distance between the bacterial SD sequence and the ATG of the eukaryotic gene in order to obtain efficient expression of the gene (for review, see Guarente et al. 1980a). Optimal placement of the promoter fragment may be achieved by the methods described in Roberts et al. (1979a,b). In brief, a unique restriction site is positioned in a plasmid within 100 bp upstream from the initiation codon. The plasmid DNA is then cleaved at that restriction site and resected to various extents with an exonuclease. A DNA fragment carrying the promoter and the SD sequence then is inserted to produce a series of plasmids that contain the SD sequence at varying distances from the initiation codon. Several DNA fragments, termed portable promoter fragments, have been designed for this purpose, including one containing the *lac*uv5 "portable promoter" and the *lacZ* SD sequence that has been used to express several prokaryotic and eukaryotic genes in *E. coli* (see Fig. 12.3) (Backman and Ptashne 1978; Roberts et al. 1979b; Guarente et al. 1980a,b; Taniguchi et al. 1980b).

A number of other vectors for producing eukaryotic proteins that are not fused to bacterial peptides are discussed below.

p*tac*12

Plasmid p*tac*12, which contains the hybrid promoter *tac* (see Fig. 12.4), can be used in the same manner as the *lac*uv5 portable promoter (E. Amann and J. Brosius, pers. comm.). However, the *tac* promoter is more efficient than the *lac*uv5 promoter and should be used in a bacterial strain containing a mutation in the *lac*I gene that causes the *lac* repressor to be overproduced (*lac*I^q). Even in such a strain, however, transcription from the *tac* and the ordinary *lac* promoters is not fully repressed when many copies of the plasmid are present (for discussion, see de Boer et al. 1982).

Because expression of some foreign proteins is detrimental to *E. coli*, it may be expedient to maximize expression of a foreign gene using the *lac* portable promoter and then to use the plasmid with the optimal SD-ATG spacing to construct a derivative carrying the stronger *tac* promoter. This is best achieved by using *Hpa*II to isolate a DNA fragment (encoding the *lac* Pribnow box, the *lac* leader, and part of the coding region of the gene) from the plasmid in which the *lac*uv5 promoter has been positioned in front of a eukaryotic gene. This fragment is then inserted into the *Cla*I site of the plasmid pEA300 (see Fig. 12.5) (E. Amann and J. Brosius, pers. comm.). The *lac* promoter is thus replaced by a *tac* promoter without altering the leader or the spacing between the SD sequence and the ATG. Plasmid pEA300 has the transcriptional termination signals from the *rrnB* operon located downstream from the site of insertion of the gene.

p*trpL*1

Plasmid p*trpL*1 carries another portable promoter fragment that can be used in the same way as the *lac* and *tac* portable promoters (see Fig. 12.6) (Edman et al. 1981).

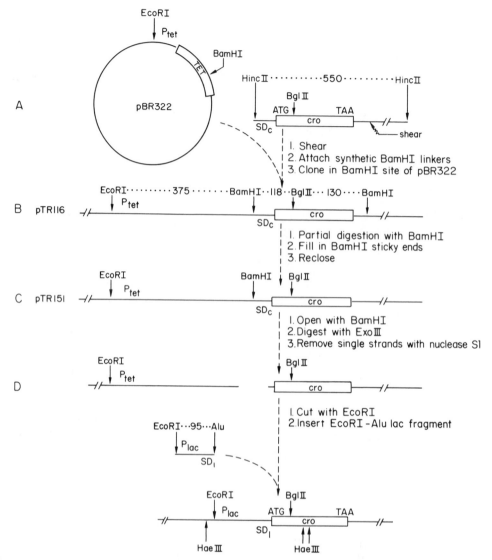

Figure 12.3

Schematic representation of the portable promoter procedure. The approximate locations of several restriction endonuclease cleavage sites are shown for the plasmid pBR322, for a DNA fragment bearing the *cro* gene of phage λ, and for a DNA fragment bearing the promoter of the *lac* operon (see Backman and Ptashne 1978 for the source of this fragment). The locations of the *tet* and *lac* promoters are indicated, as are the extent of the *tet* and *cro* genes. SD$_c$ and SD$_l$ indicate the Shine-Dalgarno sequences of the *cro* and *lacZ* genes, respectively. AUG and UAA are the start and stop signals for translation of the *cro* protein. Distances are indicated in base pairs. *(A)* The DNA fragment bearing the *cro* gene was shortened by shearing to remove certain λ control elements near the 3′ end of the gene, and the smaller fragment was inserted into the *Bam*HI site in pBR322 by using *Bam*HI linkers. *(B)* The *Bam*HI site near the carboxyl terminus of the *cro* gene was eliminated. *(C)* The plasmid was opened at the *Bam*HI site, and varying amounts of DNA were removed by treatment with exonuclease III and nuclease S1. *(D)* The partially resected plasmid was cut at the *Eco*RI site, and the *lac* promoter (bearing the uv5 mutation, rendering it independent of catabolite activator) was inserted by "sticky-end" ligation at its *Eco*RI end and by "blunt-end" ligation to the resected plasmid DNA at its *Alu* end. The efficiency of steps *C* and *D* is fairly high—about 200–400 plasmids result from each microgram of pTR151 used. (Adapted from Roberts et al. 1979a.)

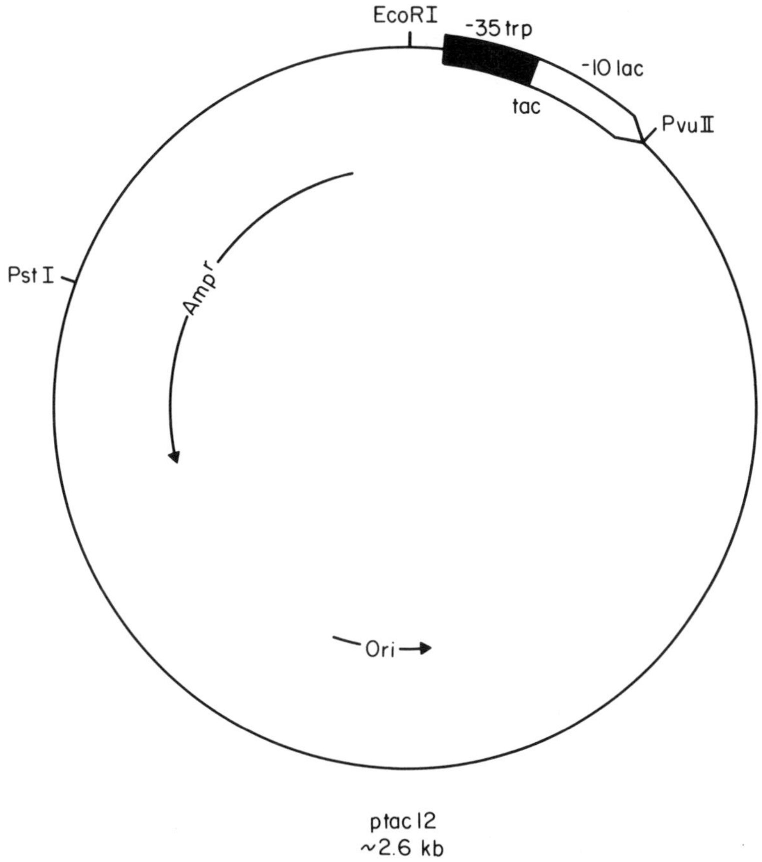

Figure 12.4

p*tac*12, a plasmid ~ 2.6 kb in length that contains the portable hybrid *trp-lac (tac)* promoter (E. Amann and J. Brosius, pers. comm.). A fragment (~ 260 bp) containing the promoter is isolated after digestion of this plasmid with *Eco*RI and *Pvu*II. This DNA encodes the *tac* promoter and the *lac* leader sequences. The *Pvu*II site is located 5 bp downstream from the SD sequence of the *lacZ* gene. The portable promoter fragment is inserted in front of the initiation codon of the gene to be expressed (as outlined in Fig. 12.3). Because the *tac* promoter is strong, recombinant plasmids containing the portable promoter should be carried in a $lacI^q$ strain to repress transcription.

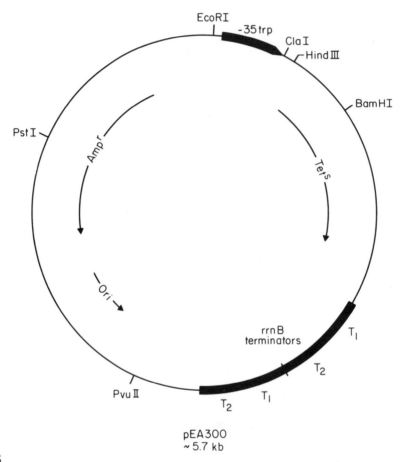

Figure 12.5

pEA300, a plasmid ~ 5.7 kb in length that carries a 192-bp fragment of *trp* DNA sequences cloned into the *Cla*I site of pBR322. By cutting this plasmid with *Cla*I and inserting an *Hpa*II fragment derived from a plasmid carrying the *lac*uv5 promoter, the *tac* promoter is constructed. In addition, pEA300 carries downstream from the promoter two 500-bp fragments, each of which contains two *rrn*B terminators, arranged in tandem (E. Amman and J. Brosius, pers. comm.).

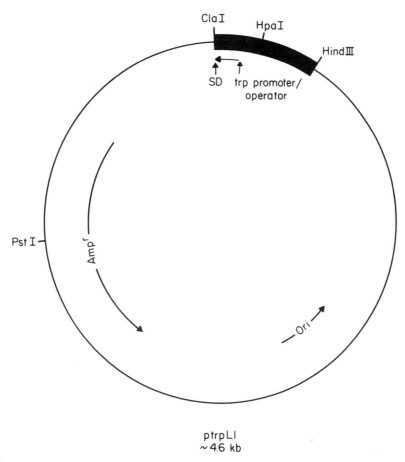

Figure 12.6

ptrpL1, a plasmid ~ 4.6 kb in length containing a sequence that could be used as a portable *trp* promoter. To use the portable promoter, the plasmid is cleaved at the *Cla*I site (3 bp downstream from the SD sequence of *trpL*). After the protruding termini have been repaired by the Klenow fragment of DNA polymerase, the plasmid is digested with *Hind*III and the promoter fragment is isolated. This fragment can be placed in front of the gene to be expressed. Alternatively, the plasmid can be used directly as an expression vector by inserting into the *Cla*I site of the ptrpL1 plasmid a DNA fragment extending from a restriction site immediately upstream from the ATG of the gene to be expressed through (or into) the coding sequence (Edman et al. 1981).

pAS1

Another system for the expression of unfused foreign genes in *E. coli* utilizes the vector pAS1 (Fig. 12.7; A. Shatzman and M. Rosenberg, pers. comm.), which provides not only a promoter (λp_L) and an SD sequence, but also an ATG codon separated from the SD by its normal distance in the λcII gene. This vector is therefore particularly useful for expressing eukaryotic coding sequences that lack an ATG. pAS1 is cleaved at the *Bam*HI site and digested with mung-bean nuclease to create a blunt end immediately downstream from the ATG. This ATG can then be joined by blunt-end ligation to a DNA fragment coding for the segment of protein to be expressed.

This method requires that a blunt-end be created immediately before a particular codon (Panayotatos and Truong 1981; A. Shatzman and M. Rosenberg, pers. comm.). This can be done as follows (see Fig. 12.8).

1. Insert two unique restriction sites upstream from the coding sequence to be expressed.

    Site 1 must generate after cleavage a protruding terminus, 4 nucleotides long; the sixth nucleotide of this site will eventually become the first nucleotide of the second codon of the protein that will be expressed.

    Site 2 must lie between site 1 and the start of the gene and must be closer to the start of the gene than to site 1.

2. Cleave the DNA at site 2 and resect with nuclease *Bal*31 to produce a population of molecules that terminate close to the desired codon.

3. Cleave the DNA at site 1.

4. Repair the DNA by the Klenow fragment of *E. coli* DNA polymerase I to create a blunt end that contains 5 of the 6 nucleotides of restriction site 1.

5. Recircularize the DNA by ligation and use the plasmid to transform bacteria to antibiotic resistance.

6. Screen individual colonies for the presence of plasmids in which restriction site 1 has been regenerated. Such regeneration occurs when, by chance, digestion with *Bal*31 generates a DNA molecule whose terminus carries the particular base pair required to complete site 1. Included in this subpopulation of plasmids will be those in which the last nucleotide of site 1 is the first nucleotide of the second codon of the gene to be expressed.

7. Cleave one of these plasmids at site 1 and digest with mung-bean nuclease to generate a blunt end directly preceding the first nucleotide of the second codon, and then cleave with a restriction enzyme that cuts downstream from the gene to obtain a DNA fragment that can be inserted after the ATG of plasmid pAS1.

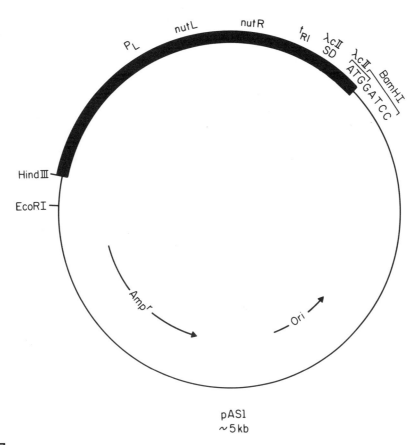

Figure 12.7

pAS1, a plasmid ~ 5 kb in length that carries the bacteriophage λ p_L promoter and a unique *Bam*HI site located at the ATG of the λ*c*II gene. This plasmid is a derivative of pKC30 (Fig. 12.2) into which the λ*c*II gene was inserted at the *Hpa*I site. The *c*II gene was then resected by exonuclease digestion until only the initiation codon ATG remained (the G of the ATG is the first nucleotide of a *Bam*HI site). To express a gene lacking an initiation codon, pAS1 is digested with *Bam*HI and then treated with mung-bean nuclease to remove the protruding, single-stranded termini. Ligation of this blunt-ended DNA to a blunt-ended DNA fragment that begins with the second codon of the gene to be expressed places that gene in-frame with the ATG. Genes inserted in this manner are regulated by introducing the recombinant plasmid into a temperature-sensitive, bacteriophage λ lysogen (*c*Its857). The cells are grown to late log phase at 32°C then shifted to 42°C to inactivate the repressor and to turn on the p_L promoter. The inserted gene can also be regulated by the action of the N protein at *nut*L and *nut*R to antiterminate at t_{RI} (A. Shatzman and M. Rosenberg, pers. comm.).

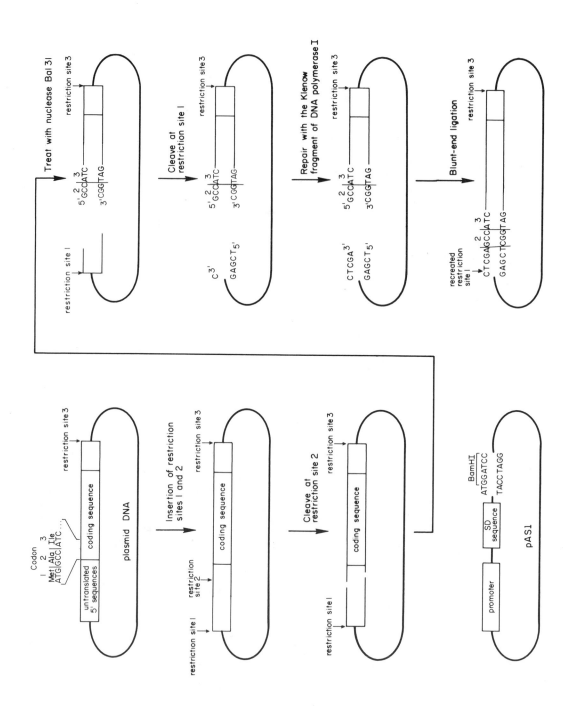

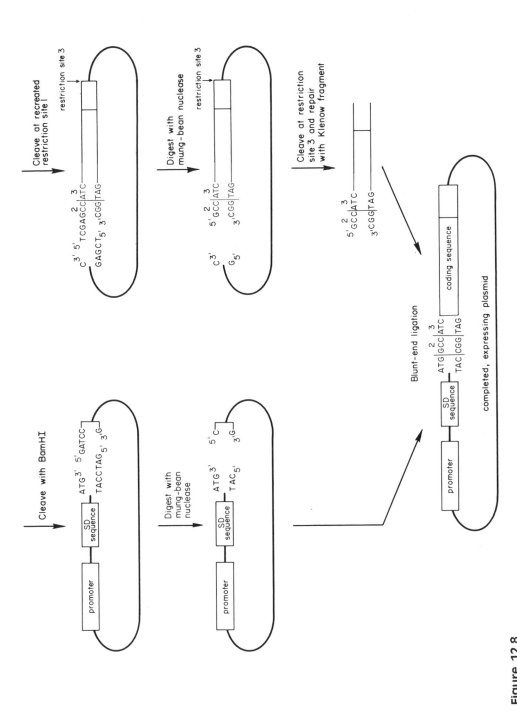

Figure 12.8
A method to create a blunt end immediately before a particular nucleotide in a segment of cloned DNA (see page 418 for details).

Vectors That Express Fused Eukaryotic Proteins

In some cases eukaryotic DNA sequences have been expressed as part of a fusion protein whose amino terminus is encoded by prokaryotic sequences and whose carboxyl terminus is encoded by eukaryotic sequences. Although usually less desirable than the unaltered eukaryotic protein, fused proteins have some useful properties:

1. Many of them are more stable in bacteria than the native eukaryotic protein (Itakura et al. 1977; Goeddel et al. 1979b; Davis et al. 1981).

2. The fusion protein may be secreted by the expressing bacteria if the eukaryotic DNA is attached to a bacterial sequence coding for a signal peptide that causes export of proteins to be translocated across membranes (Talmadge et al. 1980). However, this has been demonstrated in only one case.

3. In some cases, fusion proteins can be chemically treated to release the eukaryotic peptide in a biologically active form (Goeddel et al. 1979b; Shine et al. 1980).

4. Fusion proteins may be used as antigens (e.g., Kleid et al. 1981).

For eukaryotic sequences to be expressed, the translational reading frame must coincide with that of the prokaryotic gene to which they are fused. Some vectors are available that contain restriction cleavage sites in all three reading frames of the prokaryotic sequence (e.g., Charnay et al. 1978; Talmadge et al. 1980; Tacon et al. 1980; T. Silhavy, pers. comm.). Other vectors require the addition of synthetic linkers to allow the eukaryotic gene to be inserted in-frame (Shine et al. 1980). In either case, the DNA encoding the gene to be expressed must contain a restriction site that can be used to make the fusion; if a convenient site is not present, one must be inserted by exonuclease treatment followed by linker insertion.

pOP203-13

The plasmid pOP203-13 (Fuller 1981; see Fig. 12.9) is useful for constructing gene fusions whose products contain only a few amino acids encoded by a prokaryotic sequence (e.g., Fraser and Bruce 1978). Insertion of the coding sequence to be expressed into the *Eco*RI site of pOP203-13 allows a fusion protein to be expressed that contains the first 7 amino acids of β-galactosidase, a few amino acids encoded by *Eco*RI linker, and the amino acids of the eukaryotic polypeptide. The expression vector provides not only a *lac*uv5 promoter, which can be regulated by *lac* repressor in a *lacI*^q strain of *E. coli*, but also the ribosome-binding site of the *lacZ* gene. Because only a small number of amino acids of β-galactosidase are added to the protein to be expressed, fusion proteins synthesized using this vector may more nearly

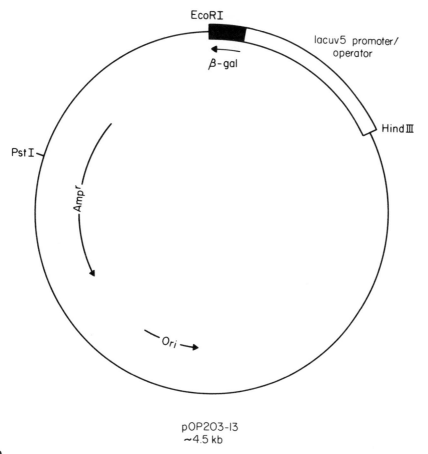

Figure 12.9

pOP203-13, a plasmid ~ 4.5 kb in length carrying the *lac*uv5 promoter and a single *Eco*RI site. Sequences inserted in-frame into this site will be translated into a fusion protein containing 7 amino acids of β-galactosidase and a few amino acids encoded by the *Eco*RI linker at their amino termini (Fuller 1981). This plasmid has been used to express a β-galactosidase–chicken-ovalbumin fusion protein (Fraser and Bruce 1978) and a fusion between β-galactosidase and human influenza virus hemagglutinin (Heiland and Gething 1981).

resemble the native protein than fusion proteins synthesized using the vectors described below. pOP203-13 has been used to express in *E. coli* a protein that consists of the N-terminal sequences of β-galactosidase fused to a large segment of influenza virus hemagglutinin (Heiland and Gething 1981).

pLC24

This plasmid (Figure 12.10) contains the strong p_L promoter of bacteriophage λ and directs the synthesis of fusion proteins consisting of 98 amino acids of MS2 polymerase and a polypeptide encoded by the foreign DNA sequence inserted at *Bam*HI or *Hin*dIII sites. Fusion proteins synthesized in this vector include MS2 polymerase–human fibroblast interferon (Derynck et al. 1980) and MS2-polymerase–foot-and-mouth-disease virus VP1 (Küpper et al. 1981). Expression of these fusion proteins may be conveniently controlled by regulating the activity of the p_L promoter in a lysogen with a λ*c*I*ts*857 prophage.

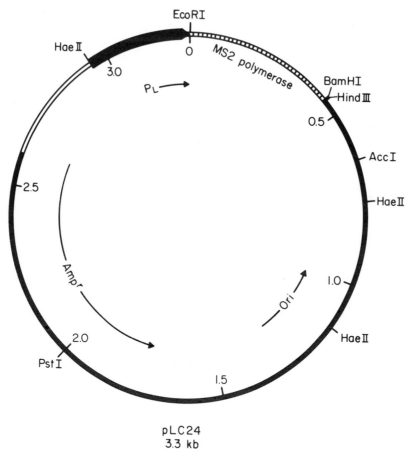

pLC24
3.3 kb

Figure 12.10

pLC24, a plasmid 3.3 kb in length that carries the bacteriophage λ p_L promoter and the ribosome-binding site of the bacteriophage MS2 polymerase gene. A fusion protein containing 98 amino acids of MS2 polymerase is expressed when a foreign gene is inserted in-frame into the *Bam*HI or *Hin*dIII sites. As in the case of other p_L promoter plasmids, inserted genes can be regulated by temperature in a λ*c*I*ts*857 lysogen (Remaut et al. 1981). pLC24 has been used for expression of fusion proteins MS2 polymerase with human fibroblast interferon (Derynck et al. 1980) and with foot-and-mouth-disease virus VP1 (Küpper et al. 1981).

p*trpED*5-1

This plasmid (Figure 12.11) was designed to express a fusion protein composed partly of *trpD* and partly of a foreign polypeptide whose coding sequences have been inserted into the *Hind*III site of *trpD* (Hallewell and Emtage 1980). p*trpED*5-1 has been modified so that the foreign sequences can be inserted into each of the three reading frames (Tacon et al. 1980). With the *trp* attenuator present, very little transcription occurs in repressed conditions, so that this plasmid is useful for cloning genes whose products are toxic to *E. coli*. A fusion protein made from this vector contains about 15% of the sequence of the *trpD* polypeptide. The truncated *trpD* fragment is itself stable in *E. coli* cells, perhaps because it associates with the *trpE*-gene product (Hallewell and Emtage 1980). The fusion of the *trpD* polypeptide and a foreign gene product may also be stabilized by a similar association with *trpE*.

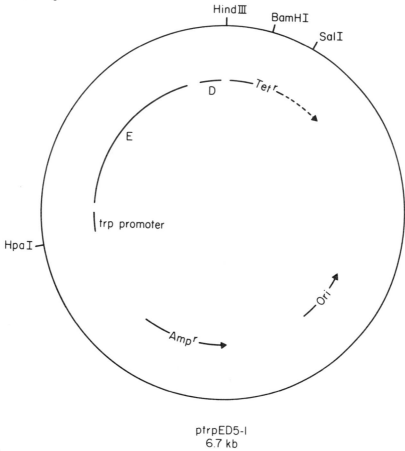

p*trpED*5-I
6.7 kb

Figure 12.11

p*trpED*5-1, a plasmid ~6.7 kb in length carrying a *trp* promoter and designed to produce fusion proteins containing the aminoterminal 15% (75 amino acids) of the *trpD* protein. Induction of the *trp* operon with 3-indolylacetic acid results in at least a 50-fold increase in *trp*-gene products. Under conditions of repression, the *trp* proteins (and any fusion proteins derived from them) are expressed at a relatively low level (the *trp* attenuator is present in this plasmid). Genes are inserted at the *Hind*III site to allow expression of fusion proteins that may be relatively stable in *E. coli* (Hallewell and Emtage 1980). A different plasmid exists for each of the three reading frames (Tacon et al. 1980).

pNCV

Two other expression vectors have been designed to allow the synthesis of large fusion proteins that also may stabilize foreign gene products. One, plasmid pNCV (see Fig. 12.12), has been used to produce stable human influenza virus hemagglutinin (Davis et al. 1981) and foot-and-mouth-disease virus VP3 antigens (Kleid et al. 1981). The stability is a consequence of fusing the *trpLE* protein encoded by *trpΔLE* 1413 with the viral protein (Goeddel et al. 1980b). This vector was constructed by removing the termination codon of *trpLE* and replacing it with synthetic DNA encoding two *Eco*RI sites and a *Pst*I site. The sequence coding for a foreign protein may be inserted at one of these sites. This vector lacks the *trp* attenuator region, and the strong *trp* promoter is therefore always partially derepressed even in the presence of excess tryptophan.

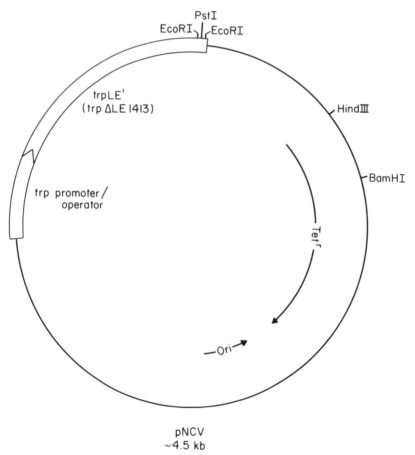

Figure 12.12

pNCV, a plasmid ~4.5 kb in length that allows synthesis of fusion proteins containing most of the *trpLE* fusion protein. The gene of interest may be inserted into the unique *Eco*RI or *Pst*I sites. The *trp* promoter lacks the attenuator region and is partially induced even in the presence of excess tryptophan. The vector has been used to produce stable fusion proteins of the *trpLE* protein with human leukocyte interferon (Goeddel et al. 1980b), with a human influenza virus hemagglutinin (Davis et al. 1981), and with foot-and-mouth-disease virus VP3 (Kleid et al. 1981).

pβ-*gal*13C

Another expression vector employing a large fusion protein for stabilizing foreign gene products is pβ-*gal*13C (see Fig. 12.13; Goeddel et al. 1979b). By cloning foreign DNA sequences into the single *Eco*RI site in *lacZ*, a fusion is made to the first 1005 amino acids of β-galactosidase (Goeddel et al. 1979b). A similar vector has been used to synthesize a hybrid β-galactosidase/β-endorphin protein (Shine et al. 1980).

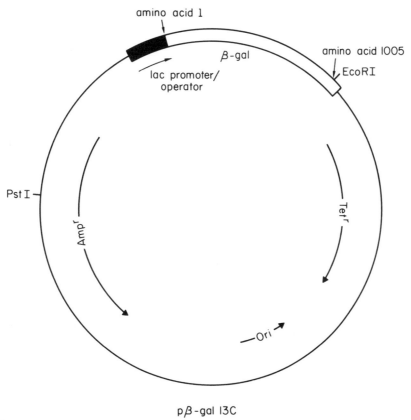

pβ-gal 13C

Figure 12.13

pβ-*gal*13C, a plasmid ∼7 kb in length that contains a large segment of the *lacZ* gene, which encodes β-galactosidase. Insertion of DNA fragments in-frame into the unique *Eco*RI site may be more stable than the native eukaryotic protein in bacteria. Vectors designed for expression of genes as fusion proteins with 1005 amino acids of β-galactosidase at their amino terminus have been used to produce fusions with human insulin peptides (Goeddel et al. 1979b), with antigenic determinants of human influenza virus hemagglutinin (Davis et al. 1981), with somatostatin (Itakura et al. 1977), and with hepatitis B virus surface antigen (Charnay et al. 1980). Several β-galactosidase fusion proteins appear to be insoluble.

pKT287

The use of fusion proteins as a means to export foreign proteins from bacteria is still under investigation. A series of vectors has been constructed that allows insertion of DNA sequences in all three reading frames at a *Pst*I site in the *bla* gene of pBR322 (Fig. 12.14; Talmadge et al. 1980). Fusion proteins containing various amounts of the amino terminus of penicillinase (4 to 27 amino acids) may be made from these vectors. The first 23 amino acids encoded by *bla* comprise the β-lactamase signal sequence. A proinsulin sequence fused to this leader resulted in expression of a protein, 50% or more of which was processed and secreted to the periplasm (Talmadge et al. 1980). The signal sequence normally found on proinsulin also functions in *E. coli* to allow the secretion of proinsulin (Talmadge et al. 1981). However, some other eukaryotic proteins that have leader sequences (e.g., human pre-β-interferon) are not processed in bacteria (Taniguchi et al. 1980b).

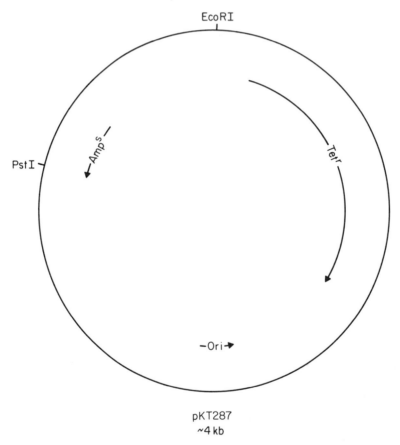

pKT287
~4 kb

Figure 12.14

pKT287, a plasmid ~ 4 kb in length containing the promoter of the β-lactamase gene. If a gene is inserted in-frame at the *Pst*I site, a fusion protein is produced containing the leader sequence. Other plasmids in which the *Pst*I site is located at various positions within or downstream from the sequences encoding the signal peptide have also been constructed. These plasmids were made by cleaving pBR322 at the *Pst*I site, digesting the DNA with *Bal*31, and inserting a *Pst*I linker (Talmadge et al. 1980).

pMH621

This vector also has the potential to cause the secretion of fusion proteins from bacteria (Fig. 12.15) (Hall and Silhavy 1981a,b; T. Silhavy, pers. comm.). When a foreign coding sequence is inserted at the *Bgl*II site, a fusion protein that has the *E. coli ompF* signal sequence is synthesized, as well as 12 amino acids of the mature *ompF* protein attached to the amino acids of the foreign polypeptide. Several polylinkers (see Chapter 1) have been inserted at the *Bgl*II site of pMH621 to allow insertion of foreign DNA using different restriction sites, to facilitate insertion in the correct reading frame, and to allow directional insertion. This vector is designed to allow the synthesis of large amounts of a fusion protein (the outer membrane protein encoded by *ompF* is present up to a level of about 100,000 copies per cell). In a bacterial strain with a cold-sensitive mutation in a positive regulatory gene (*ompR*), high levels of expression of the fusion protein occur only at high temperature (expression is minimal at 30°C). Because the hybrid proteins will contain the *ompF* signal sequence, they will probably be exported from the cytoplasm. However, it is not known whether the signal sequence will be sufficient for export. By analogy with the β-lactamase vector described above (pKT287), pMH621 may allow production of mature proteins whose leaders have been processed. This system may even export proteins to the cell surface.

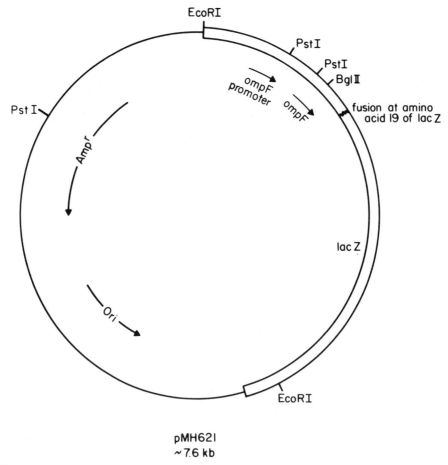

pMH62l
~7.6 kb

Figure 12.15

pMH621, a plasmid ~ 7.6 kb in length carrying the *ompF* promoter (*ompF* is a gene coding for a major outer membrane protein present at levels up to 100,000 copies per cell). Inserts in-frame into the *Bgl*II site produce a fusion protein containing the *ompF* signal peptide and 12 N-terminal amino acids of mature *ompF*. Such fusion proteins may possibly be exported to the outer membrane of *E. coli*. The promoter is under positive control and mutants have been isolated that allow expression from the *ompF* promoter to be regulated by temperature shift (T. Silhavy, pers. comm.; for background information, see Hall and Silhavy 1981a,b).

MAXIMIZING EXPRESSION OF CLONED GENES

The levels of expression of a foreign gene in *E. coli* are usually monitored by an appropriate functional assay (Struhl and Davis 1977; Chang et al. 1978), or by various immunological assays (Broome and Gilbert 1978; Heiland and Gething 1981). However, when attempting to maximize the expression of a gene product, such assays are cumbersome and can be expensive. To overcome this problem when an unfused eukaryotic gene product is to be expressed, a method in which the foreign DNA is fused to an appropriate fragment of the *lacZ* gene (Guarente et al. 1980b) is used. This method allows plasmids to be identified in which the cloned gene is efficiently transcribed and translated, even if the protein encoded by the cloned gene has no assayable activity. This technique employs a plasmid (pLG) containing *lacZ* DNA encoding an enzymatically active carboxyterminal fragment of β-galactosidase. No β-galactosidase is synthesized, however, because the promoter for the gene and the SD sequence are both absent from the plasmid. The gene to be expressed is fused in-frame to the 5′ end of the *lacZ* gene. The resulting fused gene encodes a hybrid protein that in most cases has β-galactosidase activity. However, this protein is not expressed because it still lacks a promoter and, if the upstream sequence is derived from a eukaryotic gene, an SD sequence. A portable promoter fragment can then be properly positioned in front of the fused gene as described above. Those promoter fragments that are optimally positioned should direct the synthesis of high levels of a fusion protein that has β-galactosidase activity. The plasmids with optimally placed promoter fragments can be recognized by transforming Lac⁻ bacteria and scoring for β-galactosidase activity on lactose indicator plates. These plasmids can then be used to reconstitute an unfused eukaryotic gene that is expressed at high levels (Guarente et al. 1980a,b).

INCREASED GENE DOSAGE

Once a gene has been expressed at high levels, a DNA fragment containing the entire assembly (consisting of the promoter, ribosome-binding site, and gene) may be transferred to a plasmid that can conditionally attain very high copy number in the absence of chloramphenicol amplification. The temperature-inducible "runaway replication" vector pKN402 (Uhlin et al. 1979), a small derivative of R1*drd*19, was designed for this purpose. Several derivatives of pKN402 have been constructed that provide additional cloning sites allowing one to score for insertion of foreign DNA (Bittner and Vapnek 1981). One derivative of pKN402, the vector pAS2, has proven useful for production of large amounts of protein from a cloned gene (Brent and Ptashne 1981; A. Poteete, unpubl.; A. Sancar, pers. comm.). A DNA fragment containing the *lac* or *tac* promoter and the expressed gene was cloned into pAS2. Cultures of a *lacI*^{q} strain carrying the plasmid were grown at 42°C for a few hours to increase plasmid copy number to more than 1000 copies per cell, and then isopropyl-β-D-thiogalactoside was added to inactivate the *lac* repressor. Very large increases in the production of the cloned gene product were obtained with this technique. A thermoinducible bacteriophage λ–ColE1 chimeric plasmid, pKCl6 (Rao and Rogers 1978), may be used in a similar manner. This plasmid contains genes for ampicillin and kanamycin resistance, a region of bacteriophage λ DNA (from λN through λP, including the replication origin and a λcI*ts*857 gene encoding a temperature-sensitive repressor), and a ColE1 segment including the replication and immunity regions. When the temperature of the bacterial culture is raised to 42°C, λ transcription is turned on, the λ replication system is used, and the plasmid copy number increases to about 250 copies per cell (Rao and Rogers 1978). Large increases in the production of exonuclease III were obtained by cloning the *exo*III gene into pKCl6 and then growing cells containing the recombinant plasmid at 32°C, followed by shifting the culture to 42°C (Rao and Rogers 1978).

SUMMARY

A number of expression vectors, some of which have been described here, have been designed to direct efficient transcription and translation of cloned genes. Several strong promoters have been employed, and because no single method has been used to compare their efficiencies, the choice among promoters is still somewhat arbitrary. The ability to regulate transcription is an important consideration, since the expression of gene products at high levels may be toxic to *E. coli* cells.

The signals in the DNA that direct efficient translation may be provided in several ways. If the foreign gene is to be expressed as a native, unfused protein, it may be (1) positioned so that its ATG is an optimal distance from the bacterial SD sequence by scoring for maximum expression (for review, see Guarente et al. 1980a), (2) placed downstream from an SD sequence so that its ATG is a predetermined distance from the SD sequence (e.g., Goeddel et al. 1979a), or (3) fused to an ATG present on the vector (A. Shatzman and M. Rosenberg, pers. comm.). Neither the effect of substituting nucleotides between the SD sequence and the ATG of the ribosome-binding site nor the result of changing the DNA sequence after the ATG has been examined in a systematic fashion. However, the spacing between the SD sequence and the ATG is known to affect the amount of protein synthesized (Backman and Ptashne 1978).

If the native protein is unstable in *E. coli*, expression of the gene product as part of a fusion protein may be desirable, especially if the native protein can be chemically cleaved from the purified fusion protein (e.g., Goeddel et al. 1979b) or if the protein is to be used to elicit the production of antibodies (e.g., Kleid et al. 1981). Alternatively, some proteins may be stabilized by synthesis in large amounts in *E. coli* (Shimatake and Rosenberg 1981).

Vectors that allow fusion of foreign genes to DNA encoding a signal sequence may be useful for exporting proteins out of the cytoplasm, especially if the signal peptide is cleaved during export of the protein. Export of the proteins may assist in subsequent purification and may serve to isolate them from cytoplasmic proteases. However, the factors that determine whether a given protein will be secreted when it is fused with a particular leader peptide have not been elucidated.

In most cases, the levels of expression of eukaryotic genes in *E. coli* are less than expected. In addition to factors that have been discussed, expression levels may be influenced by mRNA secondary structure and stability or by codon usage. In any case, conditions necessary for optimization of expression have not been generalized, so at present each new protein poses a different problem.

Appendices

Appendix A: Biochemical Techniques

In this appendix, we have collected a set of biochemical techniques that are frequently used in molecular cloning and that form the basis of many of the protocols presented in this manual.

GLASSWARE AND PLASTICWARE

All glassware and plasticware should be sterilized by autoclaving. It is particularly important to have available a supply of sterilized Eppendorf tubes and disposable tips for automatic pipetting devices. All of the procedures commonly used in molecular cloning can be carried out in plastic or glassware prepared in this way; there is no significant loss of material by absorption onto the surfaces of the containers. However, for certain procedures (e.g., handling very small quantities of single-stranded DNA or sequencing by the Maxam-Gilbert technique), it is advisable to use glass and plasticware that has been coated with a thin film of silicone. A simple procedure for siliconizing small items such as pipettes, tubes, beakers, and so forth is given below. For large items such as glass plates, see note ii below.

Siliconizing Glassware and Plasticware[1]

1. Place the items to be siliconized inside a large, glass desiccator.

2. Add 1 ml of dichlorodimethylsilane to a small beaker inside the desiccator.

3. Attach the jar through a trap to a vacuum and turn on the vacuum for 5 minutes.

4. Turn off the vacuum and quickly allow air to enter the desiccator. This causes uniform dispersion of the gaseous dichlorodimethylsilane.

5. Turn on the vacuum again until a vacuum is achieved inside the desiccator.

6. Close the system and leave the desiccator under vacuum for 2 hours.

7. Open the desiccator. Glassware should be baked at 180°C for at least 2 hours before use. Plasticware should be rinsed very well with water before use.

Notes

i. Dichlorodimethylsilane is toxic and highly volatile and should be used only in a chemical hood.

ii. Large items of glassware should be siliconized by soaking or rinsing in a 5% solution of dichlorodimethylsilane in chloroform. Then rinse them many times with water and bake at 180°C for 2 hours before use.

[1]B. Seed (unpubl.).

PREPARATION OF ORGANIC REAGENTS

Phenol

Many batches of commercial, liquified phenol can be used without redistillation. However, batches that are pink or yellow and all crystalline phenol must be redistilled at 160°C to remove contaminants that cause breakdown or crosslinking of RNA and DNA. Liquified and redistilled phenol should be stored at −20°C in small aliquots, preferably under nitrogen gas.

As needed, phenol is removed from the freezer, allowed to warm to room temperature, and melted at 68°. 8-Hydroxyquinoline is then added to a final concentration of 0.1%. This yellow compound is an antioxidant, a partial inhibitor of RNase, and a weak chelator of metal ions (Kirby 1956). In addition, the yellow color provides a convenient way to identify the phase.

The melted phenol is then extracted several times with an equal volume of buffer (usually 1.0 M Tris [pH 8.0], followed by 0.1 M Tris [pH 8.0] and 0.2% β-mercaptoethanol), until the pH of the aqueous phase is >7.6. The phenol solution can be stored at 4°C under equilibration buffer for periods of up to 1 month.

Caution

Phenol is highly corrosive and can cause severe burns. Safety glasses and gloves should be worn. Any areas of skin that come into contact with phenol should be rinsed with a large volume of water and washed with soap and water. Do *not* use ethanol.

Chloroform:Isoamyl Alcohol (24:1)

A mixture of chloroform and isoamyl alcohol (24:1 v/v) is used to remove proteins from preparations of nucleic acid. The chloroform denatures proteins while isoamyl alcohol reduces foaming during the extraction and facilitates the separation of the aqueous and organic phases.

Neither reagent requires treatment before use. The mixture is stable and may be stored in closed bottles at room temperature.

Ether Saturated with Water

Ether is used to extract traces of phenol or other organic substances from aqueous solutions of nucleic acid. Traces of ether remaining after the extraction may be removed easily by heating to 68°C or by blowing a gentle stream of nitrogen gas over the sample. The ether is saturated with water to prevent loss of water from the sample during extraction and to inhibit the formation of free radicals, which form during storage and may damage DNA.

Mix an equal volume of ethyl ether (anhydrous) and distilled water in a screw-capped bottle. Shake well (tightly capped) and store at room temperature in a chemical hood. The ether is the upper of the two phases.

Caution

Ether is highly volatile and extremely flammable and should be worked with and stored in an explosion-proof hood.

LIQUID MEDIA

NZCYM Medium

Per liter:

NZ amine[2]	10 g
NaCl	5 g
yeast extract	5 g
casamino acids	1 g
$MgSO_4 \cdot 7H_2O$	2 g

Adjust pH to 7.5 with sodium hydroxide.

NZYM Medium

Identical to NZCYM except that casamino acids are omitted.

LB (Luria-Bertani) Medium

500 ml

Per liter:

Bacto-tryptone	10 g
Bacto-yeast extract	5 g
NaCl	10 g

* cover when mixing

Adjust pH to 7.5 with sodium hydroxide.

7.5g of agarose

M9 Medium

Per liter:

Na_2HPO_4	6 g
KH_2PO_4	3 g
NaCl	0.5 g
NH_4Cl	1 g

Adjust pH to 7.4, autoclave, cool, and then add:

1 M $MgSO_4$	2 ml
20% glucose	10 ml
1 M $CaCl_2$	0.1 ml

The above solutions should be sterilized separately by filtration (glucose) or autoclaving.

---

[2]Type-A hydrolysate of casein from Humko Sheffield Chemical Division of Kraft, Inc., 1099 Wall St. West, Lafayette, NJ 07071.

M9CA Medium

Identical to M9 medium except that 2.0 g/l of casamino acids are included.

χ1776 Medium

Per liter:

Bacto-tryptone	25 g
Bacto-yeast extract	7.5 g
1 M Tris·Cl (pH 7.5)	20 ml

Autoclave, cool, and then add:

1 M MgCl₂	5 ml
1% diaminopimelic acid	10 ml
0.4% thymidine	10 ml
20% glucose	25 ml

Magnesium chloride should be sterilized separately by autoclaving, and the other ingredients should be sterilized separately by filtration.

SOB Medium

Per liter:

Bacto-tryptone	20 g
Yeast extract	5 g
NaCl	0.5 g

Adjust pH to 7.5 with potassium hydroxide and sterilize by autoclaving. Just before use, add 20 ml of 1 M MgSO₄, sterilized separately by autoclaving.

Media Containing Agar or Agarose

Make up liquid media according to the appropriate formula given above. Just before autoclaving, add (per liter) one of the following:

Bacto-agar	15 g (for plates)
Bacto-agar	7 g (for top agar)
Agarose (type-1, low EEO)	15 g (for plates)
Agarose	7 g (for top agarose)

200 ml

When preparing plates, media that contain agar or agarose should be sterilized by autoclaving and allowed to cool to 55°C before thermolabile substances (e.g., antibiotics) are added. Plates can then be poured directly from the flask, allowing about 30–35 ml per 85-mm petri dish. If bubbles are a problem, flame the surface of the medium with a bunsen burner before the agar hardens.

Concentrated Media

Some media (e.g., LB, NZYM, and NZCYM) can be prepared and stored as a 5× concentrate. Medium may then be prepared rapidly by diluting the 5× concentrate with the appropriate volume of sterile water.

SOLUTIONS FOR WORKING WITH BACTERIOPHAGE λ

Maltose

Maltose (0.2%) is often added to the medium during growth of bacteria that are to be used for plating bacteriophage λ.

maltose	20 g
H_2O	to 100 ml

Sterilize by filtration.
Add 1 ml of sterile 20% maltose solution for every 100 ml of medium.

SM

This medium is used for phage storage and dilution.

Per liter:

NaCl	5.8 g
$MgSO_4 \cdot 7\ H_2O$	2 g
1 M Tris·Cl (pH 7.5)	50 ml
2% gelatin	5 ml

Sterilize by autoclaving and store in 50-ml lots.

ANTIBIOTICS

Ampicillin

Stock solution. 25 mg/ml of the sodium salt of ampicillin in water. Sterilize by filtration and store in aliquots at −20°C.

Working concentration. 35–50 μg/ml.

Chloramphenicol

Stock solutions. 34 mg/ml in 100% ethanol. Store at −20°C.

Working concentration. For amplification of plasmids, 170 μg/ml; for selection of resistant bacteria, 30 μg/ml.

Kanamycin

Stock solution. 25 mg/ml in water. Sterilize by filtration and store in aliquots at −20°C.

Working concentration. 50 μg/ml.

Nalidixic Acid

Stock solution. 20 mg/ml in water. Sterilize by filtration and store in aliquots at −20°C.

Working concentration. 20 μg/ml.

Streptomycin

Stock solution. 20 mg/ml in water. Sterilize by filtration and store in aliquots at −20°C.

Working concentration. 25 μg/ml.

Tetracycline

Stock solution. 12.5 mg/ml tetracycline hydrochloride in ethanol/water (50% v/v). Store at −20°C.

Note. Because tetracycline is light-sensitive, solutions and plates containing the antibiotic should be stored in the dark.

Working concentration. 12.5–15.0 μg/ml. Magnesium ions are antagonists of tetracycline. Use media without magnesium salts (e.g., LB) for selection of bacteria resistant to tetracycline.

**TABLE A.1. CONCENTRATIONS OF ACIDS AND BASES—
COMMON COMMERCIAL STRENGTHS**

	Molecular weight	Moles/liter	Grams/liter	% (by weight)	Specific gravity
Acetic acid, glacial	60.05	17.4	1045	99.5	1.05
Acetic acid	60.05	6.27	376	36	1.045
Formic acid	46.02	23.4	1080	90	1.20
Hydrochloric acid	36.5	11.6	424	36	1.18
		2.9	105	10	1.05
Nitric acid	63.02	15.99	1008	71	1.42
		14.9	938	67	1.40
		13.3	837	61	1.37
Perchloric acid	100.5	11.65	1172	70	1.67
		9.2	923	60	1.54
Phosphoric acid	80	18.1	1445	85	1.70
Sulfuric acid	98.1	18.0	1766	96	1.84
Ammonium hydroxide	35.0	14.8	251	28	0.898
Potassium hydroxide	56.1	13.5	757	50	1.52
		1.94	109	10	1.09
Sodium hydroxide	40.0	19.1	763	50	1.53
		2.75	111	10	1.11

PREPARATION OF BUFFERS AND SOLUTIONS

The following general guidelines should be followed in preparing and using buffers and solutions.

1. Use the highest grade of reagents available.

2. Prepare all solutions with double-distilled, deionized water.

3. Wherever possible, sterilize all solutions by autoclaving or by filtration through a 0.22-μm filter.

TABLE A.2. PREPARATION OF STOCK SOLUTIONS

Solution	Method of preparation	Comments
1 M Tris	Dissolve 121.1 g Tris base in 800 ml of H_2O. Adjust the pH to the desired value by adding concentrated HCl.	If the 1 M solution has a yellow color, discard it and obtain better-quality Tris.
	Approximate amount of concentrated HCl: desired pH: pH 7.4 70 ml pH 7.6 60 ml pH 8.0 42 ml	Although many types of electrodes do not accurately measure the pH of Tris solutions, suitable electrodes can be obtained from most manufacturers.
	Allow the solution to cool to room temperature before making the final adjustments to the pH. Make up the volume of the solution to 1 liter. Dispense into aliquots and sterilize by autoclaving.	
0.5 M EDTA (pH 8.0)	Add 186.1 g of disodium ethylene diamine tetraacetate · $2H_2O$ to 800 ml of H_2O. Stir vigorously on a magnetic stirrer. Adjust the pH to 8.0 with NaOH ($\sim$ 20 g of NaOH pellets). Dispense into aliquots and sterilize by autoclaving.	The disodium salt of EDTA will not go into solution until the pH of the solution is adjusted to approximately 8.0 by the addition of NaOH.
5 M NaCl	Dissolve 292.2 g of NaCl in 800 ml of H_2O. Adjust volume to 1 liter. Dispense into aliquots and sterilize by autoclaving.	
1 M $MgCl_2$	Dissolve 203.3 g of $MgCl_2 \cdot 6H_2O$ in 800 ml of H_2O. Adjust volume to 1 liter. Dispense into aliquots and sterilize by autoclaving.	$MgCl_2$ is extremely hygroscopic. Buy small bottles (e.g., 100 g) and do not store opened bottles for long periods of time.
3 M Sodium acetate (pH 5.2)	Dissolve 408.1 g of sodium acetate · $3H_2O$ in 800 ml of H_2O. Adjust pH to 5.2 with glacial acetic acid. Adjust volume to 1 liter. Dispense into aliquots and sterilize by autoclaving.	

1 M Dithiothreitol (DTT)	Dissolve 3.09 g of DTT in 20 ml of 0.01 M sodium acetate (pH 5.2). Sterilize by filtration. Dispense into 1-ml aliquots and store at −20°C.	Do not autoclave DTT or solutions containing DTT.
β-Mercaptoethanol (BME)	Usually obtained as a 14.4 M solution. Store in a dark bottle at 4°C.	Do not autoclave BME or solutions containing BME.
10% Sodium dodecyl sulfate (SDS) (also called sodium lauryl sulfate)	Dissolve 100 g of electrophoresis-grade SDS in 900 ml of H_2O. Heat to 68°C to assist dissolution. Adjust the pH to 7.2 by adding a few drops of concentrated HCl. Adjust volume to 1 liter. Dispense into aliquots.	Wear a mask when weighing SDS. There is no need to sterilize 10% SDS.
1 M Magnesium acetate	Dissolve 214.46 g of magnesium acetate · $4H_2O$ in 800 ml of H_2O. Adjust the volume to 1 liter. Sterilize by filtration.	
5 M Ammonium acetate	Dissolve 385 g of ammonium acetate in 800 ml of H_2O. Adjust volume to 1 liter. Sterilize by filtration.	
5 M Potassium acetate	To 60 ml of 5 M potassium acetate add 11.5 ml of glacial acetic acid and 28.5 ml of H_2O. The resulting solution is 3 M with respect to potassium and 5 M with respect to acetate.	
20× SSC	Dissolve 175.3 g of NaCl and 88.2 g of sodium citrate in 800 ml of H_2O. Adjust pH to 7.0 with a few drops of a 10 N solution of NaOH. Adjust volume to 1 liter. Dispense into aliquots. Sterilize by autoclaving.	✗ misprint lower pH w/ HCl
20× SSPE	Dissolve 174 g of NaCl, 27.6 g of NaH_2PO_4 · H_2O, and 7.4 g of EDTA in 800 ml of H_2O. Adjust pH to 7.4 with NaOH (∼ 6.5 ml of a 10 N solution). Adjust volume to 1 liter. Dispense into aliquots. Sterilize by autoclaving.	
Ethidium bromide 10 mg/ml	Add 1 g of ethidium bromide to 100 ml of H_2O. Stir on a magnetic stirrer for several hours to ensure that the dye has dissolved. Wrap the container in aluminum foil or transfer to a dark bottle and store at 4°C.	Ethidium bromide is a powerful mutagen. Wear gloves and a mask when weighing it out.
Trichloroacetic acid (TCA) 100% solution	To a bottle containing 500 g of TCA, add 227 ml of H_2O. The resulting solution will contain 100% (w/v) TCA.	

Solutions

TE

pH 7.4

10 mM Tris·Cl (pH 7.4)
1 mM EDTA (pH 8.0)

pH 7.6

10 mM Tris·Cl (pH 7.6)
1 mM EDTA (pH 8.0)

pH 8.0

10 mM Tris·Cl (pH 8.0,
1 mM EDTA (pH 8.0)

STE (also called TNE)

10 mM Tris·Cl (pH 8.0)
100 mM NaCl
1 mM EDTA (pH 8.0)

Formamide (Deionized)

Mix 50 ml of formamide and 5 g of mixed-bed, ion-exchange resin (e.g., Bio-Rad AG 501-X8, 20–50 mesh). Stir for 30 minutes at room temperature. Filter twice through Whatman No. 1 filter paper. Dispense into 1-ml aliquots and store at −20°C.

Denhardt's Solution (50×)

Ficoll	5 g
polyvinylpyrrolidone	5 g
BSA (Pentax Fraction V)	5 g
H_2O	to 500 ml

Filter through a disposable Nalgene filter. Dispense into 25-ml aliquots and store at −20°C.

10% Bovine Serum Albumin (BSA)

BSA (Pentax Fraction V)	1 g
H_2O	10 ml

Dispense into aliquots. Store at −20°C.

ATP (0.1 M)

Dissolve 60 mg of ATP in 0.8 ml of H_2O. Adjust the pH to 7.0 with 0.1 M NaOH. Adjust the volume to 1.0 ml with H_2O. Dispense the solution into small aliquots and store at $-70°C$.

Ribo- and Deoxyribonucleotide Triphosphates (~10 mM)

Dissolve NTP or dNTP in water directly in the shipping bottle at an expected concentration 10 mM. Using a dilute solution (0.05 M) of Tris base, an automatic micropipettor, and pH paper, adjust the pH to 7.0. Dilute an aliquot of the neutralized NTP or dNTP appropriately and read the optical density at the wavelengths given in Table A.3. Using the values for the extinction coefficients in the table, calculate the actual concentration. Freeze away in small aliquots at $-20°C$.

TABLE A.3. OPTICAL DENSITIES OF RIBO- AND DEOXYRIBONUCLEOTIDE TRIPHOSPHATES

Base	Wavelength	Extinction coefficients for bases ϵ (M^{-1} cm^{-1})
A	259	1.54×10^4
G	253	1.37×10^4
C	271	9.1×10^3
U	262	1.0×10^4
T	260	7.4×10^3

For a cell with a 1-cm path length, absorbance = ϵ/M

TABLE A.4.   PROTEOLYTIC ENZYMES

	Stock solution	Storage temperature	Concentration in reaction	Reaction buffer	Temperature	Pretreatment
Pronase	20 mg/ml in H_2O	−20°C	1 mg/ml	0.01 M Tris (pH 7.8) 0.01 M EDTA 0.5% SDS	37°C	self-digestion for 2 hours at 37°C
Proteinase K	20 mg/ml in H_2O	−20°C	50 μg/ml	0.01 M Tris (pH 7.8) 0.005 M EDTA 0.5% SDS	37°C	none required

Preparations of pronase are often contaminated with DNase and RNase. These activities can be eliminated by incubating the stock solution for 2 hours at 37°C.

ENZYMES

Lysozyme

Stock solution. 50 mg/ml in water. Dispense into aliquots and store at $-20°C$. Discard each aliquot after use; do not refreeze.

RNase That Is Free of DNase

Dissolve pancreatic RNase (RNase A) at a concentration of 10 mg/ml in 10 mM Tris · Cl (pH 7.5) and 15 mM NaCl. Heat to 100°C for 15 minutes and allow to cool slowly to room temperature. Dispense into aliquots and store at $-20°C$.

DNase That Is Free of RNase

Unfortunately, many commercial preparations of pancreatic DNase I, even those sold as "RNase-free," are contaminated by amounts of ribonuclease that are sufficient to cause significant degradation of high-molecular-weight RNA. Two methods, given below, are available to remove the contaminating RNase activity.

Affinity chromotography on agarose-coupled 5′-(4-aminophenyl-phosphoryl) uridine 2′(3′) phosphate (Maxwell et al. 1977).

1. Equilibrate 10 ml of agarose-5′-(4-aminophenyl-phosphoryl) uridine 2′(3′) phosphate (commercially available from Miles-Yeda Laboratories) with 0.02 M sodium acetate (pH 5.2). Make a column in a 25-ml disposable syringe.

2. Dissolve 20 mg of pancreatic DNase I (DPFF, Worthington Biochemicals) in 1 ml of 0.02 M sodium acetate (pH 5.2).

3. Apply the solution of DNase I to the column and elute with 0.02 M sodium acetate (pH 5.2) at room temperature. Collect 1-ml fractions into RNase-free tubes (see pages 190, 437) until all material absorbing at 280 nm has eluted from the column.

4. Pool the fractions that contain protein. Read the OD_{280} and calculate the concentration of protein (1 $OD_{280} \simeq 1$ mg of protein). Dispense the enzyme preparation into small aliquots and store at $-20°C$.

Adsorption to macaloid. Macaloid, a clay that has been known for many years to adsorb RNase, is available from the National Lead Company, Houston, Texas. It is prepared as follows (Schaffner 1982):

a. Suspend 0.5 g of macaloid powder in 50 ml of sterile 50 mM Tris·Cl (pH 7.6). Heat to 100°C for 5 minutes with constant agitation.

b. Centrifuge at room temperature for 5 minutes at 2500g.

c. Discard the supernatant. Resuspend the sticky pellet completely in 40 ml of sterile 50 mM Tris·Cl (pH 7.6).

d. Repeat the centrifugation and washing steps twice more.

e. Centrifuge the suspension for 15 minutes at 3500g.

f. Resuspend the pellet in 30 ml of sterile 50 mM Tris·Cl (pH 7.6). The final concentration of macaloid is 16 mg/ml. The suspension may be stored indefinitely at 4°C.

The macaloid suspension is used in the following steps to remove contaminating RNase activity from DNase.

1. Dissolve 100 mg of DNase I (DPFF, Worthington Biochemicals) in 5 ml of:

   20 mM Tris·Cl (pH 7.6)
   50 mM NaCl
   1 mM dithiothreitol
   100 μg/ml BSA
   50% glycerol

2. Add 15 ml of ice-cold 50 mM Tris·Cl (pH 7.6). Mix gently.

3. Add 7.0 ml of an ice-cold, well-dispersed suspension of macaloid and mix on a rotating wheel for 30 minutes at 4°C.

4. Centrifuge for 10 minutes at 8000g at 0°C. Decant the supernatant into a fresh tube.

5. Add another 7.0 ml of macaloid suspension and mix as before.

6. Centrifuge for 15 minutes at 12,000g.

7. Carefully remove the supernatant and mix it gently with an equal volume of ice-cold, sterile glycerol.

8. Dispense into small aliquots and store at −20°C. The concentration of DNase I is approximately 3.0 mg/ml.

BUFFERS FOR RESTRICTION ENDONUCLEASE DIGESTION

Low-salt Buffer

> 10 mM Tris · Cl (pH 7.5)
> 10 mM MgCl$_2$
> 1 mM dithiothreitol

Medium-salt Buffer

> 50 mM NaCl
> 10 mM Tris · Cl (pH 7.5)
> 10 mM MgCl$_2$
> 1 mM dithiothreitol

High-salt Buffer

> 100 mM NaCl
> 50 mM Tris · Cl (pH 7.5)
> 10 mM MgCl$_2$
> 1 mM dithiothreitol

Buffer for *Sma*I

> 20 mM KCl
> 10 mM Tris · Cl (pH 8.0)
> 10 mM MgCl$_2$
> 1 mM dithiothreitol

3.M ml acetic acid
0.5M EDTA . 100 ml
1000 ml

COMMONLY USED ELECTROPHORESIS BUFFERS

Tris-Acetate (TAE)

Working solution

0.04 M Tris-acetate
0.002 M EDTA

Concentrated stock solution (50×)

Per liter:

Tris base	242 g	*(121g)*
glacial acetic acid	57.1 ml	*(28.55)*
0.5 M EDTA (pH 8.0)	100 ml	*(50 ml)*

Tris-Phosphate (TPE)

Working solution

0.08 M Tris-phosphate
0.008 M EDTA

Concentrated stock solution (10×)

Per liter

Tris base	108 g
85% phosphoric acid (1.679 mg/ml)	15.1 ml
0.5 M EDTA (pH 8.0)	40 ml

Tris-Borate (TBE)

Working solution

0.089 M Tris-borate
0.089 M boric acid

Concentrated stock solution (5×)

Per liter:

Tris base	54 g
boric acid	27.5 g
0.05 M EDTA (pH 8.0)	20 ml

0.5 M EDTA (pH 8.0) 20 ml

COMMONLY USED GEL-LOADING BUFFERS

Loading buffers are solutions of high density that enable samples to be introduced easily into gel slots. They also contain one or more tracking dyes that allow the progress of the electrophoresis to be monitored easily.

Type I

6× buffer

   0.25% bromophenol blue
   0.25% xylene cyanol
   40% (w/v) sucrose in H_2O

Store at 4°C.

Type II

10× buffer

   0.25% bromophenol blue
   0.25% xylene cyanol
   25% Ficoll (type 400) in H_2O

Store at room temperature.

Type III

6× buffer

   0.25% bromophenol blue
   0.25% xylene cyanol .
   30% glycerol in H_2O

Store at 4°C.

Type IV

6× buffer

   0.25% bromophenol blue
   40% (w/v) sucrose in H_2O

Store at 4°C.

PREPARATION OF DIALYSIS TUBING

1. Cut the tubing into pieces of convenient length (10–20 cm).

2. Boil for 10 minutes in a large volume of 2% sodium bicarbonate and 1 mM EDTA.

3. Rinse the tubing thoroughly in distilled water.

4. Boil for 10 minutes in distilled water.

5. Allow to cool and store at 5°C. Be sure that the tubing is always submersed.

6. Before use, wash the tubing inside and out with distilled water. Always handle the tubing with gloves.

Note

Instead of boiling for 10 minutes in water (step 4), the tubing can be autoclaved for 10 minutes on liquid cycle in a loosely capped jar filled with water.

DRYING DOWN ³²P-LABELED NUCLEOTIDES FROM MIXTURES OF ETHANOL AND WATER

Most commercial suppliers sell [³²P]dNTPs as concentrated, stabilized aqueous solutions that can be added directly to the appropriate reaction mixtures. However, some manufacturers still supply [³²P]dNTPs dissolved in 50% ethanol, which must be removed by evaporation before the [³²P]dNTP can be used.

1. Using an automatic micropipettor, carefully dispense the desired quantity of [³²P]dNTPs into an Eppendorf tube.

2. Plug the top of the tube with a small piece of cotton and cover with two or three layers of parafilm (Fig. A.1).

3. Poke many holes in the parafilm with a needle.

4. Place the tube securely in a beaker or rack and evaporate the [³²P]dNTPs to dryness under vacuum at room temperature or in a lyophilizer.

5. Discard the parafilm and cotton into radioactive waste. Add a small volume (5 μl) of H₂O to the tube. Vortex for 15 seconds.

6. Add the remaining ingredients of the reaction mixture to the tube. Mix by vortexing and incubate as indicated in the relevant protocol.

Notes

i. Steps 1–3 can be eliminated by using a speed-vac concentrator, which prevents bumping of the contents of the tube under vacuum.

ii. Wherever possible, manipulations involving ³²P should be carried out behind lucite screens to shield personnel from exposure to radioactivity.

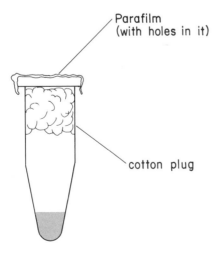

Parafilm
(with holes in it)

cotton plug

Figure A.1

PURIFICATION OF NUCLEIC ACIDS

Perhaps the most basic of all procedures in molecular cloning is the purification of nucleic acid. The key step, the removal of proteins, can often be carried out simply by extracting aqueous solutions of nucleic acids with phenol and/or chloroform. Such extractions are used whenever it is necessary to inactivate and remove enzymes that are used in one step of a cloning operation before proceeding to the next. However, additional measures are required when nucleic acids are purified from complex mixtures of molecules such as cell lysates. In these cases (see Chapter 9), it is usual to remove most of the protein by digesting with proteolytic enzymes such as pronase or proteinase K (see Table A.4), which are active against a broad spectrum of native proteins, before extracting with organic solvents.

Extraction with Phenol/Chloroform

The standard way to remove proteins from nucleic acid solutions is to extract once with phenol, once with a 1:1 mixture of phenol and chloroform, and once with chloroform. This procedure takes advantage of the fact that deproteinization is more efficient when two different organic solvents are used instead of one. Furthermore, although phenol denatures proteins efficiently, it does not completely inhibit RNase activity, and it is a solvent for RNA molecules that contain long tracts of poly(A) (Brawerman et al. 1972). Both of these problems can be circumvented by using a mixture of phenol and chloroform (1:1). Also, the final extraction with chloroform removes any lingering traces of phenol from the nucleic acid preparation.

Remember that "chloroform" means a 24:1 (v/v) mixture of chloroform and isoamyl alcohol. "Phenol" means phenol equilibrated with buffer and containing 0.1% hydroxyquinoline and 0.2% β-mercaptoethanol (see page 438).

1. Mix the DNA sample with an equal volume of phenol or phenol/chloroform in a polypropylene tube with a plastic cap.

2. Mix the contents of the tube until an emulsion forms (see note below).

3. Centrifuge for 3 minutes at 1600g or for 15 seconds in an Eppendorf centrifuge at room temperature. If the organic and aqueous phases are not well-separated, centrifuge again for a longer time or at a higher speed.

4. Use a pipette to transfer the upper, aqueous phase to a fresh polypropylene tube. For small volumes ($<$200 μl), use an automatic pipettor fitted with a disposable tip. Discard the interface and lower organic phase.

Note. To achieve the best recovery, the organic phase and interface may be "back-extracted" as follows. After the first aqueous phase has been transferred as described above, add an equal volume of TE (pH 7.8) to the organic phase and interface. Mix well. Separate the phases by centrifugation. Combine the second aqueous phase with the first and proceed to step 5.

5. Add an equal volume of a 1:1 mixture of phenol and chloroform. Repeat steps 2, 3, and 4.

6. Add an equal volume of chloroform and repeat steps 2, 3, and 4.

7. Recover the DNA by precipitation with ethanol as described on page 461.

Note

The organic and aqueous phases may be mixed by vortexing when isolating small DNAs (< 10 kb) or by gentle shaking when isolating DNAs of moderate size (10–30 kb).

More extensive precautions are necessary to avoid shearing large (> 30 kb) DNA molecules:

a. The organic and aqueous phases should be mixed by rotating the tube slowly on a wheel (20 rpm).

b. Large-bore pipettes should be used to transfer the DNA from one tube to another.

c. The DNA should not be precipitated with ethanol (step 7). Instead, traces of chloroform should be removed either by dialyzing the DNA solution extensively against large volumes of ice-cold TNE *or* by extraction with water-saturated ether.

Extraction of Phenol/Chloroform with Water-saturated Ether

Ether can be used to remove traces of phenol or chloroform from DNA solutions. Ether is highly volatile and extremely flammable and should be worked with and stored in an explosion-proof chemical hood.

1. Combine the DNA sample with an equal volume of water-saturated ether. Mix. Let the organic and aqueous phases separate by standing for 2-5 minutes.

2. Remove and discard the upper layer (ether is less dense than water).

3. Repeat steps 1 and 2.

4. Remove traces of ether by heating the DNA solution to 68°C for 5-10 minutes with gentle mixing; or by blowing a stream of nitrogen gas over the surface of the solution for 10-30 minutes.

5. Precipitate the DNA with ethanol or (in the case of high-molecular-weight DNA) dialyze extensively against ice-cold TE (pH 7.8) containing 0.1 M NaCl (STE).

CONCENTRATION OF NUCLEIC ACIDS

Precipitation with Ethanol or Isopropanol

The most widely used method for concentrating DNA is precipitation with ethanol. The precipitate of DNA, which is allowed to form at low temperature ($-20°C$ or less) in the presence of moderate concentrations of monovalent cations, is recovered by centrifugation and redissolved in an appropriate buffer at the desired concentration. The technique is rapid and is quantitative even with nanogram amounts of DNA.

1. Estimate the volume of the DNA solution.

2. Adjust the concentration of monovalent cations either by dilution with TE (pH 8.0) if the DNA solution contains a high concentration of salts or by addition of one of the salt solutions shown in Table A.5.

3. Mix well. Add exactly 2 volumes of ice-cold ethanol and mix well. Chill to $-20°C$.

4. Store at low temperature to allow the DNA precipitate to form. Usually 30–60 minutes at $-20°C$ is sufficient, but when the size of the DNA is small (< 1 kb) or when it is present in small amounts (< 0.1 μg/ml), the period of storage should be extended and the temperature should be lowered to $-70°C$.

5. Centrifuge at $0°C$. For most purposes, 10 minutes in an Eppendorf centrifuge or at $12,000g$ is sufficient. However, when low concentrations of DNA or very small fragments are being processed, more extensive centrifugation (e.g., Beckman SW50.1 at 30,000 rpm for 30 minutes) may be required.

6. Discard the supernatant. Stand the tube in an inverted position on a layer of absorbent paper to allow as much of the supernatant as possible to drain away. Use capillary pipettes to remove any drops of fluid that adhere to the walls of the tube. Traces of supernatant may be removed by brief treatment (1–2 minutes) in a vacuum desiccator or lyophilizer.

TABLE A.5. SALT SOLUTIONS

	Concentrated solution	Final solution
Sodium acetate	2.5 M (pH 5.2)	0.25 M
Sodium chloride	5.0 M	0.1 M
Ammonium acetate	10.5 M	2.0 M

7. Dissolve the DNA pellet (which is often invisible) in the desired volume of buffer. Rinse the walls of the tube well with the buffer or scrape them with a sealed pipette to aid in the recovery of the DNA. The sample can be heated to 37°C for 5 minutes to assist in dissolving the pellet.

Notes

i. Isopropanol (1 volume) may be used in place of ethanol (2 volumes) to precipitate DNA. Precipitation with isopropanol has the advantage that the volume of liquid to be centrifuged is smaller. However, isopropanol is less volatile than ethanol and it is more difficult to remove the last traces; moreover, solutes such as sucrose or sodium chloride are more easily coprecipitated with DNA when isopropanol is used, especially at $-70°C$. In general, precipitation with ethanol is preferable unless it is necessary to keep the volume of fluid to a minimum.

ii. To remove any solutes that may be trapped in the precipitate, the DNA pellet may be washed with a solution of 70% ethanol. To make certain that no DNA is lost during washing, add 70% ethanol until the tube is $2/3$ full. Vortex briefly, and recentrifuge as described above. After the 70% ethanol wash, the pellet does not adhere tightly to the wall of the tube, so great care must be taken when removing the supernatant.

iii. Very short DNA molecules (<200 bp) are precipitated inefficiently by ethanol. However, adjusting the DNA solution to 0.01 M $MgCl_2$ before addition of ethanol considerably improves the efficiency with which small DNA molecules are recovered.

iv. In general, DNA precipitated from solution by ethanol can be easily redissolved in buffers of low ionic strength such as TE. Occasional difficulties arise when buffers containing $MgCl_2$ or >0.1 M NaCl are added directly to the DNA pellet. It is therefore preferable to dissolve the DNA in a small volume of low-ionic-strength buffer and to adjust the composition of the buffer later. If the sample does not easily dissolve in a small volume, add a larger volume of buffer and repeat the precipitation with ethanol.

v. To be precipitated from solution, RNA requires slightly higher concentrations of ethanol (2.5 volumes) than does DNA.

vi. Triphosphate can be removed from DNA by two sequential precipitations with ethanol from DNA solutions containing 2 M ammonium acetate.

Concentration by Extraction with Butanol

During extraction of aqueous solutions with solvents such as secondary butyl alcohol (2-butanol) or *n*-butyl alcohol (1-butanol), some of the water molecules (but not DNA or solutes) become partitioned into the organic phase. By carrying out several cycles of extraction, the volume of a DNA solution can be reduced significantly. This method of concentrating DNA is used to reduce the volume of dilute DNA solutions to the point where the DNA can be easily recovered by precipitation with ethanol.

1. Add an equal volume of 2-butanol to the DNA sample and mix well.

   Note. Addition of too much 2-butanol can result in removal of all the water and precipitation of the DNA.

2. Centrifuge at 1600*g* for 1 minute. Remove and discard the upper (2-butanol) phase.

3. Repeat steps 1 and 2 until the desired volume is achieved.

4. Extract the sample twice with water-saturated ether to remove the 2-butanol. Remove the ether by evaporation.

Note

Because 2-butanol extraction does not remove salt, the salt concentration increases in proportion to the reduction in volume of the solution. Therefore, adjust the buffer concentration by dialysis or recover the DNA by precipitation with ethanol.

CHROMATOGRAPHY THROUGH SEPHADEX G-50

This technique, which employs gel filtration to separate high-molecular-weight DNA from smaller molecules, is used most often to segregate DNA that has been labeled by nick-translation or by filling in of recessed 3' ends from unincorporated, labeled deoxynucleotide triphosphates. However, it is also used at several stages during the synthesis of double-stranded cDNA, during addition of linkers to blunt-ended DNA, and, in general, whenever it is necessary to change the composition of the buffer in which DNA is dissolved.

Two methods are available: conventional column chromatography and centrifugation through Sephadex G-50 packed in disposable syringes.

Preparation of Sephadex G-50

Slowly add 30 g of Sephadex G-50 (medium) to 250 ml of TE (pH 8.0) in a 500-ml beaker or bottle. Make sure the powder is well dispersed. Let stand overnight at room temperature, or heat at 65°C for 1–2 hours, or autoclave for 15 minutes at 15 lb/in^2 on liquid cycle. Allow to cool to room temperature.

Decant the supernatant and replace with an equal volume of TE (pH 8.0). Store at 4°C in a screw-capped bottle.

Column Chromatography

1. Prepare a Sephadex G-50 column in a disposable 5-ml borosilicate glass pipette plugged with sterile glass wool. Wash the column with several column volumes of TE (pH 8.0).

2. Apply the DNA sample (in a volume of 200 μl or less) to the column. Wash the tube with approximately 100 μl of TE (pH 8.0) and load the washings onto the column. Connect a reservoir of TE (pH 8.0) to the column so that the flow rate is about 0.5 ml/min.

3. Collect 12–15 fractions (0.5 ml) into Eppendorf tubes. If the DNA is labeled with ^{32}P, measure the radioactivity in each of the tubes, by using either a hand-held minimonitor or by Cerenkov counting in a liquid scintillation counter.

    The DNA will be excluded from the Sephadex gel and will be found in the void volume (usually ~30% of the total column volume). The leading peak of radioactivity therefore consists of nucleotides incorporated into DNA, while the trailing peak consists of unincorporated [^{32}P]dNTPs.

4. Pool the radioactive fractions in the leading peak and store at $-20°C$.

Instead of collecting individual fractions, it is possible with practice to follow the progress of the incorporated and unincorporated $[^{32}P]dNTPs$ down the column using a hand-held minimonitor. The leading peak should be collected into a sterile polypropylene tube as it elutes from the column. The bottom of the column should then be clamped off and the buffer reservoir disconnected. The column should be discarded into the radioactive waste.

Caution

Columns should be run behind lucite screens to shield personnel from exposure to radioactivity.

Note

Column chromatography can be used with a variety of matrixes (Sephadex G-75, G-100, Sepharose CL-4B, etc.) to separate DNA from small oligonucleotides or to fractionate DNA crudely by size (see page 226). The matrixes used for particular purposes are indicated at appropriate places in the text.

Spun-column Procedure

This method is useful when several preparations of DNA are labeled simultaneously or when it is necessary to change the buffer in which DNA is dissolved.

1. Plug the bottom of a 1-ml disposable syringe with a small amount of sterile glass wool. In the syringe, prepare a column (0.9-ml bed volume) of Sephadex G-50 equilibrated in TE (pH 8.0), containing 0.1 M NaCl (STE).

2. Insert the syringe into a glass centrifuge tube, as shown in Figure A.2. Centrifuge at 1600g for 4 minutes in a bench centrifuge. Do not be alarmed by the appearance of the column. Usually the Sephadex packs down during centrifugation. Continue to add Sephadex until the packed column volume is 0.9 ml.

3. Add 0.1 ml of STE and recentrifuge at exactly the same speed and for exactly the same time as before.

4. Repeat step 3.

5. Apply the DNA sample to the column in a total volume of 0.1 ml (use STE to make up the volume).

6. Recentrifuge at exactly the same speed and for exactly the same time as before, collecting the 100 μl of effluent from the syringe in a decapped Eppendorf tube (see Fig. A.2)

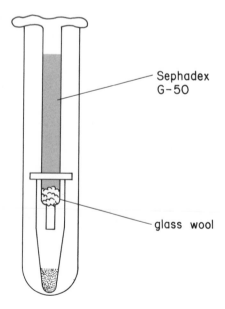

Sephadex
G-50

glass wool

Figure A.2

7. The unincorporated [^{32}P]dNTPs remain in the syringe, which should be carefully discarded. The labeled DNA is collected from the decapped Eppendorf tube.

Note

i. If the spun column is being used to change the buffer, the column should be washed 4–6 times (step 3) with the desired buffer in order to equilibrate the Sephadex G-50.

ii. Spun-column chromatography cannot be used with Sephadex G-100 since the beads are crushed by centrifugation.

QUANTITATION OF DNA AND RNA

Two methods are widely used to measure the amount of DNA or RNA in a preparation. If the sample is pure (i.e., without significant amounts of contaminants such as protein, phenol, agarose, or other nucleic acids), spectrophotometric measurement of the amount of UV irradiation absorbed by the bases is simple and accurate. If the amount of DNA or RNA is very small or if the sample contains significant quantities of impurities, the amount of nucleic acid can be estimated from the intensity of fluorescence emitted by ethidium bromide.

Spectrophotometric Determination of DNA or RNA

For quantitating the amount of DNA or RNA, readings should be taken at wavelengths of 260 nm and 280 nm. The reading at 260 nm allows calculation of the concentration of nucleic acid in the sample. An OD of 1 corresponds to approximately 50 μg/ml for double-stranded DNA, 40 μg/ml for single-stranded DNA and RNA, and 20 μg/ml for oligonucleotides. The ratio between the readings at 260 nm and 280 nm (OD_{260}/OD_{280}) provides an estimate for the purity of the nucleic acid. Pure preparations of DNA and RNA have OD_{260}/OD_{280} of 1.8 and 2.0, respectively. If there is contamination with protein or phenol, the OD_{260}/OD_{280} will be significantly less than the values given above, and accurate quantitation of the amount of nucleic acid will not be possible.

Ethidium Bromide Fluorescent Quantitation of the Amount of Double-stranded DNA

Sometimes there is not sufficient DNA (<250 ng/ml) to assay spectrophotometrically, or the DNA may be heavily contaminated with other UV-absorbing substances that impede accurate analysis. A rapid way to estimate the amount of DNA in such samples is to utilize the UV-induced fluorescence emitted by ethidium bromide molecules intercalated into the DNA. Because the amount of fluorescence is proportional to the total mass of DNA, the quantity of DNA in the sample can be estimated by comparing the fluorescent yield of the sample with that of a series of standards. As little as 1–5 ng of DNA can be detected by this method.

Plastic Wrap Method

1. Stretch a sheet of plastic wrap over a UV transilluminator or over a black sheet of paper.

2. Spot 1–5 μl of your DNA sample onto the plastic wrap.

3. Spot equal volumes of a series of DNA concentration standards (0.5–20 μg/ml) in an ordered array on the plastic wrap.

4. To each spot add an equal volume of TE containing 2 μg/ml of ethidium bromide. Mix by pipetting.

5. Photograph the spots using short-wavelength, UV illumination (see page 162). Estimate the concentration of DNA by comparing the intensity of fluorescence in the sample with that of the standard solutions.

Agarose Plate Method

Contaminants that may be present in the DNA sample can either contribute to or quench the fluorescence. To avoid these problems, the DNA samples can be spotted onto the surface of a 1% agarose slab containing 0.5 μg/ml of ethidium bromide. Allow the gel to stand at room temperature for a few hours so that small contaminating molecules have the chance to diffuse away. Photograph the gel as described above.

Minigel Method

Electrophoresis through minigels (see page 163) provides a rapid and convenient way to measure the quantity of DNA and to analyze its physical state at the same time. This is the method of choice if there is a possibility that the samples may contain significant quantities of RNA.

1. Mix 2 μl of the DNA sample with 0.4 μl of gel-loading buffer IV (bromophenol blue only; see page 455) and load into a slot in an 0.8% agarose minigel containing 0.5 μg/ml of ethidium bromide.

2. Mix 2 μl of each of a series of standard DNA solutions (0.5–50 μg/ml) with 0.4 μl of gel-loading buffer. Load the sample into the gel.

   Note. The standard DNA solution should contain a single species of DNA, approximately the same size as the unknown DNA.

3. Carry out electrophoresis until the bromophenol blue has migrated approximately 1–2 cm.

4. Destain the gel by immersing it for 5 minutes in electrophoresis buffer containing 0.01 M MgCl$_2$.

5. Photograph the gel using short-wavelength, UV irradiation. Compare the intensity of the fluorescence of the unknown DNA with that of the DNA standards and estimate the quantity of DNA in the sample.

AUTORADIOGRAPHY

Radioactive nucleic acids may be detected by autoradiography or by liquid scintillation spectroscopy. For most purposes, exact quantitation is unnecessary, and autoradiographic methods are superior: They are more sensitive, give higher resolution, do not involve destruction of the sample, and can reveal artifacts not seen by counting. Almost always, the isotope used is ^{32}P, and the instructions that follow apply only to it. For autoradiographic detection of other, weaker β-emitting isotopes, see Bonner and Laskey (1974).

1. Tape the material to be autoradiographed (see Fig. A.3) to a backing of thin mylar, cardboard, or Whatman 3MM paper.

2. Place pieces of tape marked with radioactive ink at several locations around the edge of the sample.

    Radioactive ink is made by mixing a small amount of ^{32}P with a water-proof black ink. We find it convenient to make the ink in three grades:

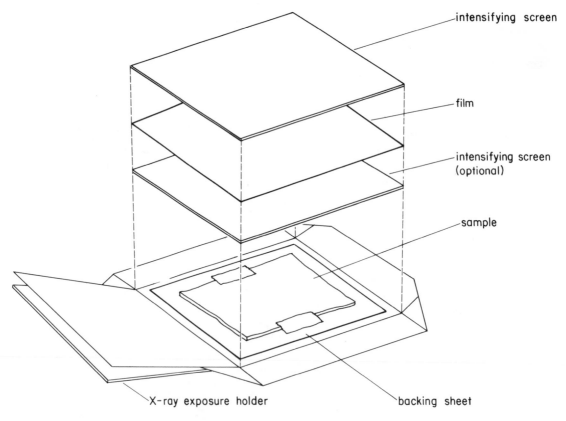

Figure A.3

very hot   >2000 cps on a minimonitor
hot   500–2000 cps on a minimonitor
cool   50–500 cps on a minimonitor

Use a fiber-tipped pen to apply ink of the desired hotness to the pieces of tape.

3. When the ink is dry, wrap the sample and backing sheet in Saran Wrap. This prevents contamination of intensifying screens and holders and prevents the sample from sticking to the film.

4. In a darkroom, place the sample in an X-ray film holder and cover it with a sheet of X-ray film. Tape the sample and film securely in place. Expose for several hours to several days. Between 1000 cpm and 5000 cpm of ^{32}P in a band 1 cm in width produces an image after an exposure of 12–16 hours.

   The most versatile film is Kodak X-Omat AR, which has a high-speed, short-exposure emulsion coated onto both sites of a colorless backing. It may be developed either in an automatic X-ray film processor or by hand:

Kodak liquid X-ray developer   5 minutes
3% acetic acid stop bath or
   water bath   1 minute
Kodak rapid fixer   10 minutes
running water   15 minutes

Dry in a warm cabinet or at room temperature.

Note. The sensitivity of the film may be increased by the use of intensifying screens (Swanstrom and Shank 1978). The best film-screen combination currently available, two calcium-tungstate-phosphor screens (Dupont Cronex Lightning-Plus) with Kodak X-Omat AR film exposed at −70°C, increases the sensitivity of detection of ^{32}P 8-fold to 10-fold. If only one screen is used, the enhancement factor for ^{32}P is 4-fold to 5-fold. The screens and films should be arranged as shown in Figure A.3. Note that the shiny side of both screens faces toward the film.

   In this system, as in conventional autoradiography, the vast majority of events recorded by the film are long-wavelength photons resulting from the fluorescence that occurs when the radiation emitted by the decay of a ^{32}P atom strikes the screen (Laskey and Mills 1977). The response of film to low intensities of light is extremely nonlinear, and the exposure is therefore carried out at −70°C in order to prolong the period of fluorescence. Preexposure (flashing) of the film (Laskey and Mills 1977) does not increase the sensitivity of detection of ^{32}P when using calcium-tungstate screens at −70°C.

a. The film holder containing the sample, film, and screens is wrapped in aluminum foil and placed in a $-70°C$ freezer for a suitable period of time. Aluminum shields should be interleaved between film holders to prevent exposure to radiation emitted by other samples. A weight should be placed on top of the stack of film holders to ensure that the samples are pressed tightly to the film. If this is not done, blurry, out-of-focus images may result.

b. The holder is removed from the freezer (use gloves) and the film is removed as quickly as possible and developed immediately. This prevents condensation on the film.

c. If it is necessary to obtain another autoradiograph, apply another film immediately and return the film holder and screens to the freezer as fast as possible. If condensation forms before you have time to apply a new film, allow the sample and screens to reach room temperature and wipe away all condensation before applying a new film.

MEASUREMENT OF RADIOACTIVITY IN NUCLEIC ACIDS

Precipitation with Trichloroacetic Acid (TCA)

1. Spot a known volume (up to 10 μl) of the sample to be assayed onto the center of a Whatman GF/C glass-fiber disc (2.4 cm diameter).

2. Add an equal volume of the sample to a tube containing 100 μl of a solution of salmon sperm DNA (500 μg/ml in 20 mM EDTA). Add 5 ml of ice-cold 10% TCA, mix, and chill on ice for 15 minutes.

3. Collect the precipitate by filtering the solution through another GF/C glass-fiber disc. Wash the filter six times with 5 ml of ice-cold 10% TCA, followed by 5 ml of 95% ethanol.

4. Dry both filters under a heat lamp. Put filters into scintillation vials. Count in a liquid scintillation counter in a toluene-based scintillation fluid such as Omnifluor (New England Nuclear). The first filter measures the total radioactivity in the sample; the second filter measures the radioactivity incorporated into nucleic acids. Nucleic acids greater than 20 nucleotides in length are quantitatively precipitated by this procedure.

Note

TCA is prepared as described on page 447.

Absorption to DE-81 Filters

1. Spot a known volume (up to 5 μl) onto the center of each of two 2.4-cm discs of Whatman DE-81 paper.

2. Wash one of the discs six times, 5 minutes per wash in 0.5 M Na_2HPO_4. Then wash the disc twice in water (1 minute per wash) and twice in 95% ethanol (1 minute per wash).

3. Dry both filters under a heat lamp. Count in a liquid scintillation counter in an aqueous scintillation fluid such as Aquasol (New England Nuclear). The unwashed filter measures the total radioactivity in the sample. The washed filter measures only the radioactivity incorporated into nucleic acids.

PREPARATION OF MULTIMERS OF PLASMIDS AS MOLECULAR-WEIGHT MARKERS[3]

1. Digest the plasmid DNA to completion with a restriction enzyme that cleaves at only one site and generates protruding termini.

2. Extract the digestion mixture with phenol/chloroform and precipitate the DNA with ethanol.

3. Dissolve the DNA at a concentration of approximately 500 μg/ml in TE (pH 7.5). Heat to 56°C for 5 minutes. Cool in ice.

4. Add 0.1 volume of 10× ligation buffer.

   10× Ligation buffer
   0.66 M Tris·Cl (pH 7.5)
   50 mM MgCl$_2$
   50 mM dithiothreitol
   10 mM ATP

5. Add approximately 1 unit (Weiss unit) of T4 DNA ligase per microgram of DNA. Incubate for 5 minutes at 12°C (for *Eco*RI termini) or at 16°C (for termini generated by other enzymes).

6. Chill the reaction to 0°C and remove an aliquot. Heat the aliquot to 68°C for 5 minutes and analyze by electrophoresis through a 0.4% agarose minigel to judge the extent of ligation. If the desired set of size markers is obtained, purify the remainder of the DNA as described above. If the amount of ligation is insufficient, warm the sample to the appropriate temperature and continue the incubation.

Note

An alternative approach is to ligate the linear DNA to completion at a high concentration and purify the concatenated DNA by gentle extraction with phenol and chloroform. The concatenated DNA is then partially digested with the restriction enzyme used for the original digestion. This aproach has the advantage that the restriction reaction is sometimes easier to control than the ligation reaction.

[3]B. Seed (unpubl.).

PROTOCOL FOR SEQUENCING BY THE MAXAM-GILBERT TECHNIQUE

Below, we describe in abbreviated form the chemical reactions developed by A. Maxam and W. Gilbert for the base-specific modification and cleavage of DNA. A more-detailed description, together with a thorough discussion of the methods used to isolate asymmetrically labeled fragments of DNA and to prepare and run sequencing gels has been published by Maxam and Gilbert in *Methods in Enzymology* (1980) **65** (part 1): 497–559.

Reagents

Dimethylsulfate (DMS) (Aldrich Chemical Co.).

Hydrazine (HZ) (Eastman Kodak).

Formic acid.

Piperidine (Fisher Scientific). Stock solution is 10 M. Dilute to 1.0 M just before use.

95% Ethanol.

70% Ethanol.

Distilled H$_2$O.

1 M Acetic acid.

0.3 M Sodium acetate (pH 5.2).

5 M NaCl.

1.2 N NaOH.

1 mM EDTA.

tRNA. Stock solution is 1 mg/ml in distilled H$_2$O.

Buffers

DMS buffer

50 mM sodium cacodylate (pH 8.0)
1 mM EDTA

It is usually unnecessary to adjust the pH.

DMS stop

1.5 M sodium acetate (pH 7.0)
1.0 M mercaptoethanol
100 μg/ml tRNA

HZ stop

0.3 M sodium acetate
0.1 mM EDTA
25 μg/ml tRNA

Loading buffer

80% (v/v) deionized or recrystallized formamide
50 mM Tris-borate (pH 8.3)
1 mM EDTA
0.1 (w/v) xylene cyanol
0.1% (w/v) bromophenol blue

Note

Filtration of the above buffers is usually not necessary. However, if a solution is turbid, filter it through a 0.45-μm nitrocellulose filter.

TABLE A.6. SUMMARY OF BASE-SPECIFIC REACTIONS FOR SEQUENCING END-LABELED DNA

	G	G & A	T & C	C	A > C
Mix	200 μl DMS buffer	10 μl H₂O	10 μl H₂O	15 μl 5 M NaCl	100 μl { 2 N NaOH / 1 mM EDTA
Chill to	5 μl [³²P]DNA / 0°C	10 μl [³²P]DNA / 0°C	10 μl [³²P]DNA / 0°C	5 μl [³²P]DNA / 0°C	5 μl [³²P]DNA
Add	1 μl DMS	50 μl formic acid	30 μl HZ	30 μl HZ	*Heat to* 90°C, 6 min. / 150 μl 1 N acetic acid / 5 μl tRNA (1 mg/ml) / 750 μl 95% ethanol
Incubate	20°C, 3–4 min.	20°C, 5 min.	20°C, 5 min.	20°C, 5 min.	
Add	50 μl DMS stop / 750 μl ethanol	180 μl HZ stop / 750 μl ethanol	200 μl HZ stop / 750 μl ethanol	200 μl HZ stop / 750 μl ethanol	
Store	−70°C, 10–15 min.	−70°C, 10–15 min.	−70°C, 10–15 min.	−70°C, 10–15 min.	−70°C, 10–15 min.
Centrifuge	10 min.	10 min.	10 min.	10 min.	10 min.
To pellet add	250 μl 0.3 M NaAc / 750 μl ethanol	250 μl 0.3 M NaAc / 750 μl ethanol	250 μl 0.3 M NaAc / 750 μl ethanol	250 μl 0.3 M NaAc / 750 μl ethanol	250 μl 0.3 M NaAc / 750 μl ethanol
Store	−70°C, 10–15 min.	−70°C, 10–15 min.	−70°C, 10–15 min.	−70°C, 10–15 min.	−70°C, 10–15 min.
Centrifuge	10 min.	10 min.	10 min.	10 min.	10 min.
Rinse pellet with	70% ethanol	70% ethanol	70% ethanol	70% ethanol	70% ethanol
Vacuum dry					
To pellet add	100 μl 1.0 M piperidine	100 μl 1.0 M piperidine	100 μl 1.0 M piperidine	100 μl 1.0 M piperidine	100 μl 1.0 M piperidine
Heat to	90°C, 30 min.	90°C, 30 min.	90°C, 30 min.	90°C, 30 min.	90°C, 30 min.
Lyophilize					
Add	10 μl H₂O	10 μl H₂O	10 μl H₂O	10 μl H₂O	10 μl H₂O
Lyophilize					
Add	10 μl H₂O	10 μl H₂O	10 μl H₂O	10 μl H₂O	10 μl H₂O
Lyophilize					
Add	10 μl loading buffer	10 μl loading buffer	10 μl loading buffer	10 μl loading buffer	10 μl loading buffer
Vortex					
Heat to	90°C, 1 min.	90°C, 1 min.	90°C, 1 min.	90°C, 1 min.	90°C, 1 min.
Chill in ice					
Load onto gel					

Reactions should be carried out in siliconized Eppendorf tubes.

Sequencing Gels

20% Acrylamide

acrylamide	96.5 g
methylene-bis-acrylamide	3.35 g
ultra-pure urea	233.5 g
5× TBE	100 ml
H_2O	to 500 ml

Urea mix

urea	233.5 g
5× TBE	100 ml
H_2O	to 500 ml

5× TBE

Tris base	54 g
boric acid	27.5 g
0.5 M EDTA (pH 8.0)	20 ml

To make an 8% sequencing gel, mix in a small flask:

20% acrylamide	20 ml
urea mix	30 ml
10% ammonium persulfate (freshly dissolved in water)	0.4 ml

Pour the solution into the barrel of a 50-ml syringe and add 50 μl of TEMED. Mix rapidly and inject the contents of the syringe (no needle should be used) into a preformed, sequencing-gel mold.

Appendix B: pBR322

NUCLEOTIDE SEQUENCE OF pBR322

The total length of pBR322 is 4362 nucleotides; in the printout of the sequence, the nucleotides are numbered in a clockwise direction from the single *Eco*RI site. Nucleotide 1 is the first thymidine residue in the *Eco*RI site:

```
    #1
     ↓
G–A–A–T–T–C
```

The nucleotide sequence was determined by Sutcliffe (1978, 1979).

```
        10              20              30              40              50
TTCTCATGTT      TGACAGCTTA      TCATCGATAA      GCTTTAATGC      GGTAGTTTAT
AAGAGTACAA      ACTGTCGAAT      AGTAGCTATT      CGAAATTACG      CCATCAAATA


        60              70              80              90             100
CACAGTTAAA      TTGCTAACGC      AGTCAGGCAC      CGTGTATGAA      ATCTAACAAT
GTGTCAATTT      AACGATTGCG      TCAGTCCGTG      GCACATACTT      TAGATTGTTA


       110             120             130             140             150
GCGCTCATCG      TCATCCTCGG      CACCGTCACC      CTGGATGCTG      TAGGCATAGG
CGCGAGTAGC      AGTAGGAGCC      GTGGCAGTGG      GACCTACGAC      ATCCGTATCC


       160             170             180             190             200
CTTGGTTATG      CCGGTACTGC      CGGGCCTCTT      GCGGGATATC      GTCCATTCCG
GAACCAATAC      GGCCATGACG      GCCCGGAGAA      CGCCCTATAG      CAGGTAAGGC


       210             220             230             240             250
ACAGCATCGC      CAGTCACTAT      GGCGTGCTGC      TAGCGCTATA      TGCGTTGATG
TGTCGTAGCG      GTCAGTGATA      CCGCACGACG      ATCGCGATAT      ACGCAACTAC


       260             270             280             290             300
CAATTTCTAT      GCGCACCCGT      TCTCGGAGCA      CTGTCCGACC      GCTTTGGCCG
GTTAAAGATA      CGCGTGGGCA      AGAGCCTCGT      GACAGGCTGG      CGAAACCGGC


       310             320             330             340             350
CCGCCCAGTC      CTGCTCGCTT      CGCTACTTGG      AGCCACTATC      GACTACGCGA
GGCGGGTCAG      GACGAGCGAA      GCGATGAACC      TCGGTGATAG      CTGATGCGCT


       360             370             380             390             400
TCATGGCGAC      CACACCCGTC      CTGTGGATCC      TCTACGCCGG      ACGCATCGTG
AGTACCGCTG      GTGTGGGCAG      GACACCTAGG      AGATGCGGCC      TGCGTAGCAC


       410             420             430             440             450
GCCGGCATCA      CCGGCGCCAC      AGGTGCGGTT      GCTGGCGCCT      ATATCGCCGA
CGGCCGTAGT      GGCCGCGGTG      TCCACGCCAA      CGACCGCGGA      TATAGCGGCT


       460             470             480             490             500
CATCACCGAT      GGGGAAGATC      GGGCTCGCCA      CTTCGGGCTC      ATGAGCGCTT
GTAGTGGCTA      CCCCTTCTAG      CCCGAGCGGT      GAAGCCCGAG      TACTCGCGAA


       510             520             530             540             550
GTTTCGGCGT      GGGTATGGTG      GCAGGCCCGT      GGCCGGGGGA      CTGTTGGGCG
CAAAGCCGCA      CCCATACCAC      CGTCCGGGCA      CCGGCCCCCT      GACAACCCGC


       560             570             580             590             600
CCATCTCCTT      GCATGCACCA      TTCCTTGCGG      CGGCGGTGCT      CAACGGCCTC
GGTAGAGGAA      CGTACGTGGT      AAGGAACGCC      GCCGCCACGA      GTTGCCGGAG
```

```
        610         620         630         640         650
   AACCTACTAC  TGGGCTGCTT  CCTAATGCAG  GAGTCGCATA  AGGGAGAGCG
   TTGGATGATG  ACCCGACGAA  GGATTACGTC  CTCAGCGTAT  TCCCTCTCGC

        660         670         680         690         700
   TCGACCGATG  CCCTTGAGAG  CCTTCAACCC  AGTCAGCTCC  TTCCGGTGGG
   AGCTGGCTAC  GGGAACTCTC  GGAAGTTGGG  TCAGTCGAGG  AAGGCCACCC

        710         720         730         740         750
   CGCGGGGCAT  GACTATCGTC  GCCGCACTTA  TGACTGTCTT  CTTTATCATG
   GCGCCCCGTA  CTGATAGCAG  CGGCGTGAAT  ACTGACAGAA  GAAATAGTAC

        760         770         780         790         800
   CAACTCGTAG  GACAGGTGCC  GGCAGCGCTC  TGGGTCATTT  TCGGCGAGGA
   GTTGAGCATC  CTGTCCACGG  CCGTCGCGAG  ACCCAGTAAA  AGCCGCTCCT

        810         820         830         840         850
   CCGCTTTCGC  TGGAGCGCGA  CGATGATCGG  CCTGTCGCTT  GCGGTATTCG
   GGCGAAAGCG  ACCTCGCGCT  GCTACTAGCC  GGACAGCGAA  CGCCATAAGC

        860         870         880         890         900
   GAATCTTGCA  CGCCCTCGCT  CAAGCCTTCG  TCACTGGTCC  CGCCACCAAA
   CTTAGAACGT  GCGGGAGCGA  GTTCGGAAGC  AGTGACCAGG  GCGGTGGTTT

        910         920         930         940         950
   CGTTTCGGCG  AGAAGCAGGC  CATTATCGCC  GGCATGGCGG  CCGACGCGCT
   GCAAAGCCGC  TCTTCGTCCG  GTAATAGCGG  CCGTACCGCC  GGCTGCGCGA

        960         970         980         990        1000
   GGGCTACGTC  TTGCTGGCGT  TCGCGACGCG  AGGCTGGATG  GCCTTCCCCA
   CCCGATGCAG  AACGACCGCA  AGCGCTGCGC  TCCGACCTAC  CGGAAGGGGT

       1010        1020        1030        1040        1050
   TTATGATTCT  TCTCGCTTCC  GGCGGCATCG  GGATGCCCGC  GTTGCAGGCC
   AATACTAAGA  AGAGCGAAGG  CCGCCGTAGC  CCTACGGGCG  CAACGTCCGG

       1060        1070        1080        1090        1100
   ATGCTGTCCA  GGCAGGTAGA  TGACGACCAT  CAGGGACAGC  TTCAAGGATC
   TACGACAGGT  CCGTCCATCT  ACTGCTGGTA  GTCCCTGTCG  AAGTTCCTAG

       1110        1120        1130        1140        1150
   GCTCGCGGCT  CTTACCAGCC  TAACTTCGAT  CACTGGACCG  CTGATCGTCA
   CGAGCGCCGA  GAATGGTCGG  ATTGAAGCTA  GTGACCTGGC  GACTAGCAGT

       1160        1170        1180        1190        1200
   CGGCGATTTA  TGCCGCCTCG  GCGAGCACAT  GGAACGGGTT  GGCATGGATT
   GCCGCTAAAT  ACGGCGGAGC  CGCTCGTGTA  CCTTGCCCAA  CCGTACCTAA
```

```
            1210             1220             1230             1240             1250
      GTAGGCGCCG       CCCTATACCT       TGTCTGCCTC       CCCGCGTTGC       GTCGCGGTGC
      CATCCGCGGC       GGGATATGGA       ACAGACGGAG       GGGCGCAACG       CAGCGCCACG


            1260             1270             1280             1290             1300
      ATGGAGCCGG       GCCACCTCGA       CCTGAATGGA       AGCCGGCGGC       ACCTCGCTAA
      TACCTCGGCC       CGGTGGAGCT       GGACTTACCT       TCGGCCGCCG       TGGAGCGATT


            1310             1320             1330             1340             1350
      CGGATTCACC       ACTCCAAGAA       TTGGAGCCAA       TCAATTCTTG       CGGAGAACTG
      GCCTAAGTGG       TGAGGTTCTT       AACCTCGGTT       AGTTAAGAAC       GCCTCTTGAC


            1360             1370             1380             1390             1400
      TGAATGCGCA       AACCAACCCT       TGGCAGAACA       TATCCATCGC       GTCCGCCATC
      ACTTACGCGT       TTGGTTGGGA       ACCGTCTTGT       ATAGGTAGCG       CAGGCGGTAG


            1410             1420             1430             1440             1450
      TCCAGCAGCC       GCACGCGGCG       CATCTCGGGC       AGCGTTGGGT       CCTGGCCACG
      AGGTCGTCGG       CGTGCGCCGC       GTAGAGCCCG       TCGCAACCCA       GGACCGGTGC


            1460             1470             1480             1490             1500
      GGTGCGCATG       ATCGTGCTCC       TGTCGTTGAG       GACCCGGCTA       GGCTGGCGGG
      CCACGCGTAC       TAGCACGAGG       ACAGCAACTC       CTGGGCCGAT       CCGACCGCCC


            1510             1520             1530             1540             1550
      GTTGCCTTAC       TGGTTAGCAG       AATGAATCAC       CGATACGCGA       GCGAACGTGA
      CAACGGAATG       ACCAATCGTC       TTACTTAGTG       GCTATGCGCT       CGCTTGCACT


            1560             1570             1580             1590             1600
      AGCGACTGCT       GCTGCAAAAC       GTCTGCGACC       TGAGCAACAA       CATGAATGGT
      TCGCTGACGA       CGACGTTTTG       CAGACGCTGG       ACTCGTTGTT       GTACTTACCA


            1610             1620             1630             1640             1650
      CTTCGGTTTC       CGTGTTTCGT       AAAGTCTGGA       AACGCGGAAG       TCAGCGCCCT
      GAAGCCAAAG       GCACAAAGCA       TTTCAGACCT       TTGCGCCTTC       AGTCGCGGGA


            1660             1670             1680             1690             1700
      GCACCATTAT       GTTCCGGATC       TGCATCGCAG       GATGCTGCTG       GCTACCCTGT
      CGTGGTAATA       CAAGGCCTAG       ACGTAGCGTC       CTACGACGAC       CGATGGGACA


            1710             1720             1730             1740             1750
      GGAACACCTA       CATCTGTATT       AACGAAGCGC       TGGCATTGAC       CCTGAGTGAT
      CCTTGTGGAT       GTAGACATAA       TTGCTTCGCG       ACCGTAACTG       GGACTCACTA


            1760             1770             1780             1790             1800
      TTTTCTCTGG       TCCCGCCGCA       TCCATACCGC       CAGTTGTTTA       CCCTCACAAC
      AAAAGAGACC       AGGGCGGCGT       AGGTATGGCG       GTCAACAAAT       GGGAGTGTTG
```

```
      1810         1820         1830         1840         1850
GTTCCAGTAA   CCGGGCATGT   TCATCATCAG   TAACCCGTAT   CGTGAGCATC
CAAGGTCATT   GGCCCGTACA   AGTAGTAGTC   ATTGGGCATA   GCACTCGTAG


      1860         1870         1880         1890         1900
CTCTCTCGTT   TCATCGGTAT   CATTACCCCC   ATGAACAGAA   ATTCCCCCTT
GAGAGAGCAA   AGTAGCCATA   GTAATGGGGG   TACTTGTCTT   TAAGGGGGAA


      1910         1920         1930         1940         1950
ACACGGAGGC   ATCAAGTGAC   CAAACAGGAA   AAAACCGCCC   TTAACATGGC
TGTGCCTCCG   TAGTTCACTG   GTTTGTCCTT   TTTTGGCGGG   AATTGTACCG


      1960         1970         1980         1990         2000
CCGCTTTATC   AGAAGCCAGA   CATTAACGCT   TCTGGAGAAA   CTCAACGAGC
GGCGAAATAG   TCTTCGGTCT   GTAATTGCGA   AGACCTCTTT   GAGTTGCTCG


      2010         2020         2030         2040         2050
TGGACGCGGA   TGAACAGGCA   GACATCTGTG   AATCGCTTCA   CGACCACGCT
ACCTGCGCCT   ACTTGTCCGT   CTGTAGACAC   TTAGCGAAGT   GCTGGTGCGA


      2060         2070         2080         2090         2100
GATGAGCTTT   ACCGCAGCTG   CCTCGCGCGT   TTCGGTGATG   ACGGTGAAAA
CTACTCGAAA   TGGCGTCGAC   GGAGCGCGCA   AAGCCACTAC   TGCCACTTTT


      2110         2120         2130         2140         2150
CCTCTGACAC   ATGCAGCTCC   CGGAGACGGT   CACAGCTTGT   CTGTAAGCGG
GGAGACTGTG   TACGTCGAGG   GCCTCTGCCA   GTGTCGAACA   GACATTCGCC


      2160         2170         2180         2190         2200
ATGCCGGGAG   CAGACAAGCC   CGTCAGGGCG   CGTCAGCGGG   TGTTGGCGGG
TACGGCCCTC   GTCTGTTCGG   GCAGTCCCGC   GCAGTCGCCC   ACAACCGCCC


      2210         2220         2230         2240         2250
TGTCGGGGCG   CAGCCATGAC   CCAGTCACGT   AGCGATAGCG   GAGTGTATAC
ACAGCCCCGC   GTCGGTACTG   GGTCAGTGCA   TCGCTATCGC   CTCACATATG


      2260         2270         2280         2290         2300
TGGCTTAACT   ATGCGGCATC   AGAGCAGATT   GTACTGAGAG   TGCACCATAT
ACCGAATTGA   TACGCCGTAG   TCTCGTCTAA   CATGACTCTC   ACGTGGTATA


      2310         2320         2330         2340         2350
GCGGTGTGAA   ATACCGCACA   GATGCGTAAG   GAGAAAATAC   CGCATCAGGC
CGCCACACTT   TATGGCGTGT   CTACGCATTC   CTCTTTTATG   GCGTAGTCCG


      2360         2370         2380         2390         2400
GCTCTTCCGC   TTCCTCGCTC   ACTGACTCGC   TGCGCTCGGT   CGTTCGGCTG
CGAGAAGGCG   AAGGAGCGAG   TGACTGAGCG   ACGCGAGCCA   GCAAGCCGAC
```

```
        2410           2420           2430           2440           2450
   CGGCGAGCGG      TATCAGCTCA      CTCAAAGGCG      GTAATACGGT      TATCCACAGA
   GCCGCTCGCC      ATAGTCGAGT      GAGTTTCCGC      CATTATGCCA      ATAGGTGTCT


        2460           2470           2480           2490           2500
   ATCAGGGGAT      AACGCAGGAA      AGAACATGTG      AGCAAAAGGC      CAGCAAAAGG
   TAGTCCCCTA      TTGCGTCCTT      TCTTGTACAC      TCGTTTTCCG      GTCGTTTTCC


        2510           2520           2530           2540           2550
   CCAGGAACCG      TAAAAAGGCC      GCGTTGCTGG      CGTTTTTCCA      TAGGCTCCGC
   GGTCCTTGGC      ATTTTTCCGG      CGCAACGACC      GCAAAAAGGT      ATCCGAGGCG


        2560           2570           2580           2590           2600
   CCCCCTGACG      AGCATCACAA      AAATCGACGC      TCAAGTCAGA      GGTGGCGAAA
   GGGGGACTGC      TCGTAGTGTT      TTTAGCTGCG      AGTTCAGTCT      CCACCGCTTT


        2610           2620           2630           2640           2650
   CCCGACAGGA      CTATAAAGAT      ACCAGGCGTT      TCCCCCTGGA      AGCTCCCTCG
   GGGCTGTCCT      GATATTTCTA      TGGTCCGCAA      AGGGGGACCT      TCGAGGGAGC


        2660           2670           2680           2690           2700
   TGCGCTCTCC      TGTTCCGACC      CTGCCGCTTA      CCGGATACCT      GTCCGCCTTT
   ACGCGAGAGG      ACAAGGCTGG      GACGGCGAAT      GGCCTATGGA      CAGGCGGAAA


        2710           2720           2730           2740           2750
   CTCCCTTCGG      GAAGCGTGGC      GCTTTCTCAA      TGCTCACGCT      GTAGGTATCT
   GAGGGAAGCC      CTTCGCACCG      CGAAAGAGTT      ACGAGTGCGA      CATCCATAGA


        2760           2770           2780           2790           2800
   CAGTTCGGTG      TAGGTCGTTC      GCTCCAAGCT      GGGCTGTGTG      CACGAACCCC
   GTCAAGCCAC      ATCCAGCAAG      CGAGGTTCGA      CCCGACACAC      GTGCTTGGGG


        2810           2820           2830           2840           2850
   CCGTTCAGCC      CGACCGCTGC      GCCTTATCCG      GTAACTATCG      TCTTGAGTCC
   GGCAAGTCGG      GCTGGCGACG      CGGAATAGGC      CATTGATAGC      AGAACTCAGG


        2860           2870           2880           2890           2900
   AACCCGGTAA      GACACGACTT      ATCGCCACTG      GCAGCAGCCA      CTGGTAACAG
   TTGGGCCATT      CTGTGCTGAA      TAGCGGTGAC      CGTCGTCGGT      GACCATTGTC


        2910           2920           2930           2940           2950
   GATTAGCAGA      GCGAGGTATG      TAGGCGGTGC      TACAGAGTTC      TTGAAGTGGT
   CTAATCGTCT      CGCTCCATAC      ATCCGCCACG      ATGTCTCAAG      AACTTCACCA


        2960           2970           2980           2990           3000
   GGCCTAACTA      CGGCTACACT      AGAAGGACAG      TATTTGGTAT      CTGCGCTCTG
   CCGGATTGAT      GCCGATGTGA      TCTTCCTGTC      ATAAACCATA      GACGCGAGAC
```

```
          3010            3020            3030            3040            3050
     CTGAAGCCAG      TTACCTTCGG      AAAAAGAGTT      GGTAGCTCTT      GATCCGGCAA
     GACTTCGGTC      AATGGAAGCC      TTTTTCTCAA      CCATCGAGAA      CTAGGCCGTT


          3060            3070            3080            3090            3100
     ACAAACCACC      GCTGGTAGCG      GTGGTTTTTT      TGTTTGCAAG      CAGCAGATTA
     TGTTTGGTGG      CGACCATCGC      CACCAAAAAA      ACAAACGTTC      GTCGTCTAAT


          3110            3120            3130            3140            3150
     CGCGCAGAAA      AAAAGGATCT      CAAGAAGATC      CTTTGATCTT      TTCTACGGGG
     GCGCGTCTTT      TTTTCCTAGA      GTTCTTCTAG      GAAACTAGAA      AAGATGCCCC


          3160            3170            3180            3190            3200
     TCTGACGCTC      AGTGGAACGA      AAACTCACGT      TAAGGGATTT      TGGTCATGAG
     AGACTGCGAG      TCACCTTGCT      TTTGAGTGCA      ATTCCCTAAA      ACCAGTACTC


          3210            3220            3230            3240            3250
     ATTATCAAAA      AGGATCTTCA      CCTAGATCCT      TTTAAATTAA      AAATGAAGTT
     TAATAGTTTT      TCCTAGAAGT      GGATCTAGGA      AAATTTAATT      TTTACTTCAA


          3260            3270            3280            3290            3300
     TTAAATCAAT      CTAAAGTATA      TATGAGTAAA      CTTGGTCTGA      CAGTTACCAA
     AATTTAGTTA      GATTTCATAT      ATACTCATTT      GAACCAGACT      GTCAATGGTT


          3310            3320            3330            3340            3350
     TGCTTAATCA      GTGAGGCACC      TATCTCAGCG      ATCTGTCTAT      TTCGTTCATC
     ACGAATTAGT      CACTCCGTGG      ATAGAGTCGC      TAGACAGATA      AAGCAAGTAG


          3360            3370            3380            3390            3400
     CATAGTTGCC      TGACTCCCCG      TCGTGTAGAT      AACTACGATA      CGGGAGGGCT
     GTATCAACGG      ACTGAGGGGC      AGCACATCTA      TTGATGCTAT      GCCCTCCCGA


          3410            3420            3430            3440            3450
     TACCATCTGG      CCCCAGTGCT      GCAATGATAC      CGCGAGACCC      ACGCTCACCG
     ATGGTAGACC      GGGGTCACGA      CGTTACTATG      GCGCTCTGGG      TGCGAGTGGC


          3460            3470            3480            3490            3500
     GCTCCAGATT      TATCAGCAAT      AAACCAGCCA      GCCGGAAGGG      CCGAGCGCAG
     CGAGGTCTAA      ATAGTCGTTA      TTTGGTCGGT      CGGCCTTCCC      GGCTCGCGTC


          3510            3520            3530            3540            3550
     AAGTGGTCCT      GCAACTTTAT      CCGCCTCCAT      CCAGTCTATT      AATTGTTGCC
     TTCACCAGGA      CGTTGAAATA      GGCGGAGGTA      GGTCAGATAA      TTAACAACGG


          3560            3570            3580            3590            3600
     GGGAAGCTAG      AGTAAGTAGT      TCGCCAGTTA      ATAGTTTGCG      CAACGTTGTT
     CCCTTCGATC      TCATTCATCA      AGCGGTCAAT      TATCAAACGC      GTTGCAACAA
```

```
            3610              3620              3630              3640              3650
     GCCATTGCTG        CAGGCATCGT        GGTGTCACGC        TCGTCGTTTG        GTATGGCTTC
     CGGTAACGAC        GTCCGTAGCA        CCACAGTGCG        AGCAGCAAAC        CATACCGAAG


            3660              3670              3680              3690              3700
     ATTCAGCTCC        GGTTCCCAAC        GATCAAGGCG        AGTTACATGA        TCCCCCATGT
     TAAGTCGAGG        CCAAGGGTTG        CTAGTTCCGC        TCAATGTACT        AGGGGGTACA


            3710              3720              3730              3740              3750
     TGTGCAAAAA        AGCGGTTAGC        TCCTTCGGTC        CTCCGATCGT        TGTCAGAAGT
     ACACGTTTTT        TCGCCAATCG        AGGAAGCCAG        GAGGCTAGCA        ACAGTCTTCA


            3760              3770              3780              3790              3800
     AAGTTGGCCG        CAGTGTTATC        ACTCATGGTT        ATGGCAGCAC        TGCATAATTC
     TTCAACCGGC        GTCACAATAG        TGAGTACCAA        TACCGTCGTG        ACGTATTAAG


            3810              3820              3830              3840              3850
     TCTTACTGTC        ATGCCATCCG        TAAGATGCTT        TTCTGTGACT        GGTGAGTACT
     AGAATGACAG        TACGGTAGGC        ATTCTACGAA        AAGACACTGA        CCACTCATGA


            3860              3870              3880              3890              3900
     CAACCAAGTC        ATTCTGAGAA        TAGTGTATGC        GGCGACCGAG        TTGCTCTTGC
     GTTGGTTCAG        TAAGACTCTT        ATCACATACG        CCGCTGGCTC        AACGAGAACG


            3910              3920              3930              3940              3950
     CCGGCGTCAA        CACGGGATAA        TACCGCGCCA        CATAGCAGAA        CTTTAAAAGT
     GGCCGCAGTT        GTGCCCTATT        ATGGCGCGGT        GTATCGTCTT        GAAATTTTCA


            3960              3970              3980              3990              4000
     GCTCATCATT        GGAAAACGTT        CTTCGGGGCG        AAAACTCTCA        AGGATCTTAC
     CGAGTAGTAA        CCTTTTGCAA        GAAGCCCCGC        TTTTGAGAGT        TCCTAGAATG


            4010              4020              4030              4040              4050
     CGCTGTTGAG        ATCCAGTTCG        ATGTAACCCA        CTCGTGCACC        CAACTGATCT
     GCGACAACTC        TAGGTCAAGC        TACATTGGGT        GAGCACGTGG        GTTGACTAGA


            4060              4070              4080              4090              4100
     TCAGCATCTT        TTACTTTCAC        CAGCGTTTCT        GGGTGAGCAA        AAACAGGAAG
     AGTCGTAGAA        AATGAAAGTG        GTCGCAAAGA        CCCACTCGTT        TTTGTCCTTC


            4110              4120              4130              4140              4150
     GCAAAATGCC        GCAAAAAAGG        GAATAAGGGC        GACACGGAAA        TGTTGAATAC
     CGTTTTACGG        CGTTTTTTCC        CTTATTCCCG        CTGTGCCTTT        ACAACTTATG


            4160              4170              4180              4190              4200
     TCATACTCTT        CCTTTTTCAA        TATTATTGAA        GCATTTATCA        GGGTTATTGT
     AGTATGAGAA        GGAAAAAGTT        ATAATAACTT        CGTAAATAGT        CCCAATAACA
```

```
        4210            4220            4230            4240            4250
CTCATGAGCG      GATACATATT      TGAATGTATT      TAGAAAAATA      AACAAATAGG
GAGTACTCGC      CTATGTATAA      ACTTACATAA      ATCTTTTTAT      TTGTTTATCC


        4260            4270            4280            4290            4300
GGTTCCGCGC      ACATTTCCCC      GAAAAGTGCC      ACCTGACGTC      TAAGAAACCA
CCAAGGCGCG      TGTAAAGGGG      CTTTTCACGG      TGGACTGCAG      ATTCTTTGGT


        4310            4320            4330            4340            4350
TTATTATCAT      GACATTAACC      TATAAAAATA      GGCGTATCAC      GAGGCCCTTT
AATAATAGTA      CTGTAATTGG      ATATTTTTAT      CCGCATAGTG      CTCCGGGAAA


        4360            4362
CGTCTTCAAG      AA
GCAGAAGTTC      TT
```

RESTRICTION SITES IN pBR322

The first column identifies restriction sites in clockwise order around the circular DNA, beginning immediately after the single *Eco*RI site. The second column identifies the number of the first nucleotide of the recognition sequence. The fourth column identifies the number of the nucleotide immediately 5' to the site of cleavage. The last column shows the nucleotide sequence of the recognition site.

Tth111II	7	CUTS?	7	TGTTTG
Alu I	15	CUTS@	16	AG^CT
Cla I	23	CUTS@	24	AT^CGAT
Taq I	24	CUTS@	24	T^CGA
Hind III	29	CUTS@	29	A^AGCTT
Alu I	30	CUTS@	31	AG^CT
HgiC I	76	CUTS@	76	G^GCACC
Hha I	101	CUTS@	103	GCG^C
Mnl I	115	CUTS?	115	CCTC
HgiC I	119	CUTS@	119	G^GCACC
Hph I	126	CUTS?	126	TCACC
ScrF I	130	CUTS?	130	CCNGG
EcoR II	130	CUTS@	129	^CCTGG
Fok I	133	CUTS?	133	GGATG
SfaN I	134	CUTS?	134	GATGC
Hpa II	161	CUTS@	161	C^CGG
Rsa I	164	CUTS@	165	GT^AC
ScrF I	170	CUTS?	170	CCNGG
Hpa II	170	CUTS@	170	C^CGG
Cau II	170	CUTS@	171	CC^GGG
Asu I	172	CUTS@	172	G^GNCC
Hae III	173	CUTS@	174	GG^CC
Mnl I	175	CUTS?	175	CCTC
EcoR V	185	CUTS@	189	GATAT^C
SfaN I	204	CUTS?	204	GCATC
Fnu4H I	226	CUTS@	227	GC^NGC
Bbv I	226	CUTS?	226	GCTGC
Hae II	232	CUTS@	236	AGCGC^T
Hha I	233	CUTS@	235	GCG^C
SfaN I	247	CUTS?	247	GATGC
Mst I	260	CUTS@	262	TGC^GCA
Hha I	261	CUTS@	263	GCG^C
HgiA I	276	CUTS@	280	GAGCA^C
Cfr I	295	CUTS@	295	T^GGCCG
Gdi II	295	CUTS@	295	T^GGCCG
Hae III	296	CUTS@	297	GG^CC
NspB II	297	CUTS@	300	GCCG^C
Fnu4H I	297	CUTS@	298	GC^NGC
NspB II	300	CUTS@	303	GCCG^C
Fnu4H I	300	CUTS@	301	GC^NGC
Taq I	339	CUTS@	339	T^CGA
FnuD II	346	CUTS@	347	CG^CG
Mbo I	349	CUTS@	348	^GATC
BamH I	375	CUTS@	375	G^GATCC
Xho II	375	CUTS@	375	G^GATCC
Mbo I	376	CUTS@	375	^GATC
Mnl I	379	CUTS?	379	CCTC
Hpa II	387	CUTS@	387	C^CGG
Hga I	390	CUTS?	390	GACGC
SfaN I	393	CUTS?	393	GCATC
Cfr I	399	CUTS@	399	T^GGCCG
Gdi II	399	CUTS@	399	T^GGCCG
Hae III	400	CUTS@	401	GG^CC
Nae I	401	CUTS@	403	GCC^GGC
Hpa II	402	CUTS@	402	C^CGG
SfaN I	405	CUTS?	405	GCATC
Hph I	408	CUTS?	408	TCACC
Hpa II	411	CUTS@	411	C^CGG
Nar I	413	CUTS@	414	GG^CGCC
Hae II	413	CUTS@	417	GGCGC^C
Acy I	413	CUTS@	414	GG^CGCC
HgiC I	413	CUTS@	413	G^GCGCC
Hha I	414	CUTS@	416	GCG^C
Nar I	434	CUTS@	435	GG^CGCC
Hae II	434	CUTS@	438	GGCGC^C
Acy I	434	CUTS@	435	GG^CGCC
HgiC I	434	CUTS@	434	G^GCGCC
Hha I	435	CUTS@	437	GCG^C
Hph I	453	CUTS?	453	TCACC
Mbo II	464	CUTS?	464	GAAGA
Mbo I	467	CUTS@	466	^GATC
HgiJ II	471	CUTS@	475	GGGCT^C

HgiJ II	485	CUTS@	489	GGGCT^C
Hae II	494	CUTS@	498	AGCGC^T
Hha I	495	CUTS@	497	GCG^C
Asu I	524	CUTS@	524	G^GNCC
Hae III	524	CUTS@	525	GG^CC
Cfr I	530	CUTS@	530	T^GGCCG
Gdi II	530	CUTS@	530	T^GGCCG
Hae III	531	CUTS@	532	GG^CC
ScrF I	533	CUTS?	533	CCNGG
Hpa II	533	CUTS@	533	C^CGG
Cau II	533	CUTS@	534	CC^GGG
Nar I	547	CUTS@	548	GG^CGCC
Hae II	547	CUTS@	551	GGCGC^C
Acy I	547	CUTS@	548	GG^CGCC
HgiC I	547	CUTS@	547	G^GCGCC
Hha I	548	CUTS@	550	GCG^C
Sph I	561	CUTS@	565	GCATG^C
NspC I	561	CUTS@	565	GCATG^C
NspB II	577	CUTS@	580	GCGG^C
Fnu4H I	577	CUTS@	578	GC^NGC
NspB II	580	CUTS@	583	GCGG^C
Fnu4H I	580	CUTS@	581	GC^NGC
HgiA I	586	CUTS@	590	GTGCT^C
Hae III	595	CUTS@	596	GG^CC
Mnl I	597	CUTS?	597	CCTC
Fnu4H I	614	CUTS@	615	GC^NGC
Bbv I	614	CUTS?	614	GCTGC
HinF I	631	CUTS@	631	G^ANTC
Hga I	648	CUTS?	648	GCGTC
Acc I	650	CUTS@	651	GT^CGAC
Hind II	650	CUTS@	652	GTC^GAC
Sal I	650	CUTS@	650	G^TCGAC
Taq I	651	CUTS@	651	T^CGA
SfaN I	657	CUTS?	657	GATGC
Alu I	685	CUTS@	686	AG^CT
Hpa II	693	CUTS@	693	C^CGG
Hha I	700	CUTS@	702	GCG^C
FnuD II	701	CUTS@	702	CG^CG
NspB II	721	CUTS@	724	GCCG^C
Fnu4H I	721	CUTS@	722	GC^NGC
Mbo I	737	CUTS?	737	TCTTC
HgiC I	765	CUTS@	765	G^GTGCC
Nae I	768	CUTS@	770	GCC^GGC
Hpa II	769	CUTS@	769	C^CGG
Bbv I	772	CUTS?	772	GCAGC
Fnu4H I	772	CUTS@	773	GC^NGC
Hae II	774	CUTS@	778	AGCGC^T
Hha I	775	CUTS@	777	GCG^C
Mnl I	796	CUTS?	796	GAGG
Asu I	798	CUTS@	798	G^GNCC
Ava II	798	CUTS@	798	G^GACC
Hha I	815	CUTS@	817	GCG^C
FnuD II	816	CUTS@	817	CG^CG
Mbo I	825	CUTS@	824	^GATC
Hae III	829	CUTS@	830	GG^CC
HinF I	851	CUTS@	851	G^ANTC
Mnl I	864	CUTS?	864	CCTC
Asu I	886	CUTS@	886	G^GNCC
Ava II	886	CUTS@	886	G^GTCC
Hae I	917	CUTS@	919	AGG^CCA
Hae III	918	CUTS@	919	GG^CC
Bgl I	928	CUTS@	934	GCCNNNN^NGGC
Nae I	928	CUTS@	930	GCC^GGC
Hpa II	929	CUTS@	929	C^CGG
NspB II	937	CUTS@	940	GCGG^C
Fnu4H I	937	CUTS@	938	GC^NGC
Xma III	938	CUTS@	938	C^GGCCG
Gdi II	938	CUTS@	938	C^GGCCG
Cfr I	938	CUTS@	938	C^GGCCG
Hae III	939	CUTS@	940	GG^CC
Hga I	943	CUTS?	943	GACGC
FnuD II	945	CUTS@	946	CG^CG

Hha I	946	CUTS@	948	GCG^C	FnuD II	1414	CUTS@	1415 CG^CG
Nru I	971	CUTS@	973	TCG^CGA	NspB II	1415	CUTS@	1418 GCGG^C
FnuD II	972	CUTS@	973	CG^CG	Fnu4H I	1415	CUTS@	1416 GC^NGC
Hga I	975	CUTS?	975	GACGC	Hha I	1418	CUTS@	1420 GCG^C
FnuD II	977	CUTS@	978	CG^CG	SfaN I	1420	CUTS?	1420 GCATC
Mnl I	980	CUTS?	980	GAGG	Ava I	1424	CUTS@	1424 C^TCGGG
Fok I	986	CUTS?	986	GGATG	Bbv I	1429	CUTS?	1429 GCAGC
Hae I	989	CUTS@	991	TGG^CCT	Fnu4H I	1429	CUTS@	1430 GC^NGC
Hae III	990	CUTS@	991	GG^CC	Asu I	1438	CUTS@	1438 G^GNCC
HinF I	1005	CUTS@	1005	G^ANTC	Ava II	1438	CUTS@	1438 G^GTCC
Mbo II	1008	CUTS?	1008	TCTTC	ScrF I	1441	CUTS?	1441 CCNGG
Hpa II	1019	CUTS@	1019	C^CGG	EcoR II	1441	CUTS@	1440 ^CCTGG
NspB II	1022	CUTS@	1025	GCGG^C	Bal I	1443	CUTS@	1445 TGG^CCA
Fnu4H I	1022	CUTS@	1023	GC^NGC	Hae I	1443	CUTS@	1445 TGG^CCA
SfaN I	1025	CUTS?	1025	GCATC	Cfr I	1443	CUTS@	1443 T^GGCCA
Fok I	1031	CUTS?	1031	GGATG	Hae III	1444	CUTS@	1445 GG^CC
SfaN I	1032	CUTS?	1032	GATGC	Mst I	1453	CUTS@	1455 TGC^GCA
FnuD II	1038	CUTS@	1039	CG^CG	Hha I	1454	CUTS@	1456 GCG^C
Hae I	1046	CUTS@	1048	AGG^CCA	Mbo I	1460	CUTS@	1459 ^GATC
Hae III	1047	CUTS@	1048	GG^CC	HgiA I	1464	CUTS@	1468 GTGCT^C
ScrF I	1058	CUTS?	1058	CCNGG	Mnl I	1478	CUTS?	1478 GAGG
EcoR II	1058	CUTS@	1057	^CCAGG	Asu I	1480	CUTS@	1480 G^GNCC
Alu I	1088	CUTS@	1089	AG^CT	Ava II	1480	CUTS@	1480 G^GACC
Mbo I	1097	CUTS@	1096	^GATC	ScrF I	1483	CUTS?	1483 CCNGG
FnuD II	1104	CUTS@	1105	CG^CG	Cau II	1483	CUTS@	1484 C^CGGG
NspB II	1105	CUTS@	1108	GCGG^C	Hpa II	1484	CUTS@	1484 C^CGG
Fnu4H I	1105	CUTS@	1106	GC^NGC	HinF I	1524	CUTS@	1524 G^ANTC
Taq I	1126	CUTS@	1126	T^CGA	Hph I	1527	CUTS?	1527 TCACC
Mbo I	1128	CUTS@	1127	^GATC	FnuD II	1536	CUTS@	1537 CG^CG
Asu I	1135	CUTS@	1135	G^GNCC	Fnu4H I	1558	CUTS@	1559 GC^NGC
Ava II	1135	CUTS@	1135	G^GACC	Bbv I	1558	CUTS?	1558 GCTGC
Mbo I	1143	CUTS@	1142	^GATC	Fnu4H I	1561	CUTS@	1562 GC^NGC
Bgl I	1162	CUTS@	1168	GCCNNNN^NGGC	Bbv I	1561	CUTS?	1561 GCTGC
NspB II	1162	CUTS@	1165	GCCG^C	Dde I	1580	CUTS@	1580 C^TNAG
Fnu4H I	1162	CUTS@	1163	GC^NGC	Mbo I	1600	CUTS?	1600 TCTTC
Mnl I	1166	CUTS?	1166	CCTC	FnuD II	1633	CUTS@	1634 CG^CG
HgiA I	1173	CUTS@	1177	GAGCA^C	Hae II	1643	CUTS@	1647 AGCGC^C
Nar I	1204	CUTS@	1205	GG^CGCC	Hha I	1644	CUTS@	1646 GCG^C
Hae II	1204	CUTS@	1208	GGCGC^C	Hpa II	1664	CUTS@	1664 C^CGG
Acy I	1204	CUTS@	1205	GG^CGCC	Xho II	1666	CUTS@	1666 G^GATCT
HgiC I	1204	CUTS@	1204	G^GCGCC	Mbo I	1667	CUTS@	1666 ^GATC
Hha I	1205	CUTS@	1207	GCG^C	SfaN I	1672	CUTS?	1672 GCATC
NspB II	1207	CUTS@	1210	GCCG^C	Fok I	1680	CUTS?	1680 GGATG
Fnu4H I	1207	CUTS@	1208	GC^NGC	SfaN I	1681	CUTS?	1681 GATGC
Mnl I	1227	CUTS?	1227	CCTC	Fnu4H I	1684	CUTS@	1685 GC^NGC
FnuD II	1233	CUTS@	1234	CG^CG	Bbv I	1684	CUTS?	1684 GCTGC
Hga I	1239	CUTS?	1239	GCGTC	Hae II	1726	CUTS@	1730 AGCGC^T
FnuD II	1243	CUTS@	1244	CG^CG	Hha I	1727	CUTS@	1729 GCG^C
ScrF I	1257	CUTS?	1257	CCNGG	Dde I	1742	CUTS@	1742 C^TNAG
Hpa II	1257	CUTS@	1257	C^CGG	Asu I	1759	CUTS@	1759 G^GNCC
Cau II	1257	CUTS@	1258	CC^GGG	Ava II	1759	CUTS@	1759 G^GTCC
Asu I	1259	CUTS@	1259	G^GNCC	NspB II	1765	CUTS@	1768 GCCG^C
Hae III	1260	CUTS@	1261	GG^CC	Fnu4H I	1765	CUTS@	1766 GC^NGC
Mnl I	1265	CUTS?	1265	CCTC	SfaN I	1768	CUTS?	1768 GCATC
Taq I	1267	CUTS@	1267	T^CGA	Mnl I	1792	CUTS?	1792 CCTC
Nae I	1282	CUTS@	1284	GCC^GGC	ScrF I	1811	CUTS?	1811 CCNGG
Hpa II	1283	CUTS@	1283	C^CGG	Hpa II	1811	CUTS@	1811 C^CGG
NspB II	1286	CUTS@	1289	GCGG^C	Cau II	1811	CUTS@	1812 CC^GGG
Fnu4H I	1286	CUTS@	1287	GC^NGC	NspC I	1815	CUTS@	1819 GCATG^T
HgiC I	1288	CUTS@	1288	G^GCACC	SfaN I	1846	CUTS?	1846 GCATC
Mnl I	1292	CUTS?	1292	CCTC	Mnl I	1850	CUTS?	1850 CCTC
HinF I	1303	CUTS@	1303	G^ANTC	Mnl I	1906	CUTS?	1906 GAGG
Hph I	1306	CUTS?	1306	TCACC	SfaN I	1909	CUTS?	1909 GCATC
Mst I	1355	CUTS@	1357	TGC^GCA	Tth111II	1921	CUTS?	1921 CAAACA
Hha I	1356	CUTS@	1358	GCG^C	Asu I	1948	CUTS@	1948 G^GNCC
FnuD II	1388	CUTS@	1389	CG^CG	Hae III	1948	CUTS@	1949 GG^CC
Hga I	1389	CUTS?	1389	GCGTC	Alu I	1998	CUTS@	1999 AG^CT
EcoP15t3	1403	CUTS?	1403	CAGCAG	Hga I	2003	CUTS?	2003 GACGC
Bbv I	1405	CUTS?	1405	GCAGC	FnuD II	2005	CUTS@	2006 CG^CG
Fnu4H I	1405	CUTS@	1406	GC^NGC	Fok I	2008	CUTS?	2008 GGATG
NspB II	1408	CUTS@	1411	GCCG^C	HinF I	2030	CUTS@	2030 G^ANTC
Fnu4H I	1408	CUTS@	1409	GC^NGC	Xmn I	2030	CUTS?	2030 GAANNNNTTC

Enzyme	Pos	Cut	Pos	Site
Alu I	2055	CUTS@	2056	AG^CT
Bbv I	2064	CUTS?	2064	GCAGC
Fnu4H I	2064	CUTS@	2065	GC^NGC
Pvu II	2065	CUTS@	2067	CAG^CTG
Alu I	2066	CUTS@	2067	AG^CT
Fnu4H I	2067	CUTS@	2068	GC^NGC
Bbv I	2067	CUTS?	2067	GCTGC
Mnl I	2071	CUTS?	2071	CCTC
FnuD II	2074	CUTS@	2075	CG^CG
Hha I	2075	CUTS@	2077	GCG^C
FnuD II	2076	CUTS@	2077	CG^CG
Hph I	2084	CUTS?	2084	GGTGA
Hph I	2093	CUTS?	2093	GGTGA
Mnl I	2101	CUTS?	2101	CCTC
NspC I	2109	CUTS@	2113	ACATG^C
Bbv I	2113	CUTS?	2113	GCAGC
Fnu4H I	2113	CUTS@	2114	GC^NGC
Alu I	2115	CUTS@	2116	AG^CT
ScrF I	2119	CUTS?	2119	CCNGG
Cau II	2119	CUTS@	2120	CC^CGG
Hpa II	2120	CUTS@	2120	C^CGG
Alu I	2134	CUTS@	2135	AG^CT
Fok I	2149	CUTS?	2149	GGATG
SfaN I	2150	CUTS?	2150	GATGC
ScrF I	2154	CUTS?	2154	CCNGG
Hpa II	2154	CUTS@	2154	C^CGG
Cau II	2154	CUTS@	2155	CC^GGG
Hha I	2178	CUTS@	2180	GCG^C
FnuD II	2179	CUTS@	2180	CG^CG
Hga I	2180	CUTS?	2180	GCGTC
Hha I	2208	CUTS@	2210	GCG^C
Bbv I	2210	CUTS?	2210	GCAGC
Fnu4H I	2210	CUTS@	2211	GC^NGC
Tth111 I	2218	CUTS@	2221	GACN^NNGTC
Acc I	2245	CUTS@	2246	GT^ATAC
Sna I	2245	CUTS?	2245	GTATAC
NspB II	2263	CUTS@	2266	GCGG^C
Fnu4H I	2263	CUTS@	2264	GC^NGC
SfaN I	2266	CUTS?	2266	GATGC
Rsa I	2281	CUTS@	2282	GT^AC
Dde I	2284	CUTS@	2284	C^TNAG
HgiA I	2290	CUTS@	2294	GTGCA^C
HgiE II	2294	CUTS?	2294	ACCNNNNNNGGT
Nde I	2296	CUTS?	2296	CATATG
SfaN I	2321	CUTS?	2321	GATGC
SfaN I	2342	CUTS?	2342	GCATC
Hae II	2348	CUTS@	2352	GGCGC^T
Hha I	2349	CUTS@	2351	GCG^C
Mbo II	2353	CUTS?	2353	TCTTC
Mnl I	2363	CUTS?	2363	CCTC
HinF I	2374	CUTS@	2374	G^ANTC
Fnu4H I	2379	CUTS@	2380	GC^NGC
Bbv I	2379	CUTS?	2379	GCTGC
Hha I	2382	CUTS@	2384	GCG^C
Fnu4H I	2397	CUTS@	2398	GC^NGC
Bbv I	2397	CUTS?	2397	GCTGC
NspB II	2400	CUTS@	2403	GCGG^C
Fnu4H I	2400	CUTS@	2401	GC^NGC
Alu I	2415	CUTS@	2416	AG^CT
HinF I	2449	CUTS@	2449	G^ANTC
NspC I	2474	CUTS@	2478	ACATG^T
Hae I	2487	CUTS@	2489	AGG^CCA
Hae III	2488	CUTS@	2489	GG^CC
Hae I	2498	CUTS@	2500	AGG^CCA
Hae III	2499	CUTS@	2500	GG^CC
ScrF I	2501	CUTS?	2501	CCNGG
EcoR II	2501	CUTS@	2500	^CCAGG
Hae III	2517	CUTS@	2518	GG^CC
NspB II	2518	CUTS@	2521	GCCG^C
Fnu4H I	2518	CUTS@	2519	GC^NGC
FnuD II	2520	CUTS@	2521	CG^CG
SfaN I	2562	CUTS?	2562	GCATC
Taq I	2574	CUTS@	2574	T^CGA
Hga I	2576	CUTS?	2576	GACGC
Mnl I	2589	CUTS?	2589	GAGG
ScrF I	2622	CUTS?	2622	CCNGG
EcoR II	2622	CUTS@	2621	^CCAGG
ScrF I	2635	CUTS?	2635	CCNGG
EcoR II	2635	CUTS@	2634	^CCTGG
Alu I	2641	CUTS@	2642	AG^CT
Mnl I	2646	CUTS?	2646	CCTC
Hha I	2652	CUTS@	2654	GCG^C
NspB II	2673	CUTS@	2676	GCCG^C
Fnu4H I	2673	CUTS@	2674	GC^NGC
Hpa II	2681	CUTS@	2681	C^CGG
Hae II	2718	CUTS@	2722	GGCGC^T
Hha I	2719	CUTS@	2721	GCG^C
Dde I	2749	CUTS@	2749	C^TNAG
Alu I	2777	CUTS@	2778	AG^CT
HgiA I	2788	CUTS@	2792	GTGCA^C
Fnu4H I	2816	CUTS@	2817	GC^NGC
Bbv I	2816	CUTS?	2816	GCTGC
Hha I	2819	CUTS@	2821	GCG^C
Hpa II	2828	CUTS@	2828	C^CGG
HinF I	2845	CUTS@	2845	G^ANTC
ScrF I	2853	CUTS?	2853	CCNGG
Cau II	2853	CUTS@	2854	CC^CGG
Hpa II	2854	CUTS@	2854	C^CGG
Bbv I	2881	CUTS?	2881	GCAGC
Fnu4H I	2881	CUTS@	2882	GC^NGC
EcoP15t3	2882	CUTS?	2882	CAGCAG
Bbv I	2884	CUTS?	2884	GCAGC
Fnu4H I	2884	CUTS@	2885	GC^NGC
Mnl I	2913	CUTS?	2913	GAGG
Hae I	2950	CUTS@	2952	TGG^CCT
Hae III	2951	CUTS@	2952	GG^CC
Hha I	2993	CUTS@	2995	GCG^C
Alu I	3034	CUTS@	3035	AG^CT
Mbo I	3041	CUTS@	3040	^GATC
Hpa II	3044	CUTS@	3044	C^CGG
Tth111II	3048	CUTS?	3048	CAAACA
HgiE II	3055	CUTS?	3055	ACCNNNNNNGGT
Tth111II	3081	CUTS?	3081	TGTTTG
Tth111II	3087	CUTS?	3103	CAAGCA
Bbv I	3090	CUTS?	3090	GCAGC
Fnu4H I	3090	CUTS@	3091	GC^NGC
EcoP15t3	3091	CUTS?	3091	CAGCAG
FnuD II	3101	CUTS@	3102	CG^CG
Hha I	3102	CUTS@	3104	GCG^C
Xho II	3115	CUTS@	3115	G^GATCT
Mbo I	3116	CUTS@	3115	^GATC
Mbo II	3124	CUTS?	3124	GAAGA
Xho II	3126	CUTS@	3126	A^GATCC
Mbo I	3127	CUTS@	3126	^GATC
Mbo I	3135	CUTS@	3134	^GATC
Hga I	3154	CUTS?	3154	GACGC
Dde I	3158	CUTS@	3158	C^TNAG
Xho II	3212	CUTS@	3212	G^GATCT
Mbo I	3213	CUTS@	3212	^GATC
Mbo II	3215	CUTS?	3215	TCTTC
Hph I	3218	CUTS?	3218	TCACC
Xho II	3224	CUTS@	3224	A^GATCC
Mbo I	3225	CUTS@	3224	^GATC
Aha III	3231	CUTS@	3233	TTT^AAA
Aha III	3250	CUTS@	3252	TTT^AAA
Mnl I	3313	CUTS?	3313	GAGG
HgiC I	3315	CUTS@	3315	G^GCACC
Dde I	3324	CUTS@	3324	C^TNAG
Mbo I	3330	CUTS@	3329	^GATC
HinF I	3362	CUTS@	3362	G^ANTC
Mnl I	3394	CUTS?	3394	GAGG
Asu I	3409	CUTS@	3409	G^GNCC
Hae III	3409	CUTS@	3410	GG^CC
Fnu4H I	3418	CUTS@	3419	GC^NGC

```
Bbv I     3418    CUTS?   3418 GCTGC
FnuD II   3431    CUTS@   3432 CG^CG
EcoP I t  3435    CUTS?   3435 AGACC
Hph I     3445    CUTS?   3445 TCACC
Hpa II    3448    CUTS@   3448 C^CGG
Bgl I     3481    CUTS@   3487 GCCNNNN^NGGC
Hpa II    3482    CUTS@   3482 C^CGG
Asu I     3488    CUTS@   3488 G^GNCC
Hae III   3489    CUTS@   3490 GG^CC
Hha I     3495    CUTS@   3497 GCG^C
Asu I     3505    CUTS@   3505 G^GNCC
Ava II    3505    CUTS@   3505 G^GTCC
Mnl I     3524    CUTS?   3524 CCTC
ScrF I    3549    CUTS?   3549 CCNGG
Hpa II    3549    CUTS@   3549 C^CGG
Cau II    3549    CUTS@   3550 CC^GGG
Alu I     3555    CUTS@   3556 AG^CT
Mst I     3587    CUTS@   3589 TGC^GCA
Hha I     3588    CUTS@   3590 GCG^C
Fnu4H I   3607    CUTS@   3608 GC^NGC
Bbv I     3607    CUTS?   3607 GCTGC
Pst I     3608    CUTS@   3612 CTGCA^G
SfaN I    3614    CUTS?   3614 GCATC
Alu I     3655    CUTS@   3656 AG^CT
Hpa II    3659    CUTS@   3659 C^CGG
Mbo I     3671    CUTS@   3670 ^GATC
Mbo I     3689    CUTS@   3688 ^GATC
Alu I     3718    CUTS@   3719 AG^CT
Asu I     3727    CUTS@   3727 G^GNCC
Ava II    3727    CUTS@   3727 G^GTCC
Mnl I     3730    CUTS?   3730 CCTC
Pvu I     3734    CUTS@   3737 CGAT^CG
Mbo I     3735    CUTS@   3734 ^GATC
Cfr I     3755    CUTS@   3755 T^GGCCG
Gdi II    3755    CUTS@   3755 T^GGCCG
Hae III   3756    CUTS@   3757 GG^CC
NspB II   3757    CUTS@   3760 GCCG^C
Fnu4H I   3757    CUTS@   3758 GC^NGC
Bbv I     3784    CUTS?   3784 GCAGC
Fnu4H I   3784    CUTS@   3785 GC^NGC
SfaN I    3824    CUTS?   3824 GATGC
Hph I     3841    CUTS?   3841 GGTGA
Rru I     3845    CUTS@   3847 AGT^ACT
Rsa I     3846    CUTS@   3847 GT^AC
Dde I     3864    CUTS@   3864 C^TNAG
NspB II   3879    CUTS@   3882 GCGG^C
Fnu4H I   3879    CUTS@   3880 GC^NGC
ScrF I    3900    CUTS?   3900 CCNGG
Cau II    3900    CUTS@   3901 CC^CGG
Hpa II    3901    CUTS@   3901 C^CGG
Acy I     3903    CUTS@   3904 GG^CGTC
Hga I     3904    CUTS?   3904 GCGTC
Hind II   3906    CUTS@   3908 GTC^AAC
FnuD II   3924    CUTS@   3925 CG^CG
Hha I     3925    CUTS@   3927 GCG^C
Aha III   3942    CUTS@   3944 TTT^AAA
HgiA I    3949    CUTS@   3953 GTGCT^C
Xmn I     3962    CUTS?   3962 GAANNNNTTC
Mbo II    3970    CUTS?   3970 TCTTC
Xho I     3992    CUTS@   3992 G^GATCT
Mbo I     3993    CUTS@   3992 ^GATC
Xho II    4009    CUTS@   4009 A^GATCC
Mbo I     4010    CUTS@   4009 ^GATC
Taq I     4018    CUTS@   4018 T^CGA
EcoK t1   4025    CUTS?   4025 AACNNNNNNGTGC
HgiA I    4034    CUTS@   4038 GTGCA^C
Mbo I     4046    CUTS@   4045 ^GATC
Mbo II    4048    CUTS?   4048 TCTTC
SfaN I    4054    CUTS?   4054 GCATC
Hph I     4067    CUTS?   4067 TCACC
Hph I     4082    CUTS?   4082 GGTGA
NspB II   4108    CUTS@   4111 GCCG^C
```

```
Fnu4H I   4108    CUTS@   4109 GC^NGC
Mbo II    4157    CUTS?   4157 TCTTC
FnuD II   4256    CUTS@   4257 CG^CG
Hha I     4257    CUTS@   4259 GCG^C
Acy I     4285    CUTS@   4286 GA^CGTC
Dde I     4290    CUTS@   4290 C^TNAG
Mnl I     4341    CUTS?   4341 GAGG
Asu I     4343    CUTS@   4343 G^GNCC
Hae III   4343    CUTS@   4344 GG^CC
Mbo II    4353    CUTS?   4353 TCTTC
EcoR I    4360    CUTS@   4360 G^AATTC
```

RESTRICTION FRAGMENTS OF pBR322 DNA

Each entry shows:

1. the number of cleavage sites for a given restriction enzyme;

2. the numbers of the nucleotides that define the beginning of each restriction fragment;

3. the length of each restriction fragment;

4. the name of the fragment.

```
              ===#=== Acc I    HAS   2 SITES ===#===
   651...  1595 BASES...B   |   2246...  2767 BASES...A
  2246...  2767 BASES...A   |    651...  1595 BASES...B

              ===#=== Acy I    HAS   6 SITES ===#===
   414...    21 BASES...F   |   1205...  2699 BASES...A
   435...   113 BASES...E   |    548...   657 BASES...B
   548...   657 BASES...B   |   4286...   490 BASES...C
  1205...  2699 BASES...A   |   3904...   382 BASES...D
  3904...   382 BASES...D   |    435...   113 BASES...E
  4286...   490 BASES...C   |    414...    21 BASES...F

              ===#=== Afl II   HAS   0 SITES ===#===

              ===#=== Aha III  HAS   3 SITES ===#===
  3233...    19 BASES...C   |   3944...  3651 BASES...A
  3252...   692 BASES...B   |   3252...   692 BASES...B
  3944...  3651 BASES...A   |   3233...    19 BASES...C

              ===#=== Alu I    HAS  16 SITES ===#===
    16...    15 BASES...O   |   1089...   910 BASES...A
    31...   655 BASES...C   |   3719...   659 BASES...B
   686...   403 BASES...E   |     31...   655 BASES...C
  1089...   910 BASES...A   |   3035...   521 BASES...D
  1999...    57 BASES...L   |    686...   403 BASES...E
  2056...    11 BASES...P   |   2135...   281 BASES...F
  2067...    49 BASES...M   |   2778...   257 BASES...G
  2116...    19 BASES...N   |   2416...   226 BASES...H
  2135...   281 BASES...F   |   2642...   136 BASES...I
  2416...   226 BASES...H   |   3556...   100 BASES...J
  2642...   136 BASES...I   |   3656...    63 BASES...K
  2778...   257 BASES...G   |   1999...    57 BASES...L
  3035...   521 BASES...D   |   2067...    49 BASES...M
  3556...   100 BASES...J   |   2116...    19 BASES...N
  3656...    63 BASES...K   |     16...    15 BASES...O
  3719...   659 BASES...B   |   2056...    11 BASES...P

              ===#=== Apa I    HAS   0 SITES ===#===

              ===#=== Asu I    HAS  15 SITES ===#===
   172...   352 BASES...C   |   1948...  1461 BASES...A
   524...   274 BASES...E   |   3727...   616 BASES...B
   798...    88 BASES...L   |    172...   352 BASES...C
   886...   249 BASES...F   |   1480...   279 BASES...D
  1135...   124 BASES...K   |    524...   274 BASES...E
  1259...   179 BASES...J   |    886...   249 BASES...F
  1438...    42 BASES...N   |   3505...   222 BASES...G
  1480...   279 BASES...D   |   4343...   191 BASES...H
  1759...   189 BASES...I   |   1759...   189 BASES...I
  1948...  1461 BASES...A   |   1259...   179 BASES...J
  3409...    79 BASES...M   |   1135...   124 BASES...K
  3488...    17 BASES...O   |    798...    88 BASES...L
  3505...   222 BASES...G   |   3409...    79 BASES...M
  3727...   616 BASES...B   |   1438...    42 BASES...N
  4343...   191 BASES...H   |   3488...    17 BASES...O

              ===#=== Asu II   HAS   0 SITES ===#===

              ===#=== Ava I    HAS   1 SITES ===#===
  1424...  4362 BASES...A   |   1424...  4362 BASES...A

              ===#=== Ava II   HAS   8 SITES ===#===
   798...    88 BASES...G   |   1759...  1746 BASES...A
   886...   249 BASES...E   |   3727...  1433 BASES...B
  1135...   303 BASES...C   |   1135...   303 BASES...C
  1438...    42 BASES...H   |   1480...   279 BASES...D
  1480...   279 BASES...D   |    886...   249 BASES...E
  1759...  1746 BASES...A   |   3505...   222 BASES...F
  3505...   222 BASES...F   |    798...    88 BASES...G
  3727...  1433 BASES...B   |   1438...    42 BASES...H
```

```
            ===#=== Ava III  HAS   0 SITES ===#===

            ===#=== Avr II   HAS   0 SITES ===#===

            ===#=== Bal I    HAS   1 SITES ===#===
 1445... 4362 BASES...A      ¦   1445... 4362 BASES...A

            ===#=== BamH I   HAS   1 SITES ===#===
  375... 4362 BASES...A      ¦    375... 4362 BASES...A

            ===#=== Bbv I    HAS  21 SITES ===#===
  226...  388 BASES...D      ¦   3784...  804 BASES...A
  614...  158 BASES...K      ¦    772...  633 BASES...B
  772...  633 BASES...B      ¦   2397...  419 BASES...C
 1405...   24 BASES...Q      ¦    226...  388 BASES...D
 1429...  129 BASES...L      ¦   1684...  380 BASES...E
 1558...    3 BASES...S      ¦   3090...  328 BASES...F
 1561...  123 BASES...M      ¦   2884...  206 BASES...G
 1684...  380 BASES...E      ¦   3418...  189 BASES...H
 2064...    3 BASES...T      ¦   3607...  177 BASES...I
 2067...   46 BASES...P      ¦   2210...  169 BASES...J
 2113...   97 BASES...N      ¦    614...  158 BASES...K
 2210...  169 BASES...J      ¦   1429...  129 BASES...L
 2379...   18 BASES...R      ¦   1561...  123 BASES...M
 2397...  419 BASES...C      ¦   2113...   97 BASES...N
 2816...   65 BASES...O      ¦   2816...   65 BASES...O
 2881...    3 BASES...U      ¦   2067...   46 BASES...P
 2884...  206 BASES...G      ¦   1405...   24 BASES...Q
 3090...  328 BASES...F      ¦   2379...   18 BASES...R
 3418...  189 BASES...H      ¦   1558...    3 BASES...S
 3607...  177 BASES...I      ¦   2064...    3 BASES...T
 3784...  804 BASES...A      ¦   2881...    3 BASES...U

            ===#=== Bcl I    HAS   0 SITES ===#===

            ===#=== Bgl I    HAS   3 SITES ===#===
  934...  234 BASES...C      ¦   1168... 2319 BASES...A
 1168... 2319 BASES...A      ¦   3487... 1809 BASES...B
 3487... 1809 BASES...B      ¦    934...  234 BASES...C

            ===#=== Bgl II   HAS   0 SITES ===#===

            ===#=== BstE II  HAS   0 SITES ===#===

            ===#=== Cau II   HAS  10 SITES ===#===
  171...  363 BASES...E      ¦    534...  724 BASES...A
  534...  724 BASES...A      ¦   2155...  699 BASES...B
 1258...  226 BASES...I      ¦   2854...  696 BASES...C
 1484...  328 BASES...G      ¦   3901...  632 BASES...D
 1812...  308 BASES...H      ¦    171...  363 BASES...E
 2120...   35 BASES...J      ¦   3550...  351 BASES...F
 2155...  699 BASES...B      ¦   1484...  328 BASES...G
 2854...  696 BASES...C      ¦   1812...  308 BASES...H
 3550...  351 BASES...F      ¦   1258...  226 BASES...I
 3901...  632 BASES...D      ¦   2120...   35 BASES...J

            ===#=== Cfr I    HAS   6 SITES ===#===
  295...  104 BASES...F      ¦   1443... 2312 BASES...A
  399...  131 BASES...E      ¦   3755...  902 BASES...B
  530...  408 BASES...D      ¦    938...  505 BASES...C
  938...  505 BASES...C      ¦    530...  408 BASES...D
 1443... 2312 BASES...A      ¦    399...  131 BASES...E
 3755...  902 BASES...B      ¦    295...  104 BASES...F

            ===#=== Cla I    HAS   1 SITES ===#===
   24... 4362 BASES...A      ¦     24... 4362 BASES...A

            ===#=== Dde I    HAS   8 SITES ===#===
 1580...  162 BASES...H      ¦   4290... 1652 BASES...A
 1742...  542 BASES...B      ¦   1742...  542 BASES...B
 2284...  465 BASES...D      ¦   3324...  540 BASES...C
```

```
2749...   409 BASES...F   |   2284...   465 BASES...D
3158...   166 BASES...G   |   3864...   426 BASES...E
3324...   540 BASES...C   |   2749...   409 BASES...F
3864...   426 BASES...E   |   3158...   166 BASES...G
4290...  1652 BASES...A   |   1580...   162 BASES...H

        ===#=== EcoR I    HAS   1 SITES ===#===
4360...  4362 BASES...A   |   4360...  4362 BASES...A

        ===#=== EcoR II   HAS   6 SITES ===#===
 129...   928 BASES...C   |   2634...  1857 BASES...A
1057...   383 BASES...D   |   1440...  1060 BASES...B
1440...  1060 BASES...B   |    129...   928 BASES...C
2500...   121 BASES...E   |   1057...   383 BASES...D
2621...    13 BASES...F   |   2500...   121 BASES...E
2634...  1857 BASES...A   |   2621...    13 BASES...F

        ===#=== EcoR V    HAS   1 SITES ===#===
 189...  4362 BASES...A   |    189...  4362 BASES...A

        ===#=== FnuD II   HAS  23 SITES ===#===
 347...   355 BASES...E   |   2521...   581 BASES...A
 702...   115 BASES...M   |   3432...   493 BASES...B
 817...   129 BASES...J   |   4257...   452 BASES...C
 946...    27 BASES...S   |   1634...   372 BASES...D
 973...     5 BASES...V   |    347...   355 BASES...E
 978...    61 BASES...R   |   2180...   341 BASES...F
1039...    66 BASES...Q   |   3925...   332 BASES...G
1105...   129 BASES...K   |   3102...   330 BASES...H
1234...    10 BASES...U   |   1244...   145 BASES...I
1244...   145 BASES...I   |    817...   129 BASES...J
1389...    26 BASES...T   |   1105...   129 BASES...K
1415...   122 BASES...L   |   1415...   122 BASES...L
1537...    97 BASES...O   |    702...   115 BASES...M
1634...   372 BASES...D   |   2077...   103 BASES...N
2006...    69 BASES...P   |   1537...    97 BASES...O
2075...     2 BASES...W   |   2006...    69 BASES...P
2077...   103 BASES...N   |   1039...    66 BASES...Q
2180...   341 BASES...F   |    978...    61 BASES...R
2521...   581 BASES...A   |    946...    27 BASES...S
3102...   330 BASES...H   |   1389...    26 BASES...T
3432...   493 BASES...B   |   1234...    10 BASES...U
3925...   332 BASES...G   |    973...     5 BASES...V
4257...   452 BASES...C   |   2075...     2 BASES...W

        ===#=== Fnu4H I   HAS  42 SITES ===#===
 227...    71 BASES...X   |   4109...   480 BASES...A
 298...     3 BASES...j   |   3091...   328 BASES...B
 301...   277 BASES...D   |   1766...   299 BASES...C
 578...     3 BASES...k   |    301...   277 BASES...D
 581...    34 BASES...e   |   3880...   229 BASES...E
 615...   107 BASES...Q   |   2885...   206 BASES...F
 722...    51 BASES...b   |   3419...   189 BASES...G
 773...   165 BASES...H   |    773...   165 BASES...H
 938...    85 BASES...T   |   2519...   155 BASES...I
1023...    83 BASES...U   |   3608...   150 BASES...J
1106...    57 BASES...z   |   2674...   143 BASES...K
1163...    45 BASES...d   |   1430...   129 BASES...L
1208...    79 BASES...W   |   1562...   123 BASES...M
1287...   119 BASES...N   |   1287...   119 BASES...N
1406...     3 BASES...l   |   2401...   118 BASES...O
1409...     7 BASES...i   |   2264...   116 BASES...P
1416...    14 BASES...h   |    615...   107 BASES...Q
1430...   129 BASES...L   |   2114...    97 BASES...R
1559...     3 BASES...m   |   3785...    95 BASES...S
1562...   123 BASES...M   |    938...    85 BASES...T
1685...    81 BASES...V   |   1023...    83 BASES...U
1766...   299 BASES...C   |   1685...    81 BASES...V
2065...     3 BASES...n   |   1208...    79 BASES...W
2068...    46 BASES...c   |    227...    71 BASES...X
2114...    97 BASES...R   |   2817...    65 BASES...Y
2211...    53 BASES...a   |   1106...    57 BASES...Z
```

```
2264... 116 BASES...P    |   2211...  53 BASES...a
2380...  18 BASES...g    |    722...  51 BASES...b
2398...   3 BASES...o    |   2068...  46 BASES...c
2401... 118 BASES...O    |   1163...  45 BASES...d
2519... 155 BASES...I    |    581...  34 BASES...e
2674... 143 BASES...K    |   3758...  27 BASES...f
2817...  65 BASES...Y    |   2380...  18 BASES...g
2882...   3 BASES...p    |   1416...  14 BASES...h
2885... 206 BASES...F    |   1409...   7 BASES...i
3091... 328 BASES...B    |    298...   3 BASES...j
3419... 189 BASES...G    |    578...   3 BASES...k
3608... 150 BASES...J    |   1406...   3 BASES...l
3758...  27 BASES...f    |   1559...   3 BASES...m
3785...  95 BASES...S    |   2065...   3 BASES...n
3880... 229 BASES...E    |   2398...   3 BASES...o
4109... 480 BASES...A    |   2882...   3 BASES...p
```

```
         ===#=== Fok I   HAS  6 SITES ===#===
 133... 853 BASES...B    |   2149... 2346 BASES...A
 986...  45 BASES...F    |    133...  853 BASES...B
1031... 649 BASES...C    |   1031...  649 BASES...C
1680... 328 BASES...D    |   1680...  328 BASES...D
2008... 141 BASES...E    |   2008...  141 BASES...E
2149... 2346 BASES...A   |    986...   45 BASES...F
```

```
         ===#=== Gdi II  HAS  5 SITES ===#===
 295... 104 BASES...E    |    938... 2817 BASES...A
 399... 131 BASES...D    |   3755...  902 BASES...B
 530... 408 BASES...C    |    530...  408 BASES...C
 938... 2817 BASES...A   |    399...  131 BASES...D
3755... 902 BASES...B    |    295...  104 BASES...E
```

```
         ===#=== Hae I   HAS  7 SITES ===#===
 919...  72 BASES...E    |   2952... 2329 BASES...A
 991...  57 BASES...F    |   1445... 1044 BASES...B
1048... 397 BASES...D    |   2500...  452 BASES...C
1445... 1044 BASES...B   |   1048...  397 BASES...D
2489...  11 BASES...G    |    919...   72 BASES...E
2500... 452 BASES...C    |    991...   57 BASES...F
2952... 2329 BASES...A   |   2489...   11 BASES...G
```

```
         ===#=== Hae II  HAS 11 SITES ===#===
 236... 181 BASES...G    |   2722... 1876 BASES...A
 417...  21 BASES...K    |   1730...  622 BASES...B
 438...  60 BASES...I    |   1208...  439 BASES...C
 498...  53 BASES...J    |    778...  430 BASES...D
 551... 227 BASES...F    |   2352...  370 BASES...E
 778... 430 BASES...D    |    551...  227 BASES...F
1208... 439 BASES...C    |    236...  181 BASES...G
1647...  83 BASES...H    |   1647...   83 BASES...H
1730... 622 BASES...B    |    438...   60 BASES...I
2352... 370 BASES...E    |    498...   53 BASES...J
2722... 1876 BASES...A   |    417...   21 BASES...K
```

```
         ===#=== Hae III HAS 22 SITES ===#===
 174... 123 BASES...L    |   3757...  587 BASES...A
 297... 104 BASES...M    |   1949...  540 BASES...B
 401... 124 BASES...K    |   1445...  504 BASES...C
 525...   7 BASES...V    |   2952...  458 BASES...D
 532...  64 BASES...P    |   2518...  434 BASES...E
 596... 234 BASES...G    |   3490...  267 BASES...F
 830...  89 BASES...N    |    596...  234 BASES...G
 919...  21 BASES...S    |   1048...  213 BASES...H
 940...  51 BASES...R    |   4344...  192 BASES...I
 991...  57 BASES...Q    |   1261...  184 BASES...J
1048... 213 BASES...H    |    401...  124 BASES...K
1261... 184 BASES...J    |    174...  123 BASES...L
1445... 504 BASES...C    |    297...  104 BASES...M
1949... 540 BASES...B    |    830...   89 BASES...N
2489...  11 BASES...U    |   3410...   80 BASES...O
2500...  18 BASES...T    |    532...   64 BASES...P
2518... 434 BASES...E    |    991...   57 BASES...Q
```

```
2952...    458 BASES...D    |    940...    51 BASES...R
3410...     80 BASES...O    |    919...    21 BASES...S
3490...    267 BASES...F    |   2500...    18 BASES...T
3757...    587 BASES...A    |   2489...    11 BASES...U
4344...    192 BASES...I    |    525...     7 BASES...V


           ===#=== Hga I   HAS  11 SITES ===#===
390...    258 BASES...H    |   3904...   848 BASES...A
648...    295 BASES...F    |   3154...   750 BASES...B
943...     32 BASES...K    |   1389...   614 BASES...C
975...    264 BASES...G    |   2576...   578 BASES...D
1239...   150 BASES...J    |   2180...   396 BASES...E
1389...   614 BASES...C    |    648...   295 BASES...F
2003...   177 BASES...I    |    975...   264 BASES...G
2180...   396 BASES...E    |    390...   258 BASES...H
2576...   578 BASES...D    |   2003...   177 BASES...I
3154...   750 BASES...B    |   1239...   150 BASES...J
3904...   848 BASES...A    |    943...    32 BASES...K


           ===#=== HgiA I  HAS   8 SITES ===#===
280...    310 BASES...F    |   2792...  1161 BASES...A
590...    587 BASES...D    |   1468...   826 BASES...B
1177...   291 BASES...G    |   4038...   604 BASES...C
1468...   826 BASES...B    |    590...   587 BASES...D
2294...   498 BASES...E    |   2294...   498 BASES...E
2792...  1161 BASES...A    |    280...   310 BASES...F
3953...    85 BASES...H    |   1177...   291 BASES...G
4038...   604 BASES...C    |   3953...    85 BASES...H


           ===#=== HgiC I  HAS   9 SITES ===#===
76...     43 BASES...H    |   1288...  2027 BASES...A
119...    294 BASES...D    |   3315...  1123 BASES...B
413...     21 BASES...I    |    765...   439 BASES...C
434...    113 BASES...F    |    119...   294 BASES...D
547...    218 BASES...E    |    547...   218 BASES...E
765...    439 BASES...C    |    434...   113 BASES...F
1204...    84 BASES...G    |   1204...    84 BASES...G
1288...  2027 BASES...A    |     76...    43 BASES...H
3315...  1123 BASES...B    |    413...    21 BASES...I


           ===#=== HgiE II HAS   2 SITES ===#===
2294...   761 BASES...B    |   3055...  3601 BASES...A
3055...  3601 BASES...A    |   2294...   761 BASES...B


           ===#=== HgiJ II HAS   2 SITES ===#===
475...     14 BASES...B    |    489...  4348 BASES...A
489...  4348 BASES...A    |    475...    14 BASES...B


           ===#=== Hha I   HAS  31 SITES ===#===
103...    132 BASES...N    |   3104...   393 BASES...A
235...     28 BASES...d    |   1729...   348 BASES...B
263...    153 BASES...J    |   3590...   337 BASES...C
416...     21 BASES...e    |   3927...   332 BASES...D
437...     60 BASES...X    |   2384...   270 BASES...E
497...     53 BASES...Y    |    948...   259 BASES...F
550...    152 BASES...K    |   4259...   206 BASES...G
702...     75 BASES...U    |   1456...   190 BASES...H
777...     40 BASES...Z    |   2821...   174 BASES...I
817...    131 BASES...O    |    263...   153 BASES...J
948...    259 BASES...F    |    550...   152 BASES...K
1207...   151 BASES...L    |   1207...   151 BASES...L
1358...    62 BASES...W    |   2210...   141 BASES...M
1420...    36 BASES...a    |    103...   132 BASES...N
1456...   190 BASES...H    |    817...   131 BASES...O
1646...    83 BASES...T    |   2995...   109 BASES...P
1729...   348 BASES...B    |   2077...   103 BASES...Q
2077...   103 BASES...Q    |   2721...   100 BASES...R
2180...    30 BASES...c    |   3497...    93 BASES...S
2210...   141 BASES...M    |   1646...    83 BASES...T
2351...    33 BASES...b    |    702...    75 BASES...U
2384...   270 BASES...E    |   2654...    67 BASES...V
2654...    67 BASES...V    |   1358...    62 BASES...W
```

```
2721... 100 BASES...R  |   437... 60 BASES...X
2821... 174 BASES...I  |   497... 53 BASES...Y
2995... 109 BASES...P  |   777... 40 BASES...Z
3104... 393 BASES...A  |  1420... 36 BASES...a
3497...  93 BASES...S  |  2351... 33 BASES...b
3590... 337 BASES...C  |  2180... 30 BASES...c
3927... 332 BASES...D  |   235... 28 BASES...d
4259... 206 BASES...G  |   416... 21 BASES...e

        ===#=== HinD II  HAS   2 SITES ===#===
 652... 3256 BASES...A  |   652... 3256 BASES...A
3908... 1106 BASES...B  |  3908... 1106 BASES...B

        ===#=== HinD III HAS   1 SITES ===#===
  29... 4362 BASES...A  |    29... 4362 BASES...A

        ===#=== HinF I   HAS  10 SITES ===#===
 631...  220 BASES...H  |  3362... 1631 BASES...A
 851...  154 BASES...I  |  2845...  517 BASES...B
1005...  298 BASES...F  |  1524...  506 BASES...C
1303...  221 BASES...G  |  2449...  396 BASES...D
1524...  506 BASES...C  |  2030...  344 BASES...E
2030...  344 BASES...E  |  1005...  298 BASES...F
2374...   75 BASES...J  |  1303...  221 BASES...G
2449...  396 BASES...D  |   631...  220 BASES...H
2845...  517 BASES...B  |   851...  154 BASES...I
3362... 1631 BASES...A  |  2374...   75 BASES...J

        ===#=== Hpa I    HAS   0 SITES ===#===

        ===#=== Hpa II   HAS  26 SITES ===#===
 161...    9 BASES...Y  |  3901...  622 BASES...A
 170...  217 BASES...G  |  2154...  527 BASES...B
 387...   15 BASES...X  |  3044...  404 BASES...C
 402...    9 BASES...Z  |  1811...  309 BASES...D
 411...  122 BASES...O  |  3659...  242 BASES...E
 533...  160 BASES...K  |  1019...  238 BASES...F
 693...   76 BASES...R  |   170...  217 BASES...G
 769...  160 BASES...L  |  1283...  201 BASES...H
 929...   90 BASES...Q  |  2854...  190 BASES...I
1019...  238 BASES...F  |  1484...  180 BASES...J
1257...   26 BASES...V  |   533...  160 BASES...K
1283...  201 BASES...H  |   769...  160 BASES...L
1484...  180 BASES...J  |  1664...  147 BASES...M
1664...  147 BASES...M  |  2681...  147 BASES...N
1811...  309 BASES...D  |   411...  122 BASES...O
2120...   34 BASES...T  |  3549...  110 BASES...P
2154...  527 BASES...B  |   929...   90 BASES...Q
2681...  147 BASES...N  |   693...   76 BASES...R
2828...   26 BASES...W  |  3482...   67 BASES...S
2854...  190 BASES...I  |  2120...   34 BASES...T
3044...  404 BASES...C  |  3448...   34 BASES...U
3448...   34 BASES...U  |  1257...   26 BASES...V
3482...   67 BASES...S  |  2828...   26 BASES...W
3549...  110 BASES...P  |   387...   15 BASES...X
3659...  242 BASES...E  |   161...    9 BASES...Y
3901...  622 BASES...A  |   402...    9 BASES...Z

        ===#=== Hph I    HAS  12 SITES ===#===
 126...  282 BASES...F  |  2093... 1125 BASES...A
 408...   45 BASES...J  |   453...  853 BASES...B
 453...  853 BASES...B  |  1527...  557 BASES...C
1306...  221 BASES...I  |  4082...  406 BASES...D
1527...  557 BASES...C  |  3445...  396 BASES...E
2084...    9 BASES...L  |   126...  282 BASES...F
2093... 1125 BASES...A  |  3218...  227 BASES...G
3218...  227 BASES...G  |  3841...  226 BASES...H
3445...  396 BASES...E  |  1306...  221 BASES...I
3841...  226 BASES...H  |   408...   45 BASES...J
4067...   15 BASES...K  |  4067...   15 BASES...K
4082...  406 BASES...D  |  2084...    9 BASES...L
```

```
                ===#=== Kpn I     HAS    0 SITES ===#===

                ===#=== Mbo I     HAS   22 SITES ===#===
    348...      27 BASES...P  |   1666... 1374 BASES...A
    375...      91 BASES...J  |   4045...  665 BASES...B
    466...     358 BASES...C  |    466...  358 BASES...C
    824...     272 BASES...F  |   3329...  341 BASES...D
   1096...      31 BASES...O  |   1142...  317 BASES...E
   1127...      15 BASES...S  |    824...  272 BASES...F
   1142...     317 BASES...E  |   3734...  258 BASES...G
   1459...     207 BASES...H  |   1459...  207 BASES...H
   1666...    1374 BASES...A  |   3224...  105 BASES...I
   3040...      75 BASES...L  |    375...   91 BASES...J
   3115...      11 BASES...U  |   3134...   78 BASES...K
   3126...       8 BASES...V  |   3040...   75 BASES...L
   3134...      78 BASES...K  |   3688...   46 BASES...M
   3212...      12 BASES...T  |   4009...   36 BASES...N
   3224...     105 BASES...I  |   1096...   31 BASES...O
   3329...     341 BASES...D  |    348...   27 BASES...P
   3670...      18 BASES...Q  |   3670...   18 BASES...Q
   3688...      46 BASES...M  |   3992...   17 BASES...R
   3734...     258 BASES...G  |   1127...   15 BASES...S
   3992...      17 BASES...R  |   3212...   12 BASES...T
   4009...      36 BASES...N  |   3115...   11 BASES...U
   4045...     665 BASES...B  |   3126...    8 BASES...V

                ===#=== Mbo II    HAS   11 SITES ===#===
    464...     273 BASES...F  |   2353...  771 BASES...A
    737...     271 BASES...G  |   3215...  755 BASES...B
   1008...     592 BASES...D  |   1600...  753 BASES...C
   1600...     753 BASES...C  |   1008...  592 BASES...D
   2353...     771 BASES...A  |   4353...  473 BASES...E
   3124...      91 BASES...J  |    464...  273 BASES...F
   3215...     755 BASES...B  |    737...  271 BASES...G
   3970...      78 BASES...K  |   4157...  196 BASES...H
   4048...     109 BASES...I  |   4048...  109 BASES...I
   4157...     196 BASES...H  |   3124...   91 BASES...J
   4353...     473 BASES...E  |   3970...   78 BASES...K

                ===#=== Mlu I     HAS    0 SITES ===#===

                ===#=== Mnl I     HAS   26 SITES ===#===
    115...      60 BASES...T  |   3730...  611 BASES...A
    175...     204 BASES...I  |   2913...  400 BASES...B
    379...     218 BASES...G  |   1478...  314 BASES...C
    597...     199 BASES...J  |   2646...  267 BASES...D
    796...      68 BASES...R  |   2101...  262 BASES...E
    864...     116 BASES...P  |   2363...  226 BASES...F
    980...     186 BASES...K  |    379...  218 BASES...G
   1166...      61 BASES...S  |   3524...  206 BASES...H
   1227...      38 BASES...X  |    175...  204 BASES...I
   1265...      27 BASES...Z  |    597...  199 BASES...J
   1292...     186 BASES...L  |    980...  186 BASES...K
   1478...     314 BASES...C  |   1292...  186 BASES...L
   1792...      58 BASES...U  |   1906...  165 BASES...M
   1850...      56 BASES...W  |   4341...  136 BASES...N
   1906...     165 BASES...M  |   3394...  130 BASES...O
   2071...      30 BASES...Y  |    864...  116 BASES...P
   2101...     262 BASES...E  |   3313...   81 BASES...Q
   2363...     226 BASES...F  |    796...   68 BASES...R
   2589...      57 BASES...V  |   1166...   61 BASES...S
   2646...     267 BASES...D  |    115...   60 BASES...T
   2913...     400 BASES...B  |   1792...   58 BASES...U
   3313...      81 BASES...Q  |   2589...   57 BASES...V
   3394...     130 BASES...O  |   1850...   56 BASES...W
   3524...     206 BASES...H  |   1227...   38 BASES...X
   3730...     611 BASES...A  |   2071...   30 BASES...Y
   4341...     136 BASES...N  |   1265...   27 BASES...Z

                ===#=== Mst I     HAS    4 SITES ===#===
    262...    1095 BASES...B  |   1455... 2134 BASES...A
   1357...      98 BASES...D  |    262... 1095 BASES...B
```

```
1455...  2134 BASES...A      |    3589...  1035 BASES...C
3589...  1035 BASES...C      |    1357...    98 BASES...D

        ===#=== Nae I   HAS   4 SITES ===#===
 403...   367 BASES...B      |    1284...  3481 BASES...A
 770...   160 BASES...D      |     403...   367 BASES...B
 930...   354 BASES...C      |     930...   354 BASES...C
1284...  3481 BASES...A      |     770...   160 BASES...D

        ===#=== Nar I   HAS   4 SITES ===#===
 414...    21 BASES...D      |    1205...  3571 BASES...A
 435...   113 BASES...C      |     548...   657 BASES...B
 548...   657 BASES...B      |     435...   113 BASES...C
1205...  3571 BASES...A      |     414...    21 BASES...D

        ===#=== Nco I   HAS   0 SITES ===#===

        ===#=== Nde I   HAS   1 SITES ===#===
2296...  4362 BASES...A      |    2296...  4362 BASES...A

        ===#=== Nru I   HAS   1 SITES ===#===
 973...  4362 BASES...A      |     973...  4362 BASES...A

        ===#=== NspB II  HAS  21 SITES ===#===
 300...     3 BASES...T      |    2676...  1084 BASES...A
 303...   277 BASES...E      |    4111...   551 BASES...B
 580...     3 BASES...U      |    1768...   498 BASES...C
 583...   141 BASES...I      |    1418...   350 BASES...D
 724...   216 BASES...G      |     303...   277 BASES...E
 940...    85 BASES...N      |    3882...   229 BASES...F
1025...    83 BASES...O      |     724...   216 BASES...G
1108...    57 BASES...Q      |    2521...   155 BASES...H
1165...    45 BASES...R      |     583...   141 BASES...I
1210...    79 BASES...P      |    2266...   137 BASES...J
1289...   122 BASES...K      |    1289...   122 BASES...K
1411...     7 BASES...S      |    3760...   122 BASES...L
1418...   350 BASES...D      |    2403...   118 BASES...M
1768...   498 BASES...C      |     940...    85 BASES...N
2266...   137 BASES...J      |    1025...    83 BASES...O
2403...   118 BASES...M      |    1210...    79 BASES...P
2521...   155 BASES...H      |    1108...    57 BASES...Q
2676...  1084 BASES...A      |    1165...    45 BASES...R
3760...   122 BASES...L      |    1411...     7 BASES...S
3882...   229 BASES...F      |     300...     3 BASES...T
4111...   551 BASES...B      |     580...     3 BASES...U

        ===#=== NspC I  HAS   4 SITES ===#===
 565...  1254 BASES...B      |    2478...  2449 BASES...A
1819...   294 BASES...D      |     565...  1254 BASES...B
2113...   365 BASES...C      |    2113...   365 BASES...C
2478...  2449 BASES...A      |    1819...   294 BASES...D

        ===#=== Pst I   HAS   1 SITES ===#===
3612...  4362 BASES...A      |    3612...  4362 BASES...A

        ===#=== Pvu I   HAS   1 SITES ===#===
3737...  4362 BASES...A      |    3737...  4362 BASES...A

        ===#=== Pvu II  HAS   1 SITES ===#===
2067...  4362 BASES...A      |    2067...  4362 BASES...A

        ===#=== Rsa I   HAS   3 SITES ===#===
 165...  2117 BASES...A      |     165...  2117 BASES...A
2282...  1565 BASES...B      |    2282...  1565 BASES...B
3847...   680 BASES...C      |    3847...   680 BASES...C

        ===#=== Sac I   HAS   0 SITES ===#===

        ===#=== Sac II  HAS   0 SITES ===#===

        ===#=== Sal I   HAS   1 SITES ===#===
 650...  4362 BASES...A      |     650...  4362 BASES...A
```

```
        ===#=== Sau I    HAS   0 SITES ===#===

        ===#=== Sca I    HAS   0 SITES ===#===

        ===#=== ScrF I   HAS  16 SITES ===#===
 130...    40 BASES...N  |  2853...   696 BASES...A
 170...   363 BASES...D  |  3900...   592 BASES...B
 533...   525 BASES...C  |   533...   525 BASES...C
1058...   199 BASES...J  |   170...   363 BASES...D
1257...   184 BASES...K  |  3549...   351 BASES...E
1441...    42 BASES...M  |  2154...   347 BASES...F
1483...   328 BASES...G  |  1483...   328 BASES...G
1811...   308 BASES...H  |  1811...   308 BASES...H
2119...    35 BASES...O  |  2635...   218 BASES...I
2154...   347 BASES...F  |  1058...   199 BASES...J
2501...   121 BASES...L  |  1257...   184 BASES...K
2622...    13 BASES...P  |  2501...   121 BASES...L
2635...   218 BASES...I  |  1441...    42 BASES...M
2853...   696 BASES...A  |   130...    40 BASES...N
3549...   351 BASES...E  |  2119...    35 BASES...O
3900...   592 BASES...B  |  2622...    13 BASES...P

        ===#=== SfaN I   HAS  22 SITES ===#===
 134...    70 BASES...O  |  2562...  1052 BASES...A
 204...    43 BASES...R  |  4054...   442 BASES...B
 247...   146 BASES...K  |  1032...   388 BASES...C
 393...    12 BASES...T  |   657...   368 BASES...D
 405...   252 BASES...E  |   405...   252 BASES...E
 657...   368 BASES...D  |  1420...   252 BASES...F
1025...     7 BASES...V  |  1909...   241 BASES...G
1032...   388 BASES...C  |  3824...   230 BASES...H
1420...   252 BASES...F  |  2342...   220 BASES...I
1672...     9 BASES...U  |  3614...   210 BASES...J
1681...    87 BASES...M  |   247...   146 BASES...K
1768...    78 BASES...N  |  2150...   116 BASES...L
1846...    63 BASES...P  |  1681...    87 BASES...M
1909...   241 BASES...G  |  1768...    78 BASES...N
2150...   116 BASES...L  |   134...    70 BASES...O
2266...    55 BASES...Q  |  1846...    63 BASES...P
2321...    21 BASES...S  |  2266...    55 BASES...Q
2342...   220 BASES...I  |   204...    43 BASES...R
2562...  1052 BASES...A  |  2321...    21 BASES...S
3614...   210 BASES...J  |   393...    12 BASES...T
3824...   230 BASES...H  |  1672...     9 BASES...U
4054...   442 BASES...B  |  1025...     7 BASES...V

        ===#=== Sna I    HAS   1 SITES ===#===
2245...  4362 BASES...A  |  2245...  4362 BASES...A

        ===#=== Sph I    HAS   1 SITES ===#===
 565...  4362 BASES...A  |   565...  4362 BASES...A

        ===#=== Stu I    HAS   0 SITES ===#===

        ===#=== Taq I    HAS   7 SITES ===#===
  24...   315 BASES...E  |  2574...  1444 BASES...A
 339...   312 BASES...F  |  1267...  1307 BASES...B
 651...   475 BASES...C  |   651...   475 BASES...C
1126...   141 BASES...G  |  4018...   368 BASES...D
1267...  1307 BASES...B  |    24...   315 BASES...E
2574...  1444 BASES...A  |   339...   312 BASES...F
4018...   368 BASES...D  |  1126...   141 BASES...G

        ===#=== Tth111 I HAS   1 SITES ===#===
2221...  4362 BASES...A  |  2221...  4362 BASES...A

        ===#=== Tth111II HAS   5 SITES ===#===
   7...  1914 BASES...A  |     7...  1914 BASES...A
1921...  1127 BASES...C  |  3103...  1266 BASES...B
3048...    33 BASES...D  |  1921...  1127 BASES...C
```

```
3103... 1266 BASES...B      |    3081...    22 BASES...E

         ===#=== Xba I    HAS    0 SITES ===#===

         ===#=== Xho I    HAS    0 SITES ===#===

         ===#=== Xho II   HAS    8 SITES ===#===
  375... 1291 BASES...B      |    1666... 1449 BASES...A
 1666... 1449 BASES...A      |     375... 1291 BASES...B
 3115...   11 BASES...H      |    3224...  768 BASES...C
 3126...   86 BASES...E      |    4009...  728 BASES...D
 3212...   12 BASES...G      |    3126...   86 BASES...E
 3224...  768 BASES...C      |    3992...   17 BASES...F
 3992...   17 BASES...F      |    3212...   12 BASES...G
 4009...  728 BASES...D      |    3115...   11 BASES...H

         ===#=== Xma I    HAS    0 SITES ===#===

         ===#=== Xma III  HAS    1 SITES ===#===
  938... 4362 BASES...A      |     938... 4362 BASES...A

         ===#=== Xmn I    HAS    2 SITES ===#===
 2030... 1932 BASES...B      |    3962... 2430 BASES...A
 3962... 2430 BASES...A      |    2030... 1932 BASES...B
```

Appendix C: Commonly Used Bacterial Strains

APPENDIX C: COMMONLY USED BACTERIAL STRAINS

Strain	Genotype	Remarks
C600	F⁻, *thi-1, thr-1, leuB6, lacY1, tonA21, supE44*, λ⁻	This strain is also known as CR34 (Appleyard 1954).
DP50, *supF*	F⁻, *tonA53, dapD8, lacY1, glnV44 (supE44)*, Δ *(gal-uvrB)47*, λ⁻, *tyrT58 (supF58), gyrA29*, Δ *(thyA57), hsdS3*	This strain, which is also known as χ2098, was originally constructed by D. Pereira in the laboratory of R. Curtiss as an EK2 host for isolation and propagation of bacteriophage λ recombinants. The use of DP50, *supF* is no longer required by the current National Institutes of Health guidelines, and it has been replaced as a host for bacteriophage λ by strains of *E. coli* that are easier to grow (Leder et al. 1977; B. Bachmann, pers. comm.).
χ1776	F⁻, *tonA53, dapD8, minA1, glnV44 (supE42)*, Δ *(gal-uvrB)40*, λ⁻, *minB2, rfb-2, gyrA25, thyA142, oms-2, metC65, oms-1, (tte-1)*, Δ *(bioH-asd)29, cycB2, cycA1, hsdR2*	This strain was originally constructed as an EK2 host for certain plasmids. Although the use of χ1776 is no longer required by the National Institutes of Health guidelines, the strain is still in current use because of its high transformation efficiency (Curtiss et al. 1977; B. Bachmann, pers. comm.; D. Hanahan, pers. comm.).
LE392	F⁻, *hsdR514* (r$_k^-$, m$_k^-$), *supE44, supF58, lacY1* or Δ *(lacIZY)6, galK2, galT22, metB1, trpR55*, λ⁻	An *su⁺* strain commonly used to propagate bacteriophage λ vectors and their recombinants. LE392 was prepared by L. Enquist and is a derivative of strain ED8654 (Brock et al. 1976; Murray et al. 1977; B. Bachmann, pers. comm.)
HB101	F⁻, *hsdS20* (r$_B^-$, m$_B^-$), *recA13, ara-14, proA2, lacY1, galK2, rpsL20* (Sm^r), *xyl-5, mtl-1, supE44*, λ⁻	This strain, an *E. coli* K-12 × *E. coli* B hybrid, is commonly used as a recipient in transformation and is a good host for large-scale growth and purification of plasmids (Boyer and Roulland-Dussoix 1969; Bolivar and Backman 1979; B. Bachmann, pers. comm.).

Strain	Genotype	Remarks
RR1	F⁻; the same as HB101 except *recA⁺*	A *rec⁺* derivative of HB101 constructed by R. Rodriguez. This strain can be transformed with high efficiency with plasmid vector annealed to cDNA by homopolymeric tails (Bolivar et al. 1977; Peacock et al. 1981; B. Bachmann, pers. comm.).
C-1A	A wild-type strain.	A clone of *E. coli* strain C wild type, maintained on minimal medium for several years. *E. coli* C is F⁻ and lacks host restriction and modification activity. It is an *su⁻* host used in complementation tests with amber mutants of bacteriophage λ (Bertani 1968; B. Bachmann, pers. comm.; F. R. Blattner, pers. comm.).
CSH18	Δ (*lac pro*), *supE*, *thi⁻*, (F' *lacZ⁻ proA⁺B⁺*)	A suppressor strain used to screen recombinants made in bacteriophage λ vectors carrying a *lac* gene. These vectors, which contain a *lac* gene in the stuffer fragment, give rise to blue plaques in the presence of the chromogenic substrate Xgal; recombinants in which the stuffer fragment has been replaced by foreign DNA give rise to white plaques (Miller 1972; Williams and Blattner 1979).
DH1	F⁻, *recA1*, *endA1*, *gyrA96*, *thi-1*, *hsdR17* (r⁻ₖ, m⁺ₖ), *supE44*, *recA1?*, λ⁻	A *recA⁻* host used for plating and growth of plasmids and cosmids; it is a *gyrA*, *recA* derivative of strain MM294 crossed with strain KL16-99 (Low 1968; Meselson and Yuan 1968; D. Hanahan, pers. comm.).
1046	*recA⁻*, *supE*, *supF*, *hsdS⁻*, *met⁻*	A *recA⁻* host used for plating and growth of cosmids (Cami and Kourilsky 1978).
Kro	*recA⁻*	A strain used for propagation of λBF101.
SK1590	*gal*, *thi*, *sbcB15*, *endA*, *hsdR4*, *hsdM⁺*	A bacterial strain with a high transformation efficiency (Kushner 1978).
AB1157	*recA99*, *thr-1*, *leu-6*, *thi-1*, *lacY1*, *galK2*, *ara-14*, *xyl-5*, *mtl-1*, *proA2*, *his-4*, *argE3*, *str-31*, *tsx-33*	A strain used in the πVX screening procedure to test for bacteriophage containing an amber suppressor gene (B. Seed, pers. comm.).
NK5486	*thyA⁻*, *rha⁻*, *Smʳ*, *lacZ*am	A strain used in the πVX screening procedure to test for bacteriophage containing an amber suppressor gene; also known as J6139 (Gross and Gross 1969; B. Seed, pers. comm.).

Strain	Genotype	Remarks
K802	*hsdR*$^+$, *hsdM*$^+$, *gal*$^-$, *met*$^-$, *supE*	An *su*$^+$ strain commonly used to propagate bacteriophage λ vectors and their recombinants (Wood 1966).
W3110 r$^-$ m$^+$	F$^-$, *hsdR*$^-$, *hsdM*$^+$	An *su*$^+$ strain, derived from strain W3110 of J. Lederberg (Campbell et al. 1978).
W3110 r$^-$ m$^+$ (p3)	F$^-$, *hsdR*$^-$, *hsdM*$^+$, p3 [*kan*^{r}, *tet*^{s}, *amp*^{s} (*tet*$^r_{am}$, *amp*$^r_{am}$)]	A derivative of W3110 r$^-$m$^+$ containing a *tra*$^-$ mutant of the doubly mutated RP1 plasmid derived from phHM$_2$ (Mindich et al. 1978).
W3110 r$^-$ m$^+$ (p3)(πVX)	F$^-$, *hsdR*$^-$, *hsdM*$^+$, p3, πVX (*kan*^{r}, *tet*^{r}, *amp*^{r})	This derivative of W3110 r$^-$m$^+$ (p3) contains the miniplasmid πVX, which carries *supF* (B. Seed, unpubl.).
Q358	*hsdR*^-_k, *hsdM*^+_k, *supF*, *φ80*^{r}	An *su*$^+$ host used for growth of the bacteriophage λ vector 1059 (Karn et al. 1980).
Q359	*hsdR*^-_k, *hsdM*^+_k, *supF*, *φ80*, P2	An *su*$^+$ host used to detect Spi$^-$ recombinants of bacteriophage λ (Karn et al. 1980).
BHB2688	(N205 *recA*$^-$ [*γimm*^{434}, *cIts*, *b2*, *red*$^-$, *E*am, *S*am]/*γ*])	A *γ* lysogen used to prepare packaging extracts (Hohn and Murray 1977; Hohn 1979).
BHB2690	(N205 *recA*$^-$ [*γimm*^{434}, *cIts*, *b2*, *red*$^-$, *D*am, *S*am/*γ*])	A *γ* lysogen used to prepare packaging extracts (Hohn and Murray 1977; Hohn 1979).
JM103	Δ (*lac pro*), *thi*, *strA*, *supE*, *endA sbcB*, *hsdR*$^-$, F′*traD36*, *proAB*, *lacI*^{q}, *Z* Δ*M15*	A host for growth of the single-stranded phage M13 and its recombinants (Messing et al. 1981).

Unless noted otherwise, all strains listed above are *E. coli* K-12.

References

Aaij, C. and P. Borst. 1972. The gel electrophoresis of DNA. *Biochim. Biophys. Acta* **269**: 192.

Agarwal, K.-L., A. Yamakazi, P.J. Fashion, and H.G. Khorana. 1972. Chemical synthesis of polynucleotides. *Agnew. Chem. Int. Ed. Engl.* **11**: 451.

Akusjärvi, G. and U. Pettersson. 1978. Nucleotide sequence at the junction between the coding region of the adenovirus 2 hexon messenger RNA and its leaser sequence. *Proc. Natl. Acad. Sci.* **75**: 5822.

Allett, B., P.G.N. Jeppesen, K.J. Katagiri, and H. Delius. 1973. Mapping the DNA fragments produced by cleavage of λ DNA with exonuclease RI. *Nature* **241**: 120.

Alwine, J.C., D.J. Kemp, and G.R. Stark. 1977. Method for defection of specific RNAs in agarose gels by transfer to diazobenzyloxymethl-paper and hybridization with DNA probes. *Proc. Natl. Acad. Sci.* **74**: 5350.

Anderson, S. 1981. Shotgun DNA sequencing using cloned DNAse I-generated fragments. *Nucleic Acids Res.* **9**: 3015.

Anderson, S., M.J. Gart, L. Mayol, and I.G. Young. 1980. A short primer for sequencing DNA cloned in the single-strand phage vector M13mp2. *Nucleic Acids Res.* **8**: 1731.

Appleyard, R.K. 1954. Segregation of new lysogenic types during growth of a doubly lysogenic strain derived from *Escherichia coli* K12. *Genetics* **39**: 440.

Armstrong, K.A., V. Hershfield, and D.R. Helinski. 1972. Gene cloning and containment properties of plasmid col E1 and its derivatives. *Science* **196**: 172.

Arnheim, M. and M. Kuehn. 1979. The genetic behavior of a cloned mouse ribosomal DNA segment mimics mouse ribosomal gene evolution. *J. Mol. Biol.* **134**: 743.

Aviv, H. and P. Leder. 1972. Purification of biologically active globin messenger RNA by chromotography on oligothymidylic acid-cellulose. *Proc. Natl. Acad. Sci.* **69**: 1408.

Backman, K. 1980. A cautionary note on the use of certain restriction endonucleases with methylated substrates. *Gene* **11**: 169.

Backman, K. and M. Ptashne. 1978. Maximizing gene expression on a plasmid using recombination in vitro. *Cell* **13**: 65.

Bahl, C.P. and R. Wu. 1978. Cloned seventeen-nucleotide-long synthetic lactose operator is biologically active. *Gene* **3**: 123.

Bahl, C.P., K.J. Marians, R. Wu, J. Slawinsky, and S.A. Narang. 1976. A. general method for inserting specific DNA sequences. *Gene* **1**: 81.

Bailey, J.M. and N. Davidson. 1976. Methylmercury as a reversible denaturing agent for agarose gel electrophoresis. *Anal. Biochem.* **70**: 75.

Barnes, W.M. 1977. Plasmid detection and sizing in single colony lysates. *Science* **195**: 393.

Becker, A. and M. Gold. 1975. Isolation of the bacteriophage lambda *A*-gene protein. *Proc. Natl. Acad. Sci.* **72**: 581.

Benton, W.D. and R.W. Davis. 1977. Screening λgt recombinant clones by hybridization to single plaques in situ. *Science* **196**: 180.

Berger, S.L. and C.S. Birkenmeier. 1979. Inhibition of intractable nucleases with ribonucleoside-vanadyl complexes: Isolation of messenger ribonucleic acid from resting lymphocytes. *Biochemistry* **18**: 5143.

Berk, A.J. and P.A. Sharp. 1977. Sizing and mapping of early adenovirus mRNAs by gel electrophoresis of S1 endonuclease digested hybrids. *Cell* **12**: 721.

———. 1978. Structure of the adenovirus 2 early mRNAs. *Cell* **14**: 695.

Berkner, K.L. and W.R. Folk. 1977. Polynucleotide kinase exchange reaction *Eco*R1 cleavage and methylation of DNAs containing modified pyrimidines in the recognition sequence. *J. Biol. Chem.* **252**: 3176.

Bernard, H.U. and D.R. Helinski. 1980. Bacterial plasmid cloning vehicles. In *Genetic engineering* (ed. J.K. Setlow and A. Hollaender), vol. 2, p. 133. Plenum Press, New York.

Bernard, H.-M., E. Remaut, M.V. Hershfield, H.K. Das, D.R. Helinski, C. Yanofsky, and N. Franklin. 1979. Construction of plasmid cloning vehicles that promote gene expression from the bacteriophage lambda p_L promoter. *Gene* **5**: 59.

Berridge, M.V. and C.D. Lane. 1976. Translation of *Xenopus* liver messenger RNA in *Xenopus* oocytes: Vitellogenin synthesis and conversion to yolk platelet proteins. *Cell* **8**: 283.

Bird, A.P. and E.M. Southern. 1978. Use of restriction enzymes to study eukaryotic DNA methylation: I. The methylation pattern in ribosomal DNA from *Xenopus laevis*. *J. Mol. Biol.* **118**: 27.

Birnboim, H.C. and J. Doly. 1979. A rapid alkaline extraction procedure for screening recombinant plasmid DNA. *Nucleic Acids Res.* **7**: 1513.

Bittner, M. and D. Vapnek. 1981. Versatile cloning vectors derived from the runaway-replication vector pKM402. *Gene* **15**: 31.

Blattner, F.R., B.G. Williams, A.E. Blechl, K. Denniston-Thompson, H.E. Faber, L.-A. Furlong, D.J. Grunwald, D.O. Kiefer, D.D. Moore, E.L. Sheldon, and O. Smithies. 1977. Charon phages: Safer derivatives of bacteriophage lambda for DNA cloning. *Science* **196**: 161.

Blin, N. and D.W. Stafford. 1976. Isolation of high-molecular-weight DNA. *Nucleic Acids Res.* **3**: 2303.

Bochner, B.R., H.C. Huang, G.L. Schieven, and B.N. Ames. 1980. Positive selection for loss of tetracycline resistance. *J. Bacteriol.* **143**: 926.

Bolivar, F. and K. Backman. 1979. Plasmids of *Escherichia coli* as cloning vectors. *Methods Enzymol.* **68**: 245.

Bolivar, F., R.L. Rodriguez, P.J. Greene, M.C. Betlach, H.L. Heynecker, H.W. Boyer, J.H. Crosa, and S. Falkow. 1977. Construction and characterization of new cloning vehicles. II. A multipurpose cloning system. *Gene* **2**: 95.

Bollum, F.J. 1974. Terminal deoxynucleotidyl transferase. *The enzymes* **10**: 145.

Bonner, W.M. and R.A. Laskey. 1974. A film detection method for tritium-labelled proteins and nucleic acids in polyacrylamide gels. *Eur. J. Biochem.* **46**: 83.

Bonner, T.I., D.J. Brenner, B.R. Neufeld, and R.J. Britten. 1973. Reduction in the rate of DNA reassociation by sequence divergence. *J. Mol. Biol.* **81**: 123.

Botchan, M., G. McKenna, and P.A. Sharp. 1974. Cleavage of mouse DNA by a restriction enzyme as a clue to the arrangement of genes. *Cold Spring Harbor Symp. Quant. Biol.* **38**: 383.

Botchan, M., W. Topp, and J. Sambrook. 1976. The arrangement of simian virus 40 sequences in the DNA of transformed cells. *Cell* **9**: 269.

Boyer, H.W. and D. Roulland-Dussoix. 1969. A complementation analysis of the restriction and modification of DNA in *Escherichia coli*. *J. Mol. Biol.* **41**: 459.

Brawerman, G., J. Mendecki, and S.Y. Lee. 1972. A procedure for the isolation of mammalian messenger ribonucleic acid. *Biochemistry* **11**: 637.

Brent, R. and P. Ptashne. 1981. Mechanism of action of the *lexA* gene product. *Proc. Natl. Acad. Sci.* **78**: 4204.

Broome, S. and W. Gilbert. 1978. Immunological screening method to detect specific translation products. *Proc. Natl. Acad. Sci.* **75**: 2746.

Buell, G.N., M.P. Wickens, F. Payvar, and R.T. Schimke. 1978. Synthesis of full length cDNAs from four partially purified oviduct mRNAs. *J. Biol. Chem.* **253**: 2471.

Cami, B. and P. Kourilsky. 1978. Screening of cloned recombinant DNA in bacteria by *in situ* colony hybridization. *Nucleic Acids Res.* **5**: 2381.

Campbell, A. 1971. Genetic structure. In *The bacteriophage lambda* (ed. A.D. Hershey), p. 13. Cold Spring Harbor Laboratory, Cold Spring Harbor, New York.

Campbell, J., C.C. Richardson, and F.W. Studier. 1978. Genetic recombination and complementation between bacteriophage T7 and cloned fragments of T7 DNA. *Proc. Natl. Acad. Sci.* **75**: 2276.

Casey, J. and N. Davidson. 1977. Rates of formation and thermal stabilities of RNA:DNA and DNA:DNA duplexes at high concentrations of formamide. *Nucleic Acids Res.* **4**: 1539.

Cattaneo, R., J. Gorski, and B. Mack. 1981. Cloning of multiple copies of immunoglobulin variable kappa genes in cosmid vectors. *Nucleic Acids Res.* **9**: 2777.

Chaconas, G. and J.H. van de Sande. 1980. 5'-^{32}P Labeling of RNA and DNA restriction fragments. *Methods Enzymol.* **65**: 75.

Chan, S.J., B.E. Noyes, K.L. Agarwal, and D.F. Steiner. 1979. Construction and selection of recombinant plasmids containing full-length complementary DNAs corresponding to rat insulins I and II. *Proc. Natl. Acad. Sci.* **76**: 5036.

Chang, A.C.Y. and S.N. Cohen. 1977. Genome construction between bacterial species in vitro: Replication and expression of *Staphylococcus* plasmid gene in *Escherichia coli*. *Proc. Natl. Acad. Sci.* **71**: 1030.

———. 1978. Construction and characterization of amplifiable multicopy DNA cloning vehicles derived from the P15A cryptic miniplasmid. *J. Bacteriol.* **134**: 1141.

Chang, A.C.Y., J.H. Nunberg, R.J. Kaufman, H.A. Erlich, R.T. Schimke, and S.N. Cohen. 1978. Phenotypic expression in *E. coli* of a DNA sequence coding for mouse dihydrofolate reductase. *Nature* **257**: 617.

Charnay, P., M. Perricaudet, F. Galibert, and P. Tiollais. 1978. Bacteriophage lambda and plasma vectors, allowing fusion of cloned genes in each of the three translational phases. *Nucleic Acids Res.* **5**: 4479.

Charnay, P., M. Gervais, A. Louise, F. Galibert, and P. Tiollais. 1980. Biosynthesis of hepatitis B virus surface antigen in *Escherichia coli*. *Nature* **286**: 893.

Chase, J.W. and C.C. Richardson. 1964. Exonuclease VII of *Escherichia coli. J. Biol. Chem.* **249**: 4545.

Chirgwin, J.M., A.E. Przybyla, R.J. MacDonald, and W.J. Rutter. 1979. Isolation of biologically active ribonucleic acid from sources enriched in ribonuclease. *Biochemistry* **18**: 5294.

Clarke, L. and J. Carbon. 1976. A colony bank containing synthetic ColE1 hybrid plasmids representative of the entire *E. coli* genome. *Cell* **9**: 91.

Clarke, L., R. Hitzeman, and J. Carbon. 1979. Selection of specific clones from colony banks by screening with radioactive antibody. *Methods Enzymol.* **68**: 436.

Clewell, D.B. 1972. Nature of Col E₁ plasmid replication in *Escherichia coli* in the presence of chloramphenicol. *J. Bacteriol.* **110**: 667.

Clewell, D.B. and D.R. Helinski. 1972. Effect of growth conditions on the formation of the relaxation complex of supercoiled ColE1 deoxribonucleic acid and protein in *Escherichia coli. J. Bacteriol.* **110**: 1135.

Cohen, S.N. and A.C.Y. Chang. 1973. Recircularization and autonomous replication of a sheared R-factor DNA segment in *Escherichia coli* transformants. *Proc. Natl. Acad. Sci.* **70**: 1293.

———. 1977. Revised interpretation of the origin of the pSC101 plasmid. *J. Bacteriol.* **132**: 734.

Cohen, S.N., A.C.Y. Chang, and L. Hsu. 1973a. Non-chromosomal antibiotic resistance in bacteria: Genetic transformation of *Escherichia coli* by R-factor DNA. *Proc. Natl. Acad. Sci.* **69**: 2110.

Cohen, S.N., A.C.Y. Chang, H.W. Royer, and R.B. Helling. 1973b. Construction of biologically functional bacterial plasmids *in vitro. Proc. Natl. Acad. Sci.* **70**: 3240.

Collins, J. and B. Hohn. 1978. Cosmids: A type of plasmid gene-cloning vector that is packageable *in vitro* in bacteriophage λ heads. *Proc. Natl. Acad. Sci.* **75**: 4242.

Colman, A. and J. Morser. 1979. Export of proteins from oocytes of *Xenopus laevis. Cell* **17**: 517.

Colman, A., C.D. Lane, R. Craig, A. Boulton, T. Mohun, and J. Morser. 1981. The influence of topology and glycosylation on the fate of heterologous secretory proteins made in *Xenopus* oocytes. *Eur. J. Biochem.* **113**: 339.

Covey, C., D. Richardson, and J. Carbon. 1976. A method for the deletion of restriction sites in bacterial plasmid deoxyribonucleic acid. *Mol. Gen. Genet.* **145**: 155.

Cox, R.A. 1968. The use of guanidinium chloride in the isolation of nucleic acids. *Methods Enzymol.* **12B**: 120.

Curtiss, R., III, M. Inoue, D. Pereira, J.C. Hsu, L. Alexander, and L. Rock. 1977. Construction in use

of safer bacterial host strains for recombinant DNA research. In *Molecular cloning of recombinant DNA* (ed. W.A. Scott and R. Werner), p. 248. Academic Press, New York.

Dagert, M. and S.D. Ehrlich. 1979. Prolonged incubation in calcium chloride improves the competence of *Escherichia coli* cells. *Gene* **6**: 23.

Davidson, E. 1976. *Gene activity in early development.* Academic Press, New York.

Davidson, J.N. 1972. *The biochemistry of nucleic acids* (7th ed.). Academic Press, New York.

Davis, A.R., D.P. Nayak, M. Ueda, A.L. Hiti, D. Dowbenko, and D.G. Kleid. 1981. Expression of antigenic determinants of the hemagglutinin gene of a human influenza virus in *Escherichia coli. Proc. Natl. Acad. Sci.* **78**: 5376.

de Boer, H.A., L.J. Comstock, D.G. Yansura, and H.L. Heynecker. 1982. Construction of a tandem *trp-lac* promoter and a hybrid *trp-lac* promoter for efficient and controlled expression of the human growth hormone gene in *Escherichia coli*. In *Promoter structure and function* (ed. R.L. Rodriguez and M.J. Chamberlain). Praegar Publishers, New York. (In press.)

de Crombrugghe, B., B. Chen, M. Gottesman, I. Pastan, H.E. Varmus, M. Emmer, and R. Perlman. 1971. Regulation of *lac* mRNA synthesis in a soluble cell-free system. *Nat. New Biol.* **230**: 37.

de Martynoff, G., E. Pays, and G. Vassart. 1980. The synthesis of a full-length DNA complementary to thyroglobulin 33S mRNA. *Biochem. Biophys. Res. Commun.* **93**: 645.

Denhardt, D.J., D. Dressler, and D.S. Ray, eds. 1978. *The single-stranded DNA phages.* Cold Spring Harbor Laboratory, Cold Spring Harbor, New York.

Derynck, R., E. Remaut, E. Saman, P. Stanssens, B. De Clercq, J. Contente, and W. Fiers. 1980. Expression of human fibroblast interferon gene in *Escherichia coli. Nature* **287**: 193.

de Wet, J.R., D.L. Daniels, J.L. Schroeder, B.G. Williams, K. Denniston-Thompson, D.D. Moore, and F.R. Blattner. 1980. Restriction maps for twenty-one charon vector phages. *J. Virol.* **33**: 401.

Dove, W.F. and N. Davidson. 1962. Cation effects on the denaturation of DNA. *J. Mol. Biol.* **5**: 467.

Dretzen, G., M. Bellard, P. Sassone-Corsi, and P. Chambon. 1981. A reliable method for the recovery of DNA fragments from agarose and acrylamide gels. *Anal. Biochem.* **112**: 295.

Drouin, J. 1980. Cloning of human mitochondrial DNA in *Escherichia coli. J. Mol. Biol.* **140**: 15.

Dugaiczyk, A., H.W. Boyer, and H.M. Goodman. 1975. Ligation of *Eco*RI endonuclease-generated DNA fragments into linear and circular structures. *J. Mol. Biol.* **96**: 171.

Dumont, J. 1972. Oogenesis in *Xenopus laevis*. *J. Morphol.* **136**: 153.

Dunn, A.R. and J. Sambrook. 1980. Mapping viral mRNAs by sandwich hybridization. *Methods Enzymol.* **65**: 468.

Dworkin, M.D. and I.B. Dawid. 1980. Use of a cloned library for the study of abundant polyA⁺ RNA during *Xenopus laevis* development. *Dev. Biol.* **76**: 449.

Edgell, M.H., S. Weaver, N. Haigwood, and C.A. Hutchison, III. 1979. Gene enrichment. In *Genetic engineering* (ed. J.K. Setlow and A. Hollander), vol. 1, p. 37. Plenum Press, New York.

Edman, J.C., R.A. Hallewell, P. Valenzuela, H.M. Goodman, and W.J. Rutter. 1981. Synthesis of hepatitis B surface and core antigens in *E. coli*. *Nature* **291**: 503.

Edmonds, M., M.H. Vaughn, Jr., and H. Nakazato. 1971. Polyadrenylic acid sequences in the heterogeneous nuclear RNA and rapidly-labeled polyribosomal RNA of HeLa cells: Possible evidence for a precursor relationship. *Proc. Natl. Acad. Sci.* **68**: 1336.

Efstratiadis, A. and L. Villa-Komaroff. 1979. Cloning of double-stranded DNA. In *Genetic engineering* (ed. J.K. Stelow and A. Hollaender), vol. 1, p. 15. Plenum Press, New York.

Efstratiadis, A., F.C. Kafatos, A.M. Maxam, and T. Maniatis. 1976. Enzymatic *in vitro* synthesis of globin genes. *Cell* **7**: 279.

Ehrlich, S.D., U. Bertazzoni, and G. Bernardi. 1973. The specificity of pancreatic deoxyribonuclease. *Eur. J. Biochem.* **40**: 143.

Faber, H.E., D. Kiefer, and F.R. Blattner. 1978. Application to the National Institutes of Health for Ek2 certification of a host-vector system for DNA cloning. Supplement X. Data on *in vitro* packaging method.

Faras, A.J. and N.A. Diebble. 1975. RNA-directed DNA synthesis by the DNA polymerase of Rous sarcoma virus: Structural and functional identification of 4S primer RNA in uninfected cells. *Proc. Natl. Acad. Sci.* **72**: 859.

Favaloro, J., R. Freisman, and R. Kamen. 1980. Transcription maps of polyoma virus-specific RNA: Analysis by two-dimensional nuclease S1 gel mapping. *Methods Enzymol.* **65**: 718.

Fedorcsak, I. and L. Ehrenberg. 1966. Effects of diethyl-pyrocarbonate and methyl methanesulfonate on nucleic acids and nucleases. *Acta Chem. Scand.* **20**: 107.

Feramisco, J.R., D.M. Helfman, J.E. Smart, K. Burridge, and G.P. Thomas. 1982. Coexistence of vinculin-like protein of higher molecular weight in smooth muscle. *J. Biol. Chem.* (in press).

Fisher, M.P. and C.W. Dingman. 1971. Role of molecular conformation in determining the electrophoretic properties of polynucleotides in agarose-acrylamide composite gels. *Biochemistry* **10**: 895.

Fraser, T.H. and B.J. Bruce. 1978. Chicken ovalbumin is synthesized and secreted by *Escherichia coli*. *Proc. Natl. Acad. Sci.* **75**: 5936.

Friedman, E.Y. and M. Rosbash. 1977. The synthesis of high yields of full-length reverse transcripts of globin mRNA. *Nucleic Acids Res.* **4**: 3455.

Frischauf, A.M., H. Garoff, and H. Lehrach. 1980. A subcloning strategy for DNA sequence analysis. *Nucleic Acids Res.* **8**: 5541.

Fritsch, E.F., R.M. Caron, and T. Maniatis. 1980. Molecular cloning and characterization of the human β-like globin gene cluster. *Cell* **19**: 959.

Fuller, F. 1981. Determination and comparative analysis of two collagen mRNA and propeptide sequences. Ph.D. thesis. Dept. of Biochemistry and Molecular Biology, Harvard University, Cambridge, Massachusetts.

Gallo, R.C., W.A. Blattner, M.S. Reitz, Jr., and T. Ito. 1982. HTLV: The virus of adult T-cell leukemia in Japan and elsewhere. *Lancet* **1**: 6830.

Gentz, R., A. Langner, A.C.Y. Chang, S.N. Cohen, and H. Bujard. 1981. Cloning and analysis of strong promoters is made possible by the downstream placement of a RNA termination signal. *Proc. Natl. Acad. Sci.* **78**: 4936.

Gergen, J.P., R.H. Stern, and P.C. Wensink. 1979. Filter replicas and permanent collections of recombinant DNA plasmids. *Nucleic Acids Res.* **7**: 2115.

Gething, M.J., J. Bye, J. Skehel, and M. Waterfield. 1980. Cloning and DNA sequence of double stranded copies of haemmagglutinin genes from H2 and H3 strains elucidates antigenic shift and drift in human influenza virus. *Nature* **287**: 301.

Ghosh, P.K., V.B. Reddy, J. Swinscoe, P. Lebowitz, and S.M. Weissman. 1978. Heterogeneity and 5′-terminal structures of the late RNAs of simian virus 40. *J. Mol. Biol.* **126**: 813.

Gilham, P.T. 1964. Synthesis of polynucleotide celluloses and their use in the fractionation of polynucleotides. *J. Am. Chem. Soc.* **86**: 4982.

Gingeras, T.R., J. Milazzo, D. Sciaky, and R.J. Roberts. 1979. Computer programs used on the sequencing of the Ad2 virus genome. *Nucleic Acids Res.* **7**: 529.

Girvitz, S.C., S. Bacchetti, A.J. Rainbow, and F.L. Graham. 1980. A rapid and efficient procedure for the purification of DNA from agarose gels. *Anal. Biochem.* **106**: 492.

Glišin, V., R. Crkvenjakov, and C. Byus. 1974. Ribonucleic acid isolated by cesium chloride centrifugation. *Biochemistry* **13**: 2633.

Godson, G.N. and D. Vapnek. 1973. A simple method

of preparing large amounts of $\phi\chi$174 RFI super-coiled DNA. *Biochim. Biophys. Acta* **299**: 516.

Goeddel, D.V., H.M. Sheppard, E. Yelverton, D. Leung, and R. Crea. 1980a. Synthesis of human fibroblast interferon by *E. coli. Nucleic Acids Res.* **8**: 4057.

Goeddel, D.V., H.L. Heyneker, T. Hozumi, R. Arentzen, K. Itakura, D.G. Yansura, M.J. Ross, G. Miozzari, R. Crea, and P.H. Seeburg. 1979a. Direct expression in *Escherichia coli* of a DNA sequence coding for human growth hormone. *Nature* **281**: 544.

Goeddel, D.V., D.G. Kleid, F. Bolivar, H.L. Heyneker, D.G. Yansura, R. Crea, T. Hirose, A. Kraszewski, K. Itakura, and A. Riggs. 1979b. Expression in *Escherichia coli* of chemically synthesized genes for human insulin. *Proc. Natl. Acad. Sci.* **76**: 106.

Goeddel, D.V., E. Yelverton, A. Ullrich, H.L. Heynecker, G. Miozzari, W. Holmes, P.H. Seeburg, T. Dull, L. May, N. Stebbing, R. Crea, S. Maeda, R. McCandliss, A. Sloma, J.M. Tabor, M. Gross, P.C. Familletti, and S. Pestka. 1980b. Human leukocyte interferon produced by *E. coli* is biologically active. *Nature* **287**: 411.

Goff, S. and P. Berg. 1978. Excision of DNA segments introduced into cloning vectors by the poly-dA·dT joining method. *Proc. Natl. Acad. Sci.* **75**: 1767.

Gold, L., D. Pribnow, T. Schneider, S. Shinedling, B.S. Singer, and G. Stromo. 1981. Translational initiation in prokaryotes. *Annu. Rev. Microbiol.* **35**: 365.

Goldberg, D.A. 1980. Isolation and partial characterization of the *Drosphila* alcohol dehydrogenase gene. *Proc. Natl. Acad. Sci.* **77**: 5794.

Goldberg, M.L., R.P. Lefton, G.R. Stark, and J.G. Williams. 1979. Isolation of specific RNAs using DNA covalently linked to diazobenzyloxymethyl-cellulose on paper. *Methods Enzymol.* **68**: 206.

Greene, P.J., M.S. Poonian, A.L. Nussbaum, L. Tobias, D.E. Garfin, H.W. Boyer, and H.M. Goodman. 1975. Restriction and modification of a self complementary substrate octa nucleotide containing the Eco RI. *J. Mol. Biol.* **99**: 237.

Gronenborn, B. and J.N. Messing. 1978. Methylation of single-stranded DNA *in vitro* introduces new restriction endonuclease cleavage sites. *Nature* **272**: 375.

Grosveld, F.G., H.-H.M. Dahl, E. deBoer, and R.A. Flavell. 1981. Isolation of β-globin-related genes from a human cosmid library. *Gene* **13**: 227.

Gross, J. and M. Gross. 1969. Genetic analysis of an *E. coli* strain with a mutation affecting DNA polymerase. *Nature* **224**: 1166.

Grunstein, M. and D. Hogness. 1975. Colony hybridi-

zation: A method for the isolation of cloned DNAs that contain a specific gene. *Proc. Natl. Acad. Sci.* **72**: 3961.

Guarente, L., T.M. Roberts, and M. Ptashne. 1980a. A technique for expressing eukaryotic genes in bacteria. *Science* **209**: 1428.

Guarente, L., G. Lauer, T.M. Roberts, and M. Ptashne. 1980b. Improved methods for maximizing expression of a cloned gene: A bacterium that synthesizes rabbit β-globin. *Cell* **20**: 543.

Gurdon, J.B. 1974. *The control of gene expression in animal development*. Clarendon Press, Oxford.

Gurdon, J.B., J.B. Lingrel, and G. Marbaiz. 1973. Message stability in injected frog oocytes: Long life of mammalian α and β globin messages. *J. Mol. Biol.* **80**: 539.

Gurdon, J.B., C.D. Lane, H.R. Woodland, and G. Maraix. 1971. Use of frog eggs in oocytes for the study of mRNA and its translation in living cells. *Nature* **233**: 177.

Hall, M.N. and T.J. Silhavy. 1981a. The *ompB* locus and the regulation of the major outer membrane porin proteins of *Escherichia coli. J. Mol. Biol.* **146**: 23.

———. 1981b. Genetic analysis of the *ompB* locus in *Escherichia coli* K-12. *J. Mol. Biol.* **151**: 1.

Hallewell, R.A. and S. Emtage. 1980. Plasmid vectors containing the tryptophan operon promoter suitable for efficient regulated expression of foreign genes. *Gene* **9**: 27.

Hanahan, D. and M. Meselson. 1980. A protocol for high density screening of plasmids in χ1776. *Gene* **10**: 63.

Hardies, S.C. and R.D. Wells. 1976. Preparative fractionation of DNA restriction fragments by reverse phase column chromatography. *Proc. Natl. Acad. Sci.* **73**: 3117.

Harpold, M.M., P.R. Dobner, R.M. Evans, and F.C. Bancroft. 1978. Construction and identification by positive hybridization translation of a bacterial plasmid containing a rat hormone structural gene sequence. *Nucleic Acids Res.* **5**: 2039.

Hattman, S., J.E. Brooks, and M. Masurekar. 1978. Sequence specificity of the P1 modification methylase (M·Eco P1) and the DNA methylase (M·Eco dam) controlled by the *Escherichia coli dam* gene. *J. Mol. Biol.* **126**: 367.

Hayashi, K. 1980. A cloning vehicle suitable strand separation. *Gene* **11**: 109.

Hayward, G.S. 1972. Gel electrophoretic separation of the complementary strands of bacteriophage DNA. *Virology* **49**: 342.

Hedgpeth, J., M. Ballivet, and H. Eisen. 1978. Lambda phage promoter used to enhance expression of a plasmid-cloned gene. *Mol. Gen. Genet.* **163**: 197.

Heiland, I. and M.-J. Gething. 1981. Cloned copy of the haemagglutinin gene codes for human influenza antigenic determinants in *E. coli. Nature* **292**: 851.

Helling, R.B., H.M. Goodman, and H.W. Boyer. 1974. Analysis of R.*Eco*RI fragments of DNA from lambdoid bacteriophages and other viruses by agarose-gel electrophoresis. *J. Virol.* **14**: 1235.

Hendrix, R.W., J.W. Roberts, F.W. Stahl, and R.A. Weisberg. 1982. *Lambda II.* Cold Spring Harbor Laboratory, Cold Spring Harbor, New York. (In press.)

Hershfield, V., H.W. Boyer, C. Yanofsky, M.A. Lovett, and D.R. Helinski. 1974. Plasmid ColE1 as a molecular vehicle for cloning and amplification of DNA. *Proc. Natl. Acad. Sci.* **71**: 3455.

Herskowitz, I. 1973. Control of gene expression in bacteriophage lambda. *Annu. Rev. Genet.* **7**: 289.

Hofstetter, H., A. Schambock, J. VandenBerg, and C. Weissman. 1976. Specific excision of the inserted DNA segment from hybrid plasmids constructed by the poly(dA)·poly(dT) method. *Biochim. Biophys. Acta* **454**: 587.

Hohn, B. 1979. *In vitro* packaging of λ and cosmid DNA. *Methods Enzymol.* **68**: 299.

Hohn, B. and J. Collins. 1980. A small cosmid for efficient cloning of large DNA fragments. *Gene* **11**: 291.

Hohn, B. and K. Murray. 1977. Packaging recombinant DNA molecules into bacteriophage particles *in vitro. Proc. Natl. Acad. Sci.* **74**: 3259.

Holmes, D.S. and M. Quigley. 1981. A rapid boiling method for the preparation of bacterial plasmids. *Anal. Biochem.* **114**: 193.

Houghton, M., A.G. Stewart, S.M. Dole, J.S. Emtage, M.A.W. Eaton, J.C. Smith, T.P. Patel, H.M. Lewis, A.G. Porter, J.R. Birth, T. Cartwright, and N.H. Carey. 1980. The aminoterminal sequence of human fibroblast interferon as deduced from reverse transcripts obtained using synthetic oligonucleotide primers. *Nucleic Acids Res.* **8**: 1913.

Hsiung, H.M., R. Brosseau, J. Michnicwicz, and S.A. Narang. 1979. Synthesis of human insulin gene. I. Development of a reversed-phase chromatography in the modified triester method. Its application in the rapid and efficient synthesis of eight deoxyribonucleotide fragments constituting human insulin A DNA. *Nucleic Acids Res.* **6**: 1371.

Hutton, J.A. 1977. Renaturation kinetics and thermostability of DNA in aqueous solutions of formamide and urea. *Nucleic Acids Res.* **4**: 3537.

Ish-Horowicz, D. and J.F. Burke. 1981. Rapid and efficient cosmid vector cloning. *Nucleic Acids Res.* **9**: 2989.

Itakura, K. and A.D. Riggs. 1980. Chemical DNA synthesis and recombinant DNA studies. *Science* **209**: 1401.

Itakura, K., T. Hirose, R. Crea, A.D. Riggs, and H.L. Heynecker, F. Bolivar, and H.W. Boyer. 1977. Expression in *Escherichia coli* of a chemically synthesized gene for the hormone somatostatin. *Science* **198**: 1056.

Jackson, D.A., R.H. Symons, and P. Berg. 1972. Biochemical method for inserting new genetic information into DNA of simian virus 40: Circular SV40 DNA molecules containing lambda phage genes and the galactose operon of *Escherichia coli. Proc. Natl. Acad. Sci.* **69**: 2904.

Jacobsen, H., H. Klenow, and K. Ovargaard-Hansen. 1974. The N-terminal amino-acid sequences of DNA polymerase I from *Escherichia coli* and of the large and the small fragments obtained by a limited proteolysis. *Eur. J. Biochem.* **45**: 623.

Jaurin, B., T. Grundström, T. Edlund, and S. Normark. 1981. The *E. coli* β-lactamase attenuator mediates growth-rate-dependent regulation. *Nature* **290**: 221.

Jeffreys, A.J. and R.A. Flavell. 1977. A physical map of the DNA regions flanking the rabbit β-globin gene. *Cell* **12**: 429.

Johnson, A.D., B.J. Meyer, and M. Ptashne. 1979. Interaction between DNA-bound repressors govern regulation by λ phage repressor. *Proc. Natl. Acad. Sci.* **76**: 5061.

Johnson, P.H. and L.I. Grossman. 1977. Electrophoresis of DNA in agarose gels. Optimizing separations of conformational isomers of double and single-stranded DNAs. *Biochemistry* **16**: 4217.

Johnson, P.H. and M. Laskowski, Sr. 1970. Mung bean nuclease I. II. Resistance of double stranded deoxyribonucleic acid and susceptibility of regions rich in adenosine and thymidine to enzymatic hydrolysis. *J. Biol. Chem.* **245**: 891.

Kacian, D.L. 1977. Methods for assaying reverse transcriptase. *Methods Virol.* **6**: 143.

Kacian, D.L., S. Spiegelman, A. Banks, M. Ferada, S. Metaforda, L. Dow, and P.A. Marks. 1972. *In vitro* synthesis of DNA components of human genes for globins. *Nat. New Biol.* **235**: 167.

Kahn, M., R. Kolter, C. Thomas, D. Figursky, R. Meyer, D. Remaut, and D.R. Helinski. 1979. Plasmid cloning vehicles derived from plasmids ColE1, F, R6K, RK2. *Methods Enzymol.* **68**: 268.

Karn, J., S. Brenner, L. Barnett, and G. Cesareni. 1980. Novel bacteriophage λ cloning vector. *Proc. Natl. Acad. Sci.* **77**: 5172.

Kelley, W.S. and K.H. Stump. 1979. A rapid procedure for isolation of large quantities of *Escherichia coli* DNA polymerase I utilizing a λ pol A transducing phage. *J. Biol. Chem.* **254**: 3206.

Kelly, R.G., N. Cozzarelli, M.P. Deutscher, I.R. Lehman, and A. Kornberg. 1970. Enzymatic synthesis of deoxiribonucleic acid. *J. Biol. Chem.* **245**: 39.

Kessler, S.W. 1981. Use of protein-A-bearing staphylococci for the immune precipitation and isolation of antigens from cells. *Methods Enzymol.* **73**: 442.

Kirby, K.S. 1956. A new method for the isolation of nucleic acids from mammalian tissues. *Biochem. J.* **64**: 405.

Kleid, D.G., D. Yansura, B. Small, D. Dowbenko, D.M. Moore, M.J. Grubman, P.D. McKercher, D.O. Morgan, B.H. Robertson, and H.L. Bachrach. 1981. Cloned viral protein vaccine for foot-and-mouth disease: Responses in cattle and swine. *Science* **214**: 1125.

Klenow, H. and I. Henningsen. 1970. Selective elimination of the exonuclease activity of the deoxyribonucleic acid polymerase from *Escherichia coli* β by limited proteolysis. *Proc. Natl. Acad. Sci.* **65**: 168.

Korman, A.J., P.J. Knudsen, J.F. Kaufman, and J.L. Strominger. 1982. cDNA clones for the heavy chain of *HLA-DR* antigens obtained after immunopurification of polysomes by monoclonal antibody. *Proc. Natl. Acad. Sci.* **79**: 1844.

Kumar, A. and U. Lindberg. 1972. Characterization of messenger ribonucleoprotein and messenger RNA from KB cells. *Proc. Natl. Acad. Sci.* **69**: 681.

Küpper, H., W. Keller, C. Kurz, S. Forss, H. Schaller, R. Franze, K. Strohmaier, O. Marquardt, V.G. Zaslovsky, and P.H. Hofschneider. 1981. Cloning of cDNA of major antigen of foot-and-mouth disease virus and expression in *E. coli. Nature* **289**: 555.

Kurtz, D.T. and C.F. Nicodemus. 1981. Cloning of α2μ globulin DNA using a high efficiency technique for the cloning of trace messenger RNAs. *Gene* **13**: 145.

Kushner, S.R. 1978. An improved method for transformation of *Escherichia coli* with ColE1-derived plasmids. In *Genetic engineering* (ed. H.B. Boyer and S. Nicosia), p. 17. Elsevier/North-Holland, Amsterdam.

Laemmli, U.K. 1970. Cleavage of structural proteins during the assembly of the head of the bacteriophage T4. *Nature* **227**: 680.

Land, H.M., H. Grez, H. Hansen, W. Lindermaier, and G. Schuetz. 1981. 5'-terminal sequences of eukaryotic mRNA can be cloned with high efficiency. *Nucleic Acids Res.* **9**: 2251.

Landy, A., C. Foeller, R. Reszelback, and D. Dudock. 1976. Preparative fractionation of DNA restriction fragments by high pressure column chromotography on RPC5. *Nucleic Acids Res.* **3**: 2575.

Lane, C. and J. Knowland. 1975. The injection of mRNA into living cells: The use of frog oocytes for the assay of mRNA in the study of the control of gene expression. In *The biochemistry of animal development* (ed. R. Weber), vol. 3, p. 145. Academic Press, New York.

Laskey, R.A. 1980. The use of intensifying screens or organic scintillators for visualizing radioactive molecules resolved by gel electrophoresis. *Methods Enzymol.* **65**: 363.

Laskey, R.A. and A.D. Mills. 1977. Enhanced autoradiographic detection of ^{32}P and ^{125}I using intensifying screens and hypersensitized film. *FEBS Lett.* **82**: 314.

Laskowski, M., Sr. 1980. Purification and properties of the mung-bean nuclease. *Methods Enzymol.* **65**: 263.

Lasky, L.A., Z. Lev, J.-H. Xin, R.J. Britten, and E.H. Davidson. 1980. Messenger RNA prevalence in sea urchin embryos measured with cloned cDNAs. *Proc. Natl. Acad. Sci.* **77**: 5317.

Lau, P.P. and H.B. Gray, Jr. 1979. Extracellular nucleases of Alteromonas espejiana Bal 31. IV. The single strand specific deoxyriboendonuclease activity as a probe for regions of altered secondary structure in negatively and positively supercoiled closed circular DNA. *Nucleic Acids Res.* **6**: 331.

Lauer, J., C.-K.J. Shen, and T. Maniatis. 1980. The chromosomal arrangement of human α-like globin genes. Sequence homology and β-globin gene deletions. *Cell* **20**: 119.

Lawn, R.M., E.F. Fritsch, R.C. Parker, G. Blake, and T. Maniatis. 1978. The isolation and characterization of a linked δ- and β-globin gene from a cloned library of human DNA. *Cell* **15**: 1157.

Leder, P., D. Tiemeier, and L. Enquist. 1977. EK2 derivatives of bacteriophage lambda useful in the cloning of DNA from higher organisms: The λgt*WES* system. *Science* **196**: 175.

Legerski, R.J., J.L. Hodnett, and H.B. Gray, Jr. 1978. Extracellular nucleases of pseudomonas *Bal31* III. Use of the double-strand deoxyriboexonuclease activity as the basis of a convenient method for the mapping of fragments of DNA produced by cleavage with restriction enzymes. *Nucleic Acids Res.* **5**: 1445.

Lehrach, H., D. Diamond, J.M. Wozney, and H. Boedtker. 1977. RNA molecular weight determinations by gel electrophoresis under denaturing conditions, a critical reexamination. *Biochemistry* **16**: 4743.

Lemischka, I.R., S. Farmer, V.R. Racaniello, and P.A. Sharp. 1981. Nucleotide sequence and evolution of a mammalian α-tubulin messenger RNA. *J. Mol. Biol.* **151**: 101.

Littauer, U.Z. and M. Sela. 1962. An ultracentrifugal study of the efficiency of some macromolecular inhibitors of ribonuclease. *Biochim. Biophys. Acta* **61**: 609.

Little, J.W., I.R. Lehman, and A.D. Kaiser. 1967. An

exonuclease induced by bacteriophage λ. *J. Biol. Chem.* **242**: 672.

Liu, C.-P., P.W. Tucker, J.F. Mushinski, and F.R. Blattner. 1980. Mapping of heavy gene chains for mouse immunoglobulins M and D. *Science* **209**: 1401.

Lobban, P.E. and A.D. Kaiser. 1973. Enzymatic end-to-end joining of DNA molecules. *J. Mol. Biol.* **78**: 453.

Loenen, W.A. and W.J. Brammar. 1980. A bacteriophage lambda vector for cloning large DNA fragments made with several restriction enzymes. *Gene* **10**: 249.

Low, B. 1968. Formation of merodiploids in matings with a class of Rec⁻ recipient strains of *Escherichia coli* K12. *Proc. Natl. Acad. Sci.* **60**: 160.

Maat, J. and A.J.H. Smith. 1978. A method for sequencing restriction fragments with dideoxynucleoside triphosphates. *Nucleic Acids Res.* **5**: 4537.

Maizel, J.V., Jr. 1971. Polyacrylamide gel electrophoresis of viral proteins. *Methods Virol.* **5**: 180.

Maloy, S.R. and W.D. Nunn. 1981. Selection for loss of tetracycline resistance by *Escherichia coli*. *J. Bacteriol.* **145**: 1110.

Mandel, M. and A. Higa. 1970. Calcium dependent bacteriophage DNA infection. *J. Mol. Biol.* **53**: 154.

Maniatis, T. 1980. Recombinant DNA. In *Cell biology* (ed. D.M. Prescott). Academic Press, New York.

Maniatis, T., A. Jeffrey, and D.G. Kleid. 1975. Nucleotide sequence of the rightward operator of phage λ. *Proc. Natl. Acad. Sci.* **72**: 1184.

Maniatis, T., S.G. Kee, A. Efstratiadis, and F.C. Kafatos. 1976. Amplification and characterization of a β-globin gene synthesized *in vitro*. *Cell* **8**: 1630.

Maniatis, T., R.C. Hardison, E. Lacy, J. Lauer, C. O'Connell, D. Quon, D.K. Sim, and A. Efstratiadis. 1978. The isolation of structural genes from libraries of eucaryotic DNA. *Cell* **15**: 687.

Marcus, S.L., M.J. Modak, and L.F. Cavaliere. 1974. Purification of avian myeloblastosis virus DNA polymerase by affinity chromotography on polycytidylate-agarose. *J. Virol.* **14**: 853.

Marinus, M.G. 1973. Location of DNA methylation genes on the *Escherichia coli* K-12 genetic map. *Mol. Gen. Genet.* **127**: 47.

Marinus, M.G. and N.R. Morris. 1973. Isolation of deoxyribonucleic acid methylase mutants of *Escherichia coli* K-12. *J. Bacteriol.* **114**: 1143.

Marmur, J. and P. Doty. 1962. Determination of the base composition of deoxyribonucleic acid from its thermal denaturation temperature. *J. Mol. Biol.* **5**: 109.

Mathews, J., J. Brown, and T. Hall. 1981. Phaseolin mRNA is translated to yield glycosilated polypeptides in *Xenopus* oocytes. *Nature* **294**: 175.

Maxam, A.M. and W. Gilbert. 1977. A new method for sequencing DNA. *Proc. Natl. Acad. Sci.* **74**: 560.

———. 1980. Sequencing end-labeled DNA with base-specific chemical cleavages. *Methods Enzymol.* **65**: 499.

Maxwell, I.H., F. Maxwell, and W.E. Hahn. 1977. Removal of RNase activity from DNase by affinity chromotography on agarose-coupled aminophenylphosphoryl-uridine 2′ (3′)-phosphate. *Nucleic Acids Res.* **4**: 241.

May, M.S. and S. Hattman. 1975. Analysis of bacteriophage deoxyribonucleic acid sequences methylated by host- and R-factor-controlled enzymes. *J. Bacteriol.* **123**: 768.

McClelland, M. 1981. Purification and characterization of two new modification methylases; M*Cla*I from *Caryophanon latum* L and M*Taq*I from *Thermus aquaticus* YTI. *Nucleic Acids Res.* **9**: 6795.

McConaughy, B.L. C.D. Laird, and B.J. McCarthy. 1969. Nucleic acid reassociation in formamide. *Biochemistry* **8**: 3289.

McDonnell, M.W., M.N. Simon, and F.W. Studier. 1977. Analysis of restriction fragments of T7 DNA and determination of molecular weights by electrophoresis in neutral and alkaline gels. *J. Mol. Biol.* **110**: 119.

McMaster, G.K. and G.G. Carmichael. 1977. Analysis of single and double-stranded nucleic acids on polyacrilamide and agarose gels by using glyoxal and acridine orange. *Proc. Natl. Acad. Sci.* **74**: 4835.

Melgar, E. and D.A. Goldthwaite. 1968. Deoxyribonucleic acid nucleases. II. The effects of metals on the mechanism of action of deoxyribonuclease I. *J. Biol. Chem.* **243**: 4409.

Meselson, M. and R. Yuan. 1968. DNA restriction enzyme from *E. coli*. *Nature* **217**: 1110.

Messing, J., R. Crea, and P.H. Seeburg. 1981. A system for shotgun DNA sequencing. *Nucleic Acids Res.* **9**: 309.

Meyerowitz, E.M., G.M. Guild, L.S. Prestidge, and D.S. Hogness. 1980. A new cosmid vector and its use. *Gene* **11**: 271.

Miller, J.H., ed. 1972. *Experiments in molecular genetics.* Cold Spring Harbor Laboratory, Cold Spring Harbor, New York.

Miller, J.H. and W.S. Reznikoff, eds. 1978. *The operon.* Cold Spring Harbor Laboratory, Cold Spring Harbor, New York.

Mindich, L., J. Cohen, and M. Weisburd. 1978. Isolation of nonsense suppressor mutants in *Pseudomonas*. *J. Bacteriol.* **126**: 177.

Miyoshi, I., M. Fujishita, H. Taguchi, Y. Ohtsuki, T. Akogi, Y.M. Morimoto, and A. Nagasaki. 1982. Caution against blood transfusions from donors seropositive to adenovirus T-cell leukaemia-associated antigens. *Lancet* **1**: 683.

Molloy, G.R., M.B. Sporn, D.W. Kelley, and R.P. Perry. 1972. Localization of polyadrenylic acid sequences in messenger ribonucleic acid of mammalian cells. *Biochemistry* **11**: 3256.

Montgomery, D.L., B.D. Hall, S. Gillam, and M. Smith. 1978. Identification and isolation of the yeast cytochrome c gene. *Cell* **14**: 673.

Morse, D.E., R.D. Mostellar, and C. Yanofsky. 1970. Dynamics of synthesis, translation, and degradation of *trp* operon messenger RNA in *E. coli*. *Cold Spring Harbor Symp. Quant. Biol.* **34**: 725.

Mount, D.W. 1971. Isolation and genetic analysis of a strain of *Escherichia coli* K-12 with an amber *recA* mutation. *J. Bacteriol.* **107**: 388.

Murray, K. 1977. Applications of bacteriophage lambda in recombinant RNA research. In *Molecular cloning of recombinant DNA* (ed. S. Werner), vol. 13, p. 133. Academic Press, New York.

Murray, N.E. and K. Murray. 1974. Manipulation of restriction targets in phage λ to form receptor chromosomes for DNA fragments. *Nature* **251**: 476.

Murray, K. and N.E. Murray. 1975. Phage lambda receptor chromosomes for DNA fragments made with restriction endonuclease III of *Haemophilus influenzae* and restriction endonuclease I of *Escherichia coli*. *J. Mol. Biol.* **98**: 551.

Murray, N.E., W.J. Brammer, and K. Murray. 1977. Lamboid phages that simplify the recovery of *in vitro* recombinants. *Mol. Gen. Genet.* **150**: 53.

Myers, J.C., F. Ramirez, D.L. Kacian, M. Flood, and S. Spiegelman. 1980. A simple purification of avian myeloblastosis virus reverse transcriptase for full-length transcription of 35S RNA. *Anal. Biochem.* **101**: 88.

Norgard, M.V., K. Keem, and J.J. Monohan. 1978. Factors affecting the transformation of *Escherichia coli* strain χ1776 by pBR322 plasmid DNA. *Gene* **3**: 279.

Novick, R.P., R.C. Clowes, S.N. Cohen, R. Curtiss III, N. Datta, and S. Falkow. 1976. Uniform nomenclature for bacterial plasmids: a proposal. *Bacteriol. Rev.* **40**: 168.

Noyes, B.E. and G.R. Stark. 1975. Nucleic acid hybridization using DNA covalently coupled to cellulose. *Cell* **5**: 301.

Noyes, B.E., M. Mevarech, R. Stein, and K.L. Agarwal. 1979. Detection and partial sequence analysis of gastrin mRNA by using an oligodeoxynucleotide probe. *Proc. Natl. Acad. Sci.* **76**: 1770.

O'Farrell, P. 1981. Replacement synthesis method of labeling DNA fragments. Bethesda Research Labs *Focus* **3**(3): 1.

O'Farrell, P.H., E. Kutter, and M. Nalcanishi. 1980. A restriction map of the bacteriophage T4 genome. *Mol. Gen. Genet.* **179**: 421.

Ohkubo, H., G. Vogeli, M. Mudryj, V.E. Avvedimento, M. Sullivan, I. Pastan, and B. deCrombrugghe. 1980. Isolation and characterization of overlapping genomic clones covering the chicken α2 type I collagen gene. *Proc. Natl. Acad. Sci.* **77**: 7059.

Okayama, H. and P. Berg. 1982. High-efficiency cloning of full-length cDNA. *Mol. Cell. Biol.* **2**: 161.

Palacios, R., R.D. Palmiter, and R.T. Schimke. 1972. Identification and isolation of ovalbumin-synthesizing polysomes. *J. Biol. Chem.* **247**: 2316.

Palmiter, R.D., A.K. Christensen, and R.T. Schimke. 1970. Organization of polysomes from pre-existing ribosomes in chick oviduct by a secondary administration of either estradiol or progesterone. *J. Biol. Chem.* **245**: 833.

Panayotatos, N. and K. Truong. 1981. Specific deletion of DNA sequences between preselected bases. *Nucleic Acids Res.* **9**: 5679.

Parker, R.C. and B. Seed. 1980. Two-dimensional agarose gel electrophoresis "Sea Plaque" agarose dimension. *Methods Enzymol.* **65**: 358.

Parnes, J.R., B. Velan, A. Felsenfeld, L. Ramanathan, U. Ferrini, E. Appella, and J.G. Sidman. 1981. Mouse β₂ microglobulin cDNA clones: A screening procedure for cDNA clones corresponding to rare mRNAs. *Proc. Natl. Acad. Sci.* **78**: 2253.

Paterson, B.M., B.E. Roberts, and E.L. Kuff. 1977. Structural gene identification and mapping by DNA·mRNA hybrid-arrested cell-free translation. *Proc. Natl. Acad. Sci.* **74**: 4370.

Peacock, S.L., C.M. McIver, and J.J. Monohan. 1981. Transformation of *E. coli* using homopolymer-linked plasmid chimeras. *Biochim. Biophys. Acta* **655**: 243.

Pirrotta, V., M. Ptashne, P. Chadwick, and R. Steinberg. 1971. The isolation of repressors. In *Procedures in nucleic acid research* (ed. G.L. Cantoni and D.R. Davies), vol. 2, p. 703. Harper and Row, New York.

Plucienniczak, A. and R.E. Streeck. 1981. An alternative procedure for the strand separation of DNA fragments. *FEBS Lett.* **124**: 72.

Pribnow, D. 1975. Nucleotide sequence of an RNA polymerase binding site at an early T7 promoter. *Proc. Natl. Acad. Sci.* **72**: 784.

Prives, C.L., H. Aviv, B.M. Paterson, B.E. Roberts, S. Rozenblatt, M. Revel, and E. Winocour. 1974.

Cell free translation of messenger RNA of simian virus 40: Synthesis of the major capsid protein. *Proc. Natl. Acad. Sci.* **71:** 3020.

Radloff, R., W. Bauer, and J. Vinograd. 1967. A dye-buoyant-density method for the detection and isolation of closed circular duplex DNA: The closed circular DNA in HeLa cells. *Proc. Natl. Acad. Sci.* **57:** 1514.

Rambach, A. and P. Tiollais. 1974. Bacteriophage λ having *Eco*RI endonuclease sites only in the nonessential region of the genome. *Proc. Natl. Acad. Sci.* **71:** 3927.

Rao, R.N. and S.G. Rogers. 1978. A thermoinducible λ phage–ColE1 plasmid chimera for the overproduction of gene products from cloned DNA segments. *Gene* **3:** 247.

———. 1979. Plasmid pKC7: A vector containing ten restriction endonuclease sites suitable for cloning DNA segments. *Gene* **7:** 79.

Reddy, V.B., P.K. Ghosh, P. Lebowitz, M. Piatek, and S.M. Weissman. 1979. Simian virus 40 early mRNAs. I. Genomic localization of 3′ and 5′ termini and two major splices in mRNA from transformed and lytically infected cells. *J. Virol.* **30:** 279.

Reddy, V.B., B. Thimmappaya, R. Dhar, K.N. Subramanian, B.S. Zain, J. Pan, P.K. Ghosh, M.L. Celma, and S.M. Weissman. 1978. The genome of simian virus 40. *Science* **200:** 494.

Remaut, E., P. Stanssens, and W. Fiers. 1981. Plasmid vectors for high-efficiency expression controlled by the p_L promoter of coliphage lambda. *Gene* **15:** 81.

Retzel, E.F., M.S. Collet, and A.J. Faras. 1980. Enzymatic synthesis of deoxyribonucleic acid by the avian retrovirus reverse transcriptase in vitro: Optimum conditions required for transcription of large ribonucleic acid templates. *Biochemistry* **19:** 513.

Ricciardi, R.P., J.S. Miller, and B.E. Roberts. 1979. Purification and mapping of specific mRNAs by hybridization selection and cell free translation. *Proc. Natl. Acad. Sci.* **76:** 4927.

Richardson, C.C. 1971. Polynucleotide kinase from *Escherichia coli* infected with bacteriophage T4. *Proc. Nucleic Acid Res.* **2:** 815.

Richardson, C.C., C.L. Schildkraut, H. Vasken Aposhian, and A. Kornberg. 1964. Enzymatic synthesis of deoxyribonucleic acid. *J. Biol. Chem.* **239:** 222.

Rigby, P.W.J., M. Dieckmann, C. Rhodes, and P. Berg. 1977. Labeling deoxyribonucleic acid to high specific activity *in vitro* by nick translation with DNA polymerase I. *J. Mol. Biol.* **113:** 237.

Rimm, D.L., D. Horness, J. Kucera, and F.R.

Blattner. 1980. Construction of coliphage lambda Charon vectors with *Bam*Hl cloning sites. *Gene* **12:** 301.

Robert-Guroff, M., Y. Nakao, K. Notake, Y. Ito, A. Sliski, and R.C. Gallo. 1982. Natural antibodies to human retrovirus HTLV in a cluster of Japanese patients with adenovirus T-cell leukemia. *Science* **215:** 975.

Roberts, R. 1982. Restriction and modification enzymes and their recognition sequences. *Nucleic Acids Res.* **10:** R117.

Roberts, T.M., R. Kacich, and M. Ptashne. 1979a. A general method for maximizing the expression of a cloned gene. *Proc. Natl. Acad. Sci.* **76:** 760.

Roberts, T.M., I. Bikel, R.R. Yocum, D.M. Livingston, and M. Ptashne. 1979b. Synthesis of simian virus 40 t antigen in *Escherichia coli*. *Proc. Natl. Acad. Sci.* **76:** 5596.

Roberts, T.M., S.L. Swanberg, A. Poteete, G. Riedel, and K. Backman. 1980. A plasmid cloning vehicle allowing a positive selection for inserted fragments. *Gene* **12:** 123.

Rogers, S.G. and B. Weiss. 1980. Exonuclease III of *Escherichia coli* K-12, an AP endonuclease. *Methods Enzymol* **65:** 201.

Rosenberg, M. and D. Court. 1979. Regulatory sequences involved in the promotion and termination of RNA transcription. *Annu. Rev. Genet.* **13:** 319.

Rothstein, R.J., L.F. Lau, C.P. Bahl, S.A. Narang, and R. Wu. 1980. Synthetic adaptors for cloning DNA. *Methods Enzymol.* **68:** 98.

Rougeon, F. and B. Mach. 1976. Stepwise biosynthesis *in vitro* of globin genes from globin mRNA by DNA polymerase of avian myeloblastosis virus. *Proc. Natl. Acad. Sci.* **73:** 3418.

Rougeon, F., P. Kourilsky and B. Mach. 1975. Insertion of a rabbit β-globin gene sequence into an *E. coli* plasmid. *Nucleic Acids Res.* **2:** 2365.

Rowekamp, W. and R.A. Firtel. 1980. Isolation of developmentally regulated genes from *Dictyostelium*. *Dev. Biol.* **79:** 409.

Royal, A., A. Garapin, B. Cami, F. Perrin, J.L. Mandel, M. LeMeur, F. Bregegigae, F. Gannon, J.P. LePennec, P. Chambon, and P. Kourelsky. 1979. The ovalbumin gene region: Common features in the organization of three genes expressed in chicken oviduct under hormonal control. *Nature* **279:** 125.

Roychoudhury, R., E. Jay, and R. Wu. 1976. Terminal labeling and addition of homopolymer tracts to duplex DNA fragments by terminal deoxynucleotidyl transferase. *Nucleic Acids Res.* **3:** 101.

Sanger, F. and A.R. Coulson. 1975. A rapid method for determining sequences in DNA by primed synthesis with DNA polymerase. *J. Mol. Biol.* **94**: 441.

Sanger, F., S. Nicklen, and A.R. Coulson. 1977. DNA sequencing with chain-terminating inhibitors. *Proc. Natl. Acad. Sci.* **74**: 5463.

Scalenghe, F., M. Buscaglia, C. Steinheil, and M. Crippa. 1978. Large scale isolation of nuclei and nucleoli from vitellogenic oocytes of *Xenopus laevis. Chromosoma* **66**: 299.

Schaffner, W. 1982. Purification of DNase I from RNase by macaloid treatment. *Anal. Biochem.* (in press).

Schechter, I. 1973. Biologically and chemically pure mRNA coding for a mouse immunoglobulin L-chain prepared with the aid of antibodies and immobilized oligothymidine. *Proc. Natl. Acad. Sci.* **70**: 2256.

Scheel, G. and P. Blackburn. 1979. Role of mammalian RNAse inhibitor in cell-free protein synthesis. *Proc. Natl. Acad. Sci.* **46**: 4898.

Scheller, R.H., R.E. Dickerson, H.W. Boyer, A.D. Riggs, and K. Itakus. 1977. Chemical synthesis of restriction enzyme recognition sites useful for cloning. *Science* **196**: 177.

Schutz, G., S. Kieval, B. Groner, A.E. Sippel, D.T. Kurtz, and P. Feigelson. 1977. Isolation of specific messenger RNA by adsorption to matrix-bound antibody. *Nucleic Acids Res.* **4**: 71.

Seeburg, P.H., J. Shine, J.A. Marshall, J.D. Baxter, and H.M. Goodman. 1977. Nucleotide sequence and amplification in bacteria of structural gene for rat growth hormone. *Nature* **220**: 486.

Seed, B. 1982. Diazotizable arylamine cellulose papers for the coupling and hybridization of nucleic acids. *Nucleic Acids Res.* **10**: 1799.

Seed, B., R. Parker, and N. Davidson. 1982. Representation of DNA in recombinant DNA partial digest libraries. *Gene* (in press).

Sela, M., C.B. Anfinsen, and W.F. Harrington. 1957. The correlation of ribonuclease activity with specific aspects of teritary structure. *Biochim. Biophys. Acta* **26**: 506.

Selker, E., K. Brown, and C. Yanofsky. 1977. Mitomycin C-induced expression of *trpA* of *Salmonella typhimurium* inserted into the plasmid ColE1. *J. Bacteriol.* **129**: 388.

Shapiro, S.Z. and J.R. Young. 1981. An immunochemical method for mRNA purification. *J. Biol. Chem.* **256**: 1495.

Sharp, P.A., B. Sugden, and J. Sambrook. 1973. Detection of two restriction endonuclease activities in *Haemophilus parainfluenzae* using analytical agarose. *Biochemistry* **12**: 3055.

Sheller, R.H., R.E. Dickerson, H.W. Boyer, A.D. Riggs, and K. Itakura. 1977. Chemical synthesis

of restriction enzyme recognition sites useful for cloning. *Science* **196**: 177.

Sheppard, H.M., E. Yelverton, and D.V. Goeddel. 1982. Increased synthesis in *E. coli* of fibroblast and leukocyte interferons through alterations in ribosome-binding sites. *DNA* **1**: 125.

Shimatake, H. and M. Rosenberg. 1981. Purified λ regulatory protein *c*II positively activates promoters for lysogenic development. *Nature* **292**: 128.

Shine, J. and L. Dalgarno. 1975. Determinant of cistron specificity in bacterial ribosomes. *Nature* **254**: 34.

Shine, J., I. Fettes, N.C.Y. Lan, J.L. Roberts, and J.D. Baxter. 1980. Expression of cloned β-endorphin gene sequences by *Escherichia coli. Nature* **285**: 456.

Sim, G.K., S.C. Kafatos, C.W. Jones, M.D. Koehler, A. Efstratiadis, and T. Maniatis. 1979. Use of a cDNA library for studies on evolution and developmental expression of chorion multigene family. *Cell* **18**: 1303.

Sippel, A.E. 1973. Purification and characterization of adenosine triphosphate: Ribonucleic acid adenyltransferase from *Esherichia coli. Eur. J. Biochem.* **37**: 31.

Slutsky, A.M., P.M. Rabinovich, L.Z. Yakubov, I.V. Sineokaya, A.I. Stepanov, and V.K. Gordeyev. 1980. Direct selection of DNA inserts in plasmic gene of kanamycin resistance. *Mol. Gen. Genet.* **180**: 487.

Smith, H.O. 1980. Recovery of DNA from gels. *Methods Enzymol.* **65**: 371.

Smith, H.O. and M.L. Bernstiel. 1976. A simple method for DNA restriction site mapping. *Nucleic Acids Res.* **3**: 2387.

Smithies, O., A.E. Blechl, K. Denniston-Thompson, N. Newell, J.E. Richards, J.L. Slightom, B.W. Tucker, and F.R. Blattner. 1978. Cloning human fetal γ-globin and mouse α-type globin DNA: Characterization and partial sequencing. *Science* **202**: 1284.

Southern, E. 1975. Detection of specific sequences among DNA fragments separated by gel electrophoresis. *J. Mol. Biol.* **98**: 503.

———. 1980. Gel electrophoresis of restriction fragments. *Methods Enzymol.* **69**: 152.

Stahl, F.W. 1979. Special sites in generalized recombination. *Annu. Rev. Genet.* **13**: 7.

Stahl, F.W., J.M. Crasemann, and M.M. Stahl. 1975. Rec-mediated recombinational hot spot activity in bacteriophage lambda. III. Chi mutations are site mutations stimulating rec-mediated recombination. *J. Mol. Biol.* **94**: 203.

Steitz, J.A. 1979. Genetic signals and nucleotide sequences in messenger RNA. In *Biological regu-*

lation and development (ed. R.F. Goldberger), vol. 1, p. 349. Plenum Press, New York.

Stephano, J.E. and J.D. Gralla. 1982. Spacer mutations in the *lacP*[S] promoter. *Proc. Natl. Acad. Sci.* **79**: 1069.

Sternberg, N., D. Tiemeier, and L. Enquist. 1977. *In vitro* packaging of a λ dam vector containing *Eco*RI DNA fragments of *Escherichia coli* and phage P1. *Gene* **1**: 255.

St. John, T.P. and R.W. Davis. 1979. Isolation of galactose-inducible DNA sequences from *Saccaromyces cerevisiae* by differential plaque filter hybridization. *Cell* **16**: 443.

Struhl, K. and R.W. Davis. 1977. Production of a functional eukaryotic enzyme in *Escherichia coli:* Cloning and expression of the yeast structural gene for imidazole glycerolphosphate dehydratase (*his3*). *Proc. Natl. Acad. Sci.* **74**: 5255.

Struhl, K., J.R. Cameron, and R.W. Davis. 1976. Functional genetic expression of eukaryotic DNA in *Escherichia coli. Proc. Natl. Acad. Sci.* **73**: 1471.

Suggs, S.V., R.B. Wallace, T. Hirose, E.H. Kawashima, and K. Itakura. 1981. Use of synthetic oligonucleotides as hybridization probes: Isolation of cloned cDNA sequences for human β_2-microglobulin. *Proc. Natl. Acad. Sci.* **78**: 6613.

Sugino, A., H.M. Goodman, H.L. Heyneker, J. Shine, H.W. Boyer, and N.R. Cozzarelli. 1977. Interaction of bacteriophage T4 RNA and DNA ligases in joining of duplex DNA at base -paired ends. *J. Biol. Chem.* **252**: 3987.

Sutcliffe, J.G. 1978. pBR322 restriction map marked from the DNA sequence: Accurate DNA size markers up to 4361 nucleotide pairs long. *Nucleic Acids Res.* **5**: 2721.

————. 1979. Complete nucleotide sequence of the *Escherichia coli* plasmid pBR322. *Cold Spring Harbor Symp. Quant. Biol.* **43**: 77.

Swanstrom, R. and P.R. Shank. 1978. X-ray intensifying screens greatly enhance the detection by autoradiography of the radioactive isotopes ^{32}P and ^{125}I. *Anal. Biochem.* **86**: 184.

Szalay, A.A., K. Grohmann, and R.L. Sinsheimer. 1977. Separation of the complementary strands of DNA fragments on polyacrylamide gels. *Nucleic Acids Res.* **4**: 1569.

Tacon, W., N. Carey, and S. Emtage. 1980. The construction and characterization of plasmid vectors suitable for the expression of all DNA phases under the control of the *E. coli.* tryptophan promoter. *Mol. Gen. Genet.* **177**: 427.

Talmadge, K., J. Brosius, and W. Gilbert. 1981. An "internal" signal sequence directs secretion and processing of proinsulin in bacteria. *Nature* **294**: 176.

Talmadge, K., S. Stahl, and W. Gilbert. 1980. Eukaryotic signal sequence transports insulin antigen in *Escherichia coli. Proc. Natl. Acad. Sci.* **77**: 3369.

Taniguchi, T. and C. Weissman. 1978. Site-directed mutations in the initiator region of the bacteriophage Qβ coat cistron and their effect on ribosome binding. *J. Mol. Biol.* **118**: 533.

Taniguchi, T., Y. Fujii-Kuriyama, and M. Muramatsu. 1980a. Molecular cloning of human interferon cDNA. *Proc. Natl. Acad. Sci.* **77**: 4003.

Taniguchi, T., L. Guarente, T.M. Roberts, D. Kimelman, J. Douhan, III, and M. Ptashne. 1980b. Expression of the human fibroblast interferon gene in *Escherichia coli. Proc. Natl. Acad. Sci.* **77**: 5230.

Taylor, J.M., R. Illmensee, and J. Summers. 1976. Efficient transcription of RNA into DNA by avian sarcoma virus polymerase. *Biochim. Biophys. Acta* **442**: 324.

Thomas, M. and R.W. Davis. 1975. Studies on the cleavage of bacteriophage λ DNA with *Eco*RI restriction endonuclease. *J. Mol. Biol.* **91**: 315.

Thomas, M., J.R. Cameron, and R.W. Davis. 1974. Viable molecular hybrids of bacteriophage lambda and eukaryotic DNA. *Proc. Natl. Acad. Sci.* **71**: 4579.

Thomas, P.S. 1980. Hybridization of denatured RNA and small DNA fragments transferred to nitrocellulose. *Proc. Natl. Acad. Sci.* **77**: 5201.

Thorne, H.V. 1966. Electrophoretic separation of polyoma virus DNA from host cell DNA. *Virology* **29**: 234.

————. 1967. Electrophoretic characterization and fractionation of polyoma virus DNA. *J. Mol. Biol.* **24**: 203.

Tibbetts, C., U. Pettersson, K. Johansson, and L. Philpson. 1974. Relationship of mRNA from productively infected cells to the complementary strands of adenovirus type 2 DNA. *J. Virol.* **13**: 370.

Tilghman, S.M., D.C. Tiemeer, F. Polsky, M.H. Edgell, J.G. Seidman, and A. Leder. 1977. Cloning specific segments of the mammalian genome: Bacteriophage λ containing mouse globin and surrounding sequences. *Proc. Natl. Acad. Sci.* **74**: 4406.

Timberlake, W.E. 1980. Developmental gene regulation in *Aspergillus nidulans. Dev. Biol.* **78**: 497.

Tonegawa, S., C. Brack, N. Hozumi, and R. Schuller. 1977. Cloning on the immunoglobulin variable region gene from mouse embryo. *Proc. Natl. Acad. Sci.* **74**: 3518.

Treisman, R. 1980. Characterization of polyoma late mRNA leader sequences by molecular cloning and DNA sequence analysis. *Nucleic Acids Res.* **8**: 4867.

Tu, C.-P.D. and S.N. Cohen. 1980. 3-End labeling of DNA with [α-^{32}P]cordycepin-5′-triphosphate. *Gene* **10**: 177.

Twigg, A.J. and D. Sherratt. 1980. Trans-complementable copy-number mutants of plasmid ColE1. *Nature* **283**: 216.

Uhlin, B.E., S. Molin, P. Gustafsson, and K. Nordstrom. 1979. Plasmids with temperature-dependent copy number for amplification of cloned genes and their products. *Gene* **6**: 91.

Ullrich, A., J. Shine, J. Chirgwin, R. Pictet, E. Tischer, W.J. Rutter, and H.M. Goodman. 1977. Rat insulin genes: Construction of plasmids containing the coding sequences. *Science* **196**: 1313.

van der Ploeg, L.H.T. and R.A. Flavell. 1980. DNA methylation in the human γδβ-globin locus in erthyroid and nonerythroid tissues. *Cell* **19**: 947.

Vande Woude, G.F., M. Oskarsson, L.W. Enquist, S. Nomura, and P.J. Fischinger. 1979. Cloning of integrated Moloney sarcoma proviral DNA sequences in bacteriophage λ. *Proc. Natl. Acad. Sci.* **76**: 4464.

Verma, I.M. 1977. The reverse transcriptase. *Biochim. Biophys. Acta* **473**: 1.

Villa-Komaroff, L., A. Efstratiadis, S. Broome, P. Lomedico, R. Tizard, S.P. Naker, W.L. Chick, and W. Gilbert. 1978. A bacterial clone synthesizing proinsulin. *Proc. Natl. Acad. Sci.* **75**: 3727.

Vogeli, G., E.V. Avedimento, M. Sullivan, J.V. Maizel, Jr., G. Lazano, S.L. Adams, L. Pastan, and B. deCrombrugge. 1980. Isolation and characterization of genomic DNA coding for a2 type I collagen. *Nucleic Acids Res.* **8**: 1823.

Vogt, V.M. 1973. Purification and further properties of single strand specific nuclease from *Aspergillus oryzae*. *Eur. J. Biochem.* **33**: 192.

Wahl, G.M., M. Stern, and G.R. Stark. 1979. Efficient transfer of large DNA fragments from agarose gels to diazobenzloxymethal-paper and rapid hybridization by using dextran sulfate. *Proc. Natl. Acad. Sci.* **76**: 3683.

Wallace, R.B., N.J. Johnson, T. Hirose, M. Miyake, E.H. Kawashima, and K. Itakura. 1981. The use of synthetic oligonucleotides as hybridization probes. II. Hybridization of oligonucleotides of mixed sequence to rabbit β-globin DNA. *Nucleic Acids Res.* **9**: 879.

Wallace, R.B., J. Schaffer, R.F. Murphy, J. Bonner, T. Hirose, and K. Itakura. 1979. Hybridization of synthetic oligodeoxyribonucleotides to Φ × 1974 DNA: The effect of single base pair mismatch. *Nucleic Acids Res.* **6**: 3543.

Wartell, R.M. and W.S. Reznikoff. 1980. Cloning DNA restriction endonuclease fragments with protruding, single-stranded ends. *Gene* **9**: 307.

Weislander, L. 1979. A simple method to recover intact high molecular weight RNA and DNA after electrophoretic separation in low gelling temperature agarose gels. *Anal. Biochem.* **98**: 305.

Weiss, B. 1976. Endonuclease II of *Escherichia coli* is exonuclease III. *J. Biol. Chem.* **251**: 1896.

Weiss, B., and C.C. Richardson. 1967. Enzymatic breakage and joining of deoxyribonucleic acid. III. An enzyme-adenylate intermediate in the polynucleotide ligase reaction. *J. Biol. Chem.* **242**: 427.

Weiss, B., A. Jacquemin-Sablon, T.R. Live, G.C. Fareed, and C.C. Richardson. 1968. Enzymatic breakage and joining of deoxyribonucleic acid. VI. Further purification and properties of polynucleotide ligase from *Escherichia coli* infected with bacteriophage T4. *J. Biol. Chem.* **243**: 4543.

Wensink, P.C., D.J. Finnegan, J.E. Donelson, and D.S. Hogness. 1974. A system for mapping DNA sequences in the chromosomes of *Drosophila melanogaster*. *Cell* **3**: 315.

Wickens, M.P., G.N. Buell, and R.T. Schimke. 1978. Synthesis of double-stranded DNA complementary to lysozyme, ovomucord, and ovalbumin mRNAs. *J. Biol. Chem.* **253**: 2483.

Williams, B.G. and F.R. Blattner. 1979. Construction and characterization of the hybrid bacteriophage lambda charon vectors for DNA cloning. *J. Virol.* **29**: 555.

———. 1980. Bacteriophage λ vectors for DNA cloning. In *Genetic engineering* (ed. J.K. Setlow and A. Hollaender), vol. 2, p. 201. Plenum Press, New York.

Williams, J.G. 1981. The preparation and screening of a cDNA clone bank. In *Genetic engineering* (ed. R. Williamson), vol. 1, p. 2. Academic Press, New York.

Williams, J.G. and M.M. Lloyd. 1979. Changes in the abundance of polyadenylated RNA during slime mold development measured using cloned molecular hybridization probes. *J. Mol. Biol.* **129**: 19.

Woo, S.L.C. 1979. A sensitive and rapid method for recombinant phage screening. *Methods Enzymol.* **68**: 389.

Woo, S.L.C., A. Dugaiczyk, M.-J. Tsai, E.C. Lai, J.F. Catterall, and B.W. O'Malley. 1978. The ovalbumin gene: Cloning of the natural gene. *Proc. Natl. Acad. Sci.* **75**: 3688.

Wood, K.O. and J.C. Lee. 1966. Integration of synthetic globin genes into an *E. coli* plasmid. *Nucleic Acids Res.* **3**: 1961.

Wood, W.B. 1966. Host specificity of DNA produced by *Escherichia coli*: Bacterial mutations affected the restriction and modification of DNA. *J. Mol. Biol.* **16**: 118.

Wozney, J., D. Hanahan, R. Morimoto, H. Boedtker, and P. Doty. 1981. Fine structural analysis of the chicken pro α2 collagen gene. *Proc. Natl. Acad. Sci.* **78**: 712.

Wu, R. 1972. Nucleotide sequence analysis of DNA. *Nat. New Biol.* **236**: 198.

Wu, R., E. Jay, and R. Roychoudburg. 1976. Nucleotide sequence analysis of DNA. *Methods Cancer Res.* **12**: 87.

Yamamoto, K.R., B.M. Alberts, R. Benzinger, L. Lawhorne, and G. Treiber. 1970. Rapid bacteriophage sedimentation in the presence of polyethylene glycol and its application to large-scale virus purification. *Virology* **40**: 734.

Zain, S., J. Sambrook, R.J. Roberts, W. Keller, M. Fried, and A.R. Dunn. 1979. Nucleotide sequence analysis of the leader segments in a cloned copy of of adenovirus-2 fiber mRNA. *Cell* **16**: 851.

Zasloff, M., G.D. Ginder, and G. Felsenfeld. 1979. A new method for the purification and identification of covalently closed circular DNA molecules. *Nucleic Acids Res.* **5**: 1139.

Zehavi-Willner, T. and C. Lane. 1977. Subcellular compartmentation of albumin and globin made in oocytes under the direction of injected messenger RNA. *Cell* **11**: 683.

Zimmerman, C.R., W.C. Orr, R.F. Leclerc, E.C. Barnard, and W.E. Timberlake. 1980. Molecular cloning and selection of genes regulated in *Aspergillus* development. *Cell* **21**: 709.

Zissler, J., E. Singer, and F. Schaefer. 1971. The role of recombination in growth of bacteriophage lambda. I. The gamma gene. In *The bacteriophage lambda* (ed. A.D. Hershey), p. 455. Cold Spring Harbor Laboratory, Cold Spring Harbor, New York.

Index